A CENTURY OF PROGRESS

A CENTURY OF PROGRESS

Edited by CHARLES A. BEARD

Essay Index Reprint Series

 BOOKS FOR LIBRARIES PRESS
FREEPORT, NEW YORK

Copyright © 1932 by Charles A. Beard

Copyright © renewed 1960 by Harper & Brothers
All rights reserved

Reprinted 1970 by arrangement with
Harper & Row, Publishers, Inc.

INTERNATIONAL STANDARD BOOK NUMBER:
0-8369-1903-3

LIBRARY OF CONGRESS CATALOG CARD NUMBER:
79-128205

PRINTED IN THE UNITED STATES OF AMERICA

CONTENTS

CHAPTER	PAGE
EDITOR'S PREFACE	vii
I. THE IDEA OF PROGRESS *by* Charles A. Beard	3
II. INVENTION AS A SOCIAL MANIFESTATION *by* Waldemar Kaempffert	21
III. INDUSTRY *by* Henry Ford, *in collaboration with* Samuel Crowther	66
IV. TRANSPORTATION AND COMMUNICATION *by* Edward Hungerford	85
V. AGRICULTURE *by* Frank O. Lowden	122
VI. LABOR *by* William Green	148
VII. BANKING AND FINANCE *by* H. Parker Willis	171
VIII. GOVERNMENT AND LAW *by* Charles A. Beard	201
IX. THE PROCESS OF SOCIAL TRANSFORMATION *by* Jane Addams	233
X. THE CHANGING POSITION OF WOMEN *by* Grace Abbott	253

CONTENTS

CHAPTER		PAGE
XI.	THE ADVANCEMENT OF NATURAL SCIENCE	291
	by Watson Davis	
XII.	MEDICINE	325
	by Fielding H. Garrison	
XIII.	EDUCATION	357
	by Charles H. Judd	
XIV.	THE ARTS	380
	by Fiske Kimball	
XV.	LITERATURE	399
	by John Erskine	

EDITOR'S PREFACE

THE plan of this book is simple. The authors of the several chapters were asked by the editor to summarize for the lay public, as far as possible in non-technical language, the outstanding events and achievements in their respective fields during the past century of American history. Within this general limitation each writer was given a free hand to report the findings required by a perspective of a hundred years. Concerning the competence of the various contributors to speak with authority on the subjects assigned to them it is not necessary to make any comments.

Special acknowledgments are due to Dr. Rudolf A. Clemen for excellent advice respecting the general scheme of the book and to Mr. John Dickinson and Mr. Felix Frankfurter for illuminating suggestions on the development of American government and law—without placing the slightest responsibility upon them for the outcome. For the brief introductory passages in small type at the head of the various chapters the editor is to be held accountable.

CHARLES A. BEARD.

New Milford, Connecticut,

A CENTURY OF PROGRESS

CHAPTER ONE

The Idea of Progress

By Charles A. Beard

ALTHOUGH hailed in some circles of conceit as a glorious symbol of more speed and bigger machines, and in others as a covering for cruel materialism, the concept of progress is one of the most profound and germinal ideas at work in the modern age. It is at the same time an interpretation of the long history of mankind and a philosophy of action in this world of bewildering choices. It gives a clue of meaning to the rise of civilization out of the crudities of primitive barbarism and offers a guide to the immense impending future. Briefly defined, it implies that mankind, by making use of science and invention, can progressively emancipate itself from plagues, famines, and social disasters, and subjugate the materials and forces of the earth to the purposes of the good life—here and now. In essence the idea of progress belongs to our own times, for it was unknown to the ancients and to the thinkers of the Middle Ages. It is associated, therefore, with every phase of the vast intellectual, economic, and rational movement which has transformed the classical and medieval heritage into what is called, for the sake of convenience, Western civilization.

Hence it is closely affiliated with democracy, natural science, technology, and social amelioration, and shares with them the strength of universality. It is more than a theory. It has achievements to its credit on every hand—diseases stamped out, pain silenced or assuaged, the span of life lengthened, famine made obsolete, comforts and conveniences established, sanitation supplied to multitudes, knowledge made popular through amazing instrumentalities of transmission and reproduction. And it suggests a faith of power, faith that the world, as Emerson said, "is all gates, all opportunities, strings of tension to be struck." Rejecting resignation as a philosophy of life, it confronts obstacles with assurance. Where the pessimist sees the worst, it proposes a search for the best and advances toward perfection by increments. The suffering, ignorance, and folly which drive the timid to the Nirvana of doubt and oblivion are, under the light of progress, calls to action, to research, to planning, and to conquest. Touched by the genius of universal emancipation, the idea cuts across the barriers of caste, class, race, and nationality, breaks through rigid boundaries, and regards the substances and forces of nature as potential instruments of humane purposes. Everywhere it makes its way, dissolving the feudal institutions of Europe, disturbing the slumbers of the Orient, arousing lethargic Russia, and finding a naked avowal in the United States of America: the earth may be subdued to the security, welfare, and delight of them that dwell therein.

Like religion, which may be used as a cloak for pious frauds, and patriotism, which may garb the profits of munition-makers, the idea of progress may be and indeed is employed to cloud issues, evade evident responsibilities, and justify cruelties. In the hands of the demagogue or noisy promoter it may be manipulated to avoid questions and obscure doubts. And yet the idea survives its friends as well as its enemies. Being a synthesis of all explorations, scientific, economic, and social, of all energies devoted to subduing matter and force to ordered human ends, it offers a philosophy of individual and collective action. Tendering no scheme of finality, it escapes the illusion of finality—the doom of all little systems. With natural science and the prodigious art of technology at its command, with indubitable achievements already on the credit side, it is no mere dream, but has demonstrated that symmetry and efficiency can be carried into modern life. If this is true only in part and in outline, the idea of progress, in any event, deserves exploration to its uttermost boundaries and illustration in particulars.

At the outset, the explorer confronts four fundamental questions which have perplexed thinkers since civilization began on this planet: Do nations, like human beings, pass through youth, middle life, and old age to death? Or do they revolve endlessly, as some ancient writers thought, in a cycle—despotism, kingship, tyranny, aristocracy, oligarchy, democracy, and mob-rule—or some such succession of forms? Or is it possible for a nation to stand still through countless ages, preserving what it believes to be an ideal arrangement of things? Or is there discernible, under the surface ebb and flow through the centuries, some stream of tendency, some organizing principle indicating the course of nations, and giving to their peoples some guiding rule by which to shape their activities and mold their lives and their institutions?

These questions run deeply into our religious beliefs, our philosophies, our fundamental attitudes toward life and conduct. Individuals may avoid them if they will, may move from one thing to another under immediate impulses or "the instant need of things," and make one little decision after another, trusting to luck or fate or the immortal gods as far as all larger patterns and tendencies are concerned. But no great statecraft, art, letters, or program of economy can be founded on hand-to-mouth concepts of living and working. Within the universal scheme, small projects may be constructed and executed, no doubt, sometimes with outward appearances of success, but even they are subject to laws and forces which constrain them on all sides, are themselves parts of a larger whole. No one can think long and hard about the issues of private life or public affairs without confronting and attempting to answer these basic questions respecting the nature and course of the whole. And judging by the methods of the leaders of thought and action, by the achievements that endure through time, it is only in coming as nearly as possible to the central scheme of things that the worthiest and most lasting work can be accomplished. Perhaps this is only another way of saying that the wider the horizon, the more catholic the thought, the surer will be the insight of those who attempt great enterprises. How could it be otherwise?

Now among the fundamental notions competing for the allegiance of mankind in our age, as indicated in the beginning, none is more widely discussed, warmly defended, and hotly attacked than the idea of progress. In substance, it is a theory that the lot of mankind on this earth can be continually improved by the attainment of exact knowledge and the subjugation of the material world to the requirements of human welfare. Associated with it are many subsidiary concepts. Its controlling interest is in this earth, in our own time, not in a remote heaven to be attained after death. It assumes an indefinite future and plans for greater security, health, comfort, and beauty in the coming years. While a philosophy of history, it is also a gospel of futurism. It is founded on the belief that civilization is on the threshold of time and it is characterized by the buoyancy of youth, not the skepticism and morbidity of old age. If it lays emphasis on the material benefits of civilization, it makes no assumptions that are more materialistic than those of less earthly philosophies. It does not admit that nations move from youth to death or through endless cycles, but contends that mankind is advancing, in spite of calamities, errors, and disasters, and on the whole in a desirable direction. If the truth of this allegation be questioned, its defender may reply, as did the mathematician, Poincaré, when the validity of Euclid was challenged: whether true or not, it is convenient and is at all events one of the supreme products of intellectual history, the highest of all world courts.

Thus broadly conceived, the idea of progress runs counter, of course, to the doctrine of fatalism which has possessed large sections of humanity for long ages, especially in the Orient. The fatalist sees nations decimated by plagues, famines, floods, blights, diseases, and wars, and insists that "nothing can be done about it," that the more changes we have the more we have of the same thing. Those who make a philosophy of such fatalism, leave the world to its folly and withdraw within themselves to contemplate. Seeing many horrors wrought by physical nature and human nature, they conclude that "nothing really matters"; resignation, not effort and thought, is the best, if not the only, recourse. That such an attitude is fitting for a civilization in which science and invention have created no instrumentalities for eliminating or reducing calamitous forces must be conceded, but what justification can be made for accepting undoubted evils that can be eliminated by understanding and labor? By what criterion of values is it better to endure evils than to remove them? Let the philosophy of fatalism answer that question.

In a similar way the idea of progress is opposed to certain views of life which may fairly be said to have been dominant in Europe in the Middle Ages. With exceptions, of course, medieval thinkers looked upon this world as a mere vestibule to heaven. Man, ran the current theory, is a poor and miserable sinner, born to trouble as the sparks fly upward; faith and conduct are to be shaped with reference to eternal bliss hereafter rather than to a pleasant and comfortable living on this mundane sphere. The ideal was not to refashion the world after some concept of earthly needs, but to accept most of it, as it came, and pass on to joys beyond. Riches, and the delightful life which they provided, were not unknown, to be sure, but were objects of suspicion. "There is not to be found in the writers of the early Middle Ages," declares a competent scholar, "that is to say, from the eighth to the thirteenth century, a trace of any attention to what we at the present day would designate economic questions. . . . The writings of this period, therefore, betray no sign of any interest in economic affairs." An exhaustive examination of the works of the outstanding thinkers of the time reveals not "a single passage to suggest that any of these authors suspected that the pursuit of riches, which they despised, occupied a sufficiently large place in national as well as individual life, to offer the philosopher a subject fruitful in reflections and results."[1] Such was the Christian heritage near the close of the thirteenth century.

[1] George O'Brien, *An Essay on Mediæval Economic Teaching*, pp. 13 ff.

Although later in the Middle Ages religious philosophers were compelled to give attention to the economy of life and did work out systems of thought respecting it, they related their schemes to theology—to conduct acceptable in the sight of God, to sin, hell, salvation, and heaven. And their systems of thought were based upon the idea of a fixed order of society in which the established relations of classes were to be maintained and sanctioned by morality. Kings, bishops, priests, landlords, serfs, peasants, and artisans had their rights and duties—nothing fundamental was to be changed by the daring experimentation of the mind. The best thought of the time was concentrated on salvation in the hereafter rather than on the progressive reordering of earthly affairs. The great end of life—the purpose of history—was to secure the external welfare of that portion of humanity which could pass the gates of heaven. As for this world, well, it would be utterly destroyed sometime or at best was to be viewed as a place of temporary sojourn. While the Lord's Prayer mentioned a possible Kingdom of Heaven here below, medieval thinkers certainly did not concentrate on that aspect of their theology.

To the fatalism of the Orient and the other-worldliness of the Christian Middle Ages must be added a second idea opposed to the concept of progress—that is, utopianism. This idea takes two forms. In the minds of some thinkers it is related to the past: there has been a golden age, in the "good old days of the fathers" or in some remote period of the early evolution of mankind. In seeking to escape the evils of the present, we must return to the perfection of long ago when people lived in peace, happiness, innocence, and plenty. But, in other minds, utopianism is related to the future: by doing this or that we can establish a static order of bliss—a fixed scheme of things so nearly perfect that they will never have to be changed. A variant on these aspects of dreaming may be called the utopianism of whitewash: the present order is so nearly perfect that it is almost profane to inquire into its evils or to propose modifications, for the possibility of doing harm is always greater than the chances of doing good. Historians, with all their searching, have not been able to find the golden age in the past, and skeptics doubt the perfection of the present. Still the illusion of utopianism shadows all human thought about public and private affairs, challenging the idea of progress.

Unknown to the ancients, foreign to the theology of the Middle Ages, the idea of progress was slow in taking form and winning its way as a dominant concept of life. In reality it was a kind of gigantic intellectual outcropping—the product of the great commercial revolution ushered in by the discovery of America, the circumnavigation of the globe, and the development of natural science. As J. B. Bury points out,[2] certain conditions were necessary to the flowering of the idea. First of all, there had to be respect for, and interest in, the common business of labor and industry—a respect which the slave-owners of ancient Athens and the landlords of the Middle Ages could not acquire. In the next place, since the idea of progress had to do with this world, it was necessary to shake off the dominance of other-worldliness and to think in secular terms; the recovery of ancient learning in the renaissance and the commercial revolution to which reference has been made favored this shift from heaven to earth. Finally, the idea of progress could not flourish until thinkers had cast overboard their slavish adherence to ancient books; natural science, with its emphasis on experimentation, the observation of common things, and invention, was necessary to clear the way for the emancipation of the mind from the despotism of theology and the classics. By the end of the seventeenth century, when all the American colonies except one, had been fairly started on their course, the ground was prepared for the rise and growth of the idea of progress—the steady improvement of the lot of mankind in this world as a good in itself, as a value in itself, without any reference whatever to a possible life after death.

At last the idea of progress, long in germination, already dimly foreshadowed by a few thinkers, finally came out in

[2] *The Idea of Progress* (Edition of 1932 with an introduction by Charles A. Beard).

positive form. If a single name must be associated with its origin, it may well be that of Abbé de Saint-Pierre who gave to the world in 1737 his epoch-making work entitled *Observations on the Continuous Progress of Universal Reason.* "Here," says Bury, correctly, "we have for the first time, expressed in definite terms, the vista of an immensely long progressive life in front of humanity. Civilization is only in its infancy. Bacon, like Pascal, had conceived it to be in its old age. . . . The Abbé was the first to fix his eyes on the remote destinies of the race and name immense periods of time." By shaking off its inertia, by taking thought, by devoting its talents to the enterprise, wrote the Abbé, in substance, mankind can do more in a relatively short time to establish peace and improve its lot on earth than it has done in a thousand years under the régime of resignation, indifference, and complacency.

Once announced, the new philosophy ran swiftly through the minds of the French thinkers who were preparing the way for the Revolution that was to shake Church and State in the Old World and make room for secular supremacy. It was a dominant note in the great French *Encyclopædia of Universal Knowledge.* It was implicit in Adam Smith's *Wealth of Nations* published in 1776. In many ways the titanic labors of the French Revolution were guided by the idea of progress. The constitutional, economic, educational, and law-reform policies which accompanied that upheaval were secular, mundane, and directed to the improvement of the common fortunes of mankind. If, in immediate consequences, the Revolution was bourgeois in character, its achievements and ideas far outran the purposes of its directors, breaking the path for the reconstruction of government and economy. If the bourgeois could be lifted into power, security, and well-being, why not the whole order of society? If the bourgeois could set up for themselves the goal of earthly advancement, why not the humblest laborers in the land? The genie was clearly out of the bottle and no human power could compress the spirit again into a class mold.

THE IDEA OF PROGRESS 11

During these momentous years, while the idea of progress was taking form, spreading, branching, and working its way into the remotest divisions of European thought, the English colonies in America grew to maturity and burst upon the world stage as a united and independent power. Here the natural resources, intellectual climate, and social order were highly favorable to the growth of the new concept. Here nature had provided an enormous and diversified material endowment which could be used to establish a high level of life and sustain the continuous advancement of standards, if intelligently and efficiently used. Here the population was ready for secular enterprise. While many had migrated to America in search of religious freedom, the great majority who came voluntarily had come for mundane reasons—the improvement of their condition here and now—and even those who fled for religious reasons expected, as a rule, to find a decent living somehow. All the factors which had contributed to the germination of the idea of progress in Europe were even more prominent in America—respect for industry and labor, a preoccupation with secular enterprise, and a spirit of experimentation and invention.

In these circumstances, the leading thinkers of the New World, especially Franklin and Paine, carried the idea of progress more or less consciously into the plans they formulated for American culture. "It is impossible to imagine," wrote the former, "the height to which may be carried, in a thousand years, the power of man over matter. We may perhaps learn to deprive large masses of their gravity and give them levity, for the sake of easy transportation. Agriculture may diminish its labor and double its produce; all diseases may by sure means be prevented or cured, not excepting that of old age, and our lives lengthened at pleasure even beyond the antediluvian standard. O that moral science were in a fair way of improvement, that men would cease to be wolves to one another and that human beings would at length learn what they now improperly call humanity!" Thomas Paine, in his *Rights of Man,* written in answer to Edmund Burke's

Reflections on the French Revolution—a plea for historic conservatism—sketched an outline of political economy that embraced universal education, the abolition of poverty, reform of the criminal law, pensions for the aged, the reduction of armaments, and international peace.

All through the nineteenth century the idea of progress continued to work as a powerful ferment in the opinions of the world. In America, the extension of the suffrage beyond the boundaries of the propertied classes, the adoption of universal education, and the growth of a leveling freedom in the agrarian West helped to widen its scope to include the whole population, to democratize it, in a word, and make it a guiding principle for a civilization. In previous times and in other circumstances, privileged classes and individuals could lift themselves to a position of comfort, security, and prosperity by law and economic advantage and thus enjoy the benefits and delights of culture; now at last in a vast natural theater, it was thought, a whole people could, through progressive development, enjoy the blessings of science, industry, and art, and become civilized. The hewers of wood and drawers of water were to rise above the level of serfdom and sit at the banquet prepared by applied science. Here civilization was conceived not as a beautiful fairyland of delight surrounded by brutalizing labor, illiteracy, and margin-of-subsistence living for the masses. The actualities of American life, it was easy to show, were far from the ideal held up to the faithful, but the concept of progress, once let loose in our democracy, continued to act as a dynamic force, transforming every aspect of American civilization.

With inescapable fatality the mass production made possible by machinery and nourished by our unparalleled natural resources accelerated the leveling democracy implied in the idea of progress. Gigantic industries could not flourish without an immense market. And where was that market to be found? In a small privileged class enriched by the profits of capitalism? Only one answer was possible. The few craftsmen of the Middle Ages might sell the choice products of loom, forge, kiln, and chisel to lords, ladies, bishops, princes, and kings, but masters of huge industries turning out commodities by the ton and the million could thrive in no such limited area of demand. Markets for mass production simply could not be found unless the masses themselves rose above the historic margin of subsistence and were able to buy by the ton and the million. Only when the standard of life for the multitude is constantly rising and buying capacity is expanding can widening outlets be found for the goods which pour in swelling streams from the vast industries made possible by science and machinery. If the American bourgeois were as indifferent, on moral grounds, to the lot of the masses as the French nobility of the eighteenth century to the plight of their laborious peasants, still their enterprises could not develop without a continuous enlargement of the popular market—without a steady growth in the capacity of the masses to buy and enjoy goods once confined to the classes.

Herein, no doubt, lies one of the main sources of the European criticism which is directed against the idea of progress as powerfully expressed in American civilization. Every quest for the inner nature of that criticism and for the roots of its inspiration leads immediately to an opposition of class ideals. True culture, we are told, is inevitably confined to "the superior minority," and cannot exist when boundaries are widened to include millions of nameless and unknown. This is the theme of one school of writers which had its origins in ancient Greece and survives in the latest hour, finding new spokesmen as the old are forgotten. Consciously or unconsciously, it is dominated by one secret wish or conviction: Democracy operating under the idea of progress is incompatible with "culture."

This concept and the antithesis were clearly and eloquently set forth long ago in the writings of Amiel. "In society," he remarks, "people are expected to behave as if they lived on ambrosia and concerned themselves with no interests except such as are noble. Care, need, passion, do not exist. All realism is suppressed as brutal. In a word, what is called *le grand monde* gives itself for the moment the flattering illusion that it is moving in an ethereal atmosphere and breathing the air of the gods. For this reason all vehemence, any cry of nature, all real suffering, all heedless familiarity, any genuine sign of passion, are startling and distasteful in this delicate *milieu* and at once destroy the collective work, the cloud-palace, the imposing architectural creation raised by common consent. It is like the shrill cock-crow which breaks the spell of all enchantments and puts the fairies to flight. These select gatherings produce without intending it a sort of concert for the eye and ear, an improvised work of art. By the instinctive collaboration of everybody concerned, wit and taste hold festival, and the associations of reality are exchanged for the associations of imagination. So understood, society is a form of poetry; the cultivated classes deliberately recompose the idyll of the past, and the buried world of Astræa. Paradox or not, I believe that these fugitive attempts to reconstruct a dream, whose only end is beauty, represent confused reminiscences of an age of gold haunting the human heart; or rather, aspirations toward a harmony of things which everyday reality denies to us, and of which art alone gives us a glimpse." Undoubtedly this is a fair statement of the idealized case; although a student of the world's social memoirs may be inclined to believe that such a *grand monde* never existed, save perhaps in the Tokugawa era of Japan at the height of its glory.

Having drawn his perfect picture of *le grand monde* supplied by Europe, Amiel presents the contrast afforded by the United States: "For the Americans, life means devouring, incessant activity. They must win gold, predominance, power; they must crush rivals, subdue nature. They have their hearts set on the means and never for an instant think of the end. They confound being with individual being, and the expansion of self with happiness. This means that they do not live by the soul, that they ignore the immutable and eternal, bustle at the circumference of their existence, because they cannot penetrate to its center. They are restless, eager, positive, because they are superficial. To what end all this stir, noise, greed, struggle? It is all a mere being, stunned and deafened." In short, without stopping now to dispute the correctness of Amiel's contentions, Americans do not live on ambrosia, dispense with care, move in an ethereal atmosphere, breathe the air of the gods, escape from the world, and reconstruct a dream whose only end is beauty; they are incessantly engaged in subduing nature and in seeking to develop an ordered economy which will establish security, continuity in high productive output, and the widest possible distribution of the benefits flowing from efficient industry.

When once the antithesis presented by Amiel is clearly recognized and its implications understood, the issue of civilization before us becomes perfectly evident. Whether and how long European countries will continue to maintain superior minorities concerned only with "noble" interests, with cloud palaces and associations of the imagination, is an appropriate matter for speculation. Assuming their virtues to be all that their advocates claim, it may be appropriately asked, "At what price glory?" Bent backs, knotted hands, and numbed minds must pay for parties at which such wit and taste hold festival and the idyll of the past is recovered for the delight of the participants. If, when the balance sheet is struck, the credits outweigh the debits, still it may be surmised that the knowledge released by science, the demands of industry for markets, the awakening insistence of the multitude on sharing the fruits of the earth, have made forever obsolete *le grand monde* of the lotos-eaters. Esthetes may regret it, but there is something Promethean in the vast upward thrust of the masses under the banner of progress, and those who have occasion to think, teach, or direct in the coming years will have to reckon with that invincible fact. Iron gates are closing on the dreams of privilege, and those who cherish the ideals of that order will have to look beyond this world for their lost Atlantis. This seems to inhere in the nature of things, even though poignant Americans will long continue to pay large honoraria to Europeans for the privilege of listening to deprecatory estimates respecting the very heart and dynamic of civilization in the United States.

If critics of progress fail to grasp its cosmic nature, friends of the idea often make it appear petty and ridiculous by the undiscriminating zeal with which they espouse it. As in the case of every other fruitful concept, a lunatic fringe is associated with the idea. To these short-sighted spectators at the great show, all movement is progress, means are ends, and the worth of a personality is to be measured by the number of motor-cars, telephones, radios, and bathtubs he possesses. The idea of progress thus becomes purely numerical. J. P. Morgan has more things than Dante; therefore he is superior. Jim Fiske had more diamond rings than Francis of Assisi; accordingly, his rating in civilization must be higher. Zenith has more miles of paved streets than Athens, a single apartment house in New York will hold the entire population of that ancient city; evidently America transcends in achievement the best of the Greeks. Thus a noble concept of humanity is made both absurd and contemptible, obnoxious in the house of its friends, and a shining target for abuse at the hands of its opponents.

Yet when the critics and scoffers, writing under soft lamps or lecturing for fees to well-fed audiences, in comfortable rooms electrically lighted, venture to speak of an alternative, they can only offer a return to agriculture and handicrafts. Overlooking the fact that they can themselves go at any time to any one of a thousand waste places awaiting the plow or the hoe, they prefer to advise others to incur the risk. When asked for a bill of particulars, they become hazy and vague. Are we merely to surrender the tractor and return to the steel plow? Why not to the wooden plow? Or better still, to the forked stick hardened by fire? Each advance on the most primitive instrument is a gain in efficiency, a transfer of labor from man to a tool. In the process of retreat are surgery and dentistry to go into the discard? Sanitation, antiseptics, and anesthetics? Each of these gains has marked a step in progress or rather a long series of steps, and each art steadily advances

THE IDEA OF PROGRESS

in our own time as masters of the test tube and microscope penetrate deeper and deeper into the mysteries of nature. Fundamentally the machine differs from the tool in degree, not in kind, and the chemist works in materials no less than did the most primitive woman herbalist. His knowledge is wider, his skill is greater, but his ends may well be fundamentally the same—the relief of human suffering. Where then is the line to be drawn? To what point in the long upward progress of mankind is the return to be made? To ask these questions is to answer them. The severest critic of progress is forced to admit, when cornered, that the problem is not one of retreat, but of ends and methods, of choices and uses.

If in the hands of its superficial champions the idea of progress seems to emphasize means rather than ends, an examination into the history and nature of the concept shows that this notion is without basis. Although selfish men have seized upon the instrumentalities of progress and have left in the train of their exploits hideous industrial cities, slums, poverty, and misery, that upshot is no more to be attributed to the idea itself than the cruelties of the Inquisition to the teachings of Jesus. An inquiry into the writings of those who originated and developed the theory of progress shows at the center of their thought the concept of the good life as the end of progressive endeavor, the genius which is to preside over the searches and labors of explorers and experimenters. The good life for the multitude, not for a superior minority living in a land of illusion on the sweat of the "ignoble"—this is the kernel germinating in the heart of the concept of progress. To see life whole and to see it steadily, to sound its deeps, to illuminate its possibilities, and to make the noblest and wisest use of material resources in realizing its purposes, this is the sum total of the idea of progress—a grand end, conceived in the light of universality, appealing to a mankind seeking high destiny and striving for mastery over the instrumentalities to be employed by the way. Anything less than this is a caricature of the idea.

Wrongly identified with capitalism, communism, or particular systems of economy, though standing at the very threshold of the great analysis and inquest, the idea of progress nevertheless clearly reveals the method by which ends are to be attained. Its method is that of science and technology—rationality, in short. And that method implies many things. It implies an open-eyed and open-minded attitude toward tasks in hand and problems to be solved. Working with concrete materials under positive law, technology is as indifferent to the emotional idiosyncrasies of individuals and classes as the elements themselves. Universal in its reach, as transcendent as the gods, it cannot be monopolized by any nation, period, class, government, or race. Its catholicity surpasses that of all religions. Essentially objective in its manipulations, dealing with materials, quantities, and known laws, technology is leveling and democratic in its effects; it is not a closed cult handed down by a few masters to a few students in cloistered universities. Rational in nature, corresponding to the mathematics of physical things and forces, this method is necessarily planful. It cannot begin anything without a goal, project, or purpose. To proceed at all it must stake out a field of work, a problem to be solved, and then it must proceed according to plan, on the assumption of predictable results, to predetermined ends. Inexorably, therefore, it cuts across the wild welter of unreasoned actions, irrelevant sentiments, and emotional starts and fits which have so long characterized human life in historic politics, industry, agriculture, and esthetics. Rational and planful, working in the unity of all things, this method is centripetal, drawing all arts, economies, and sciences inward toward the unity of the world—with implications so vast, so in harmony with mankind's noblest dreams, that the imagination is staggered by them.

Since the rationality of progress imposes limitations on inner impulses and cuts across external arrangements, it inevitably involves all departments of human activity—pure science, invention, industry, transportation, agriculture, government, finance, medicine, social adjustments, the work of women, education, arts, and letters. As the first carved gates of ancient Egypt celebrated the purpose of the ruling monarch, so the latest skyscraper in New York reflects the functions of its inhabitants. All branches of civilization mirror the dominant idea. If the escape of negation be sought, it will be found blocked at the exit. All arts, sciences, and crafts are drawn into the movement of regnant thought and practice. And when the thought of the thinker, the dream of the artist, and the aspiration of the practitioner draw together under a common principle of unification, the light and heat required for heroic endeavor are generated, giving to each the power of the whole, suffusing all with a sense of elevation and movement, supplying energy to the weak, and providing for the strong and willful who make history that social dynamic without which even Napoleon himself might have been a Corsican lawyer or Genoese scrivener.

CHAPTER TWO

INVENTION AS A SOCIAL MANIFESTATION

Look backward a hundred years. Half a continent yet remains to tempt a conquering people, equipped merely with canal boats, ox carts, horse-drawn wagons, steamboats, shovels, scythes, axes and flails. It takes from May to November to travel from the Atlantic to the Pacific in a prairie schooner dragged by six to twelve animals. After the conquest of California, a mail line is established, but it costs five dollars to send a letter, weighing less than half an ounce, from New York to San Francisco. There are no trains; waters and dirt roads are the only connection between farms and towns. Automobiles, gas-engines and dynamos are not even on the horizon. Thought can be transmitted no faster than a horse or a steamboat can carry it. Newspapers are printed on hand-presses at the rate of about two hundred copies an hour, circulated locally, and sold at prices above the reach of the masses. Industry is largely in the handicraft stage, associated with rural life for the most part, and unconscious of the possibilities of serving metropolitan multitudes. But the will that determines to conquer a continent also determines to transform economy. It ransacks Europe for machines, adapts and improves them. It contrives new devices for production and marketing. It makes trading combinations on a mammoth scale. The productive power of men and women is multiplied more than a hundredfold. Giant cities arise in response to the unexampled output of farms and industrial plants. A novel relation between town and country develops. As a result of the general transference of muscular labor to machines, American culture becomes distinctive. In fact, the story of the marvelous inventions—the tools of American life—is one of the great historical romances and the tale is here sketched in bold, firm strokes.

Invention as a Social Manifestation

By Waldemar Kaempffert

WHEN we speak of a culture we conjure up a picture of group behavior—of a race influenced by common instincts, passions, or motives. A social tension is evident, a tension which compels men to act, dress, and think more or less alike. In the Middle Ages that social tension expresses itself so strongly in religion that there are 110 holidays in the year (literally "holy days"), that a new architecture is evolved, that the whole of Europe rises to the spiritual need of wresting Jerusalem from infidel control. If today we ride in airplanes that can travel 200 miles an hour, talk to one another across the Atlantic Ocean, read two-cent newspapers, determine the chemical and physical constitution of a star by light that left it when dinosaurs shook the earth, it is not because the human mind is intrinsically any better than it was ten thousand years ago, but because it has acquired different interests under social tension.

Tension of any kind seeks relief. To a socially tense people relief comes through art, philosophy, religion, arms, or science, depending on the crucial need of the moment. Hence Dante, Shakespeare, Voltaire, Bach, Newton, Watt, Morse, Bell, Edison, and Marconi must be regarded as fuses that blow and that enable society to short-circuit itself by following the lines of least resistance. The leader expresses the massed, unconscious aspirations of the race and responds to forces of which even he may not be aware.

A change came over our group aspirations about the middle of the eighteenth century. What we call the objective, scientific point of view began to be taken toward the universe, which means that we succeeded in partially divorcing our personal feelings and traditions from our perceptions and proceeded to interpret sticks, stones, and stars dispassionately. In other words, we began to tear the universe apart to find out of what it was made, regardless of what Aristotle may have said about it, and to let it speak for itself. We studied bodies in motion and noted how they acted when they were subjected to forces under varying conditions. We even controlled the conditions. To be sure Archimedes, Galileo and a few others had this impersonal point of view, but it did not possess society as a whole until about the middle of the eighteenth century. It was then that the steam-engine appeared. Patent laws were enacted to encourage invention—additional evidence of a changed viewpoint. A dozen factors thus contributed toward bringing about what economists call the Industrial Revolution, by which they mean the beginning of mass production and the factory system and the introduction of machinery to do the world's work.

The wave of mechanical invention, which sets our own time apart, was confined to no one country. It swept western Europe, and since the New World was settled by western Europeans it necessarily included the United States. We must deal with it not nationally, but racially. There is something childish about Englishmen, Frenchmen, Germans, and Americans who claim for one of their countrymen the honor of having invented the "first" electric lamp or the "first" dynamo, or the "first" telegraph. There are very few "first" inventions, and among these few are the wheel, the bow and arrow, the bow-drill, fire-kindling instruments, all of which must be credited to unknown, prehistoric, savage Edisons. Every invention is rooted in old principles, and every inventor has a technical heritage without which he could not progress. Edison certainly would not have devised the phonograph or the incandescent lamp if he had been a Cro-Magnon man in southern France fifty thousand years ago. Without Galvani, Volta, Faraday, and Henry the telegraph of Morse would never have been conceived.

INVENTION AS A SOCIAL MANIFESTATION

If we consider the course of invention in the United States in this racial and cultural aspect we find that Americans are no more originative than other peoples. On the other hand the social and economic effect of invention is nowhere so apparent. For nowhere are more daring experiments in engineering made; nowhere is the machine more in evidence; nowhere has the industrialization of old crafts been carried so far; and nowhere is the future state of a mechanized society so clearly foreshadowed. In the wilderness of what was destined to become the United States, with unlimited resources, the inventive genius of western European settlers, untrammeled by tradition, found free play.

No one invention or group of inventions made industrial America what it is today. Certain basic inventions, whether they were American in origin or not, had an especially profound effect. And the general effect of these it is our purpose to examine here. They cannot be discussed individually in so brief a treatment, but they may be thus classified and swiftly identified:

The Generation of Energy

Steam-Engine. Invented in its modern reciprocating and condensing form by James Watt, a Scotsman, in 1765. High-pressure or non-condensing engines are due to Oliver Evans (American, 1787) and Richard Trevithick (British). Corliss (American) invented his valve-gear in 1848. The modern steam-turbine was invented by Sir Charles Parsons (patented in 1884)

Internal-combustion Engine. Lénoir (1860) and Otto (1876), inventors of the internal-combustion engine, were, respectively, French and German. The Diesel engine was devised by Rudolf Diesel (German) in 1893.

Electric Generator. Pacinotti (Italian) invented the first dynamo to give a continuous current. Reinvented by Gramme (Belgian) in 1870 and improved by Siemens (German), Edison (American), Brush (American), Thomson (British), and others in various countries.

Transportation

Locomotive. Reduced to its modern essentials by George and Robert Stephenson, two British engineers, in 1829.
Automobile. The modern gasoline type was independently conceived

and reduced to practice about 1883-1885 by Daimler and Benz, both Germans. American engineers modified the construction to make mass production possible.

Steamboat. John Fitch, American, deserves credit for his pioneering. His eighteenth-century European contemporaries were equally successful. Fulton's contribution was quite lacking in originality.

Airplane. Cayley, early in the nineteenth century, glided successfully in England. Lilienthal and Pilcher also glided about 1893-1896. Langley (American) flew his quarter-size steam-driven model, 1896. But the first successful man-carrying airplane was that of the Wrights, who demonstrated the soundness of its principle in 1903. Every element in the machine was old, but the combination was nevertheless a great invention.

Elevator. Vertical transportation is old. Mine elevators were used in Europe before the Declaration of Independence was signed. Passenger elevators were suggested as early as 1687. The Otis elevator (American, 1852) was distinguished by its automatic braking mechanism. It made the skyscraper possible.

COMMUNICATION

Printing from Type. European in origin, even to mechanical typesetting. American inventions in printing are entirely intensive.

Typewriter. Writing-machines go back as far as 1714, in which year Henry Mill (British) patented his invention. Keyboard machines are older than Christopher Sholes (American), who devised the modern typewriter (1867).

Motion Pictures. The step-by-step presentation of photographs in proper sequence at such a rate as to create the illusion of continuous movement by retinal persistence of images was discovered by various Europeans early in the nineteenth century. Coleman Sellers (American) patented the first projector that resembled the modern type (1861). The use of film was proposed by Stampfer (1833) and Devignes (1860).

Electric Telegraphy. Electric telegraphs were in successful operation in Europe before S. F. B. Morse invented his instrument (1832-1838). All the elements of the Morse telegraph were old, but the combination was new. The Morse code was a distinct contribution.

Telephony. Invented simultaneously and independently by Bell and Gray (Americans), both of whom filed patent applications on the same day (March 7, 1876). There is reason to believe that Reis (German) had a telephone (1861) which was not as mute as American histories of invention assert. At least it could transmit music. In view

INVENTION AS A SOCIAL MANIFESTATION 25

of the work of Bell and his American successors the telephone must be regarded as an American innovation.

Radio Signaling. Hertz (German) discovered experimentally how to generate, transmit, and receive radio waves (1887). Marconi (Italian) devised the essentials of a radio telegraph (1896) and first signaled intelligibly through space. Radio telephony was largely the result of work done by Poulsen (Dane) and Fessenden (American) in 1903. America's most striking contribution to radio and possibly to electro-mechanics as a whole was the vacuum-tube, invented by De Forest in 1906, with the Fleming valve (British) as a basis.

METALLURGY

Steel. Modern steel-making must be credited to the Englishman, Bessemer (1856), although he was anticipated by the American, Kelly (1851), who failed to push his priority aggressively. Most steel today is made by the Siemens-Martin process. Siemens, a German, invented the necessary regenerative furnace (1861), and his French licensees, the Martin brothers, made great improvements. The blast furnace goes back to Darby (British, 1713).

High-speed Tool Steel. The work of Frederick W. Taylor and Maunsel White (Americans) in the late nineteenth and early twentieth centuries resulted in one of the greatest contributions to industry. To them we owe the modern, high-speed chromium, tungsten, and vanadium alloy steels which made the cheap production of automobiles and other machines possible.

METAL-WORKING MACHINES

Rolling Mill. Invented by Darby (British) in 1713.

Machine Tools. Bramah and Maudslay, two Englishmen, invented the slide-rest about 1792—the greatest improvement in the lathe ever made. The slide-rest is an iron fist that is screwed along to hold the tool against the work. Independently invented by David Wilkinson in America (patented 1798). Eli Whitney (American) built the first milling machine in which the tool rotates while the work is fed to it. The principle of directing several tools successively against the work was invented by Stephen Fitch (American) in 1845. Fitch's turret lathe is the greatest improvement made in machine tools since the invention of the slide-rest. With the turret lathe began the series of "automatics" in which tool after tool mechanically comes into play.

ILLUMINATION

Gas. Natural coal gas was known as early as 1691 in England. Murdock (British) used manufactured coal gas for illumination in 1792. Windsor lit some London streets with gas early in the nineteenth century. Water gas is made by a process independently invented by Lowe (American) and du Motay (French) about 1874.

Arc Light. Discovered by Davy (British) in 1801.

Electric Incandescent Lamp. Invented in 1879 by Edison. Also independently by Goebel (German-American) and Swan (British). Because he planned the central-station system without which the lamp could not be effective, Edison deserves the credit as an innovator that he has received.

Miscellaneous Labor-saving Machines and Devices

Planting and Reaping Machinery. All European in origin. The Scotch reaper (1794), the Salmon reaper (1807), and the Bell scissors reaper (1826) were in use before either McCormick or Hussey invented their American reapers about 1833. McCormick, combining old elements ingeniously, produced the reaper of today.

Textile Machinery. Samuel Slater arrived in New York in 1789 and brought with him all that England knew about carding, spinning, and weaving on a factory scale. Modern automatic looms are largely of American invention.

Cotton Gin. Invented by Eli Whitney in 1792. Entirely American.

Sewing-machine. Thimmonier's sewing-machine (patented in 1830) was so much of a success in Paris that it was destroyed by seamstresses and tailors. Howe (American) received his first patent in 1846. His eye-pointed needle had previously been invented by Hunt (American) in 1832.

Interchangeable Parts. Fitfully used in Europe toward the end of the eighteenth century. The principle was reinvented by Eli Whitney to carry out a government contract for the manufacture of muskets to be used in the war of 1812. It must be regarded as a great mechanical discovery, one without which mass production of machinery would be impossible.

It is impossible, in a sketch of technical progress, to discuss each of these inventions and its effect. We must select a few and with their aid paint a picture of America transforming itself from a wilderness to the most highly mechanized social organism in the world today.

POWER

Until the coming of the steam-engine it mattered little industrially whether a country possessed coal deposits. After

INVENTION AS A SOCIAL MANIFESTATION

Watt, coal was potential energy. Nations were willing to bargain for it and to fight for it. It follows, almost as a matter of course, that because the United States is the largest coal-owning country the steam-engine proved to be enormously important in the conquest of the wilderness. In truth not until the steam-engine was introduced, not until American coal could be converted into energy in American factories, not until American inventors had steam at their command to operate the many mechanical substitutes for the human hand that they had devised, did the United States take its place in the front rank of industrial nations.

Yet the earlier contributions of the United States to the development of the steam-engine have not been conspicuous for great originality. In colonial times there was brilliant, erratic Oliver Evans, who received one of the three patents granted by the United States in the first year of its existence. Evans probably invented the first high-pressure engine—an engine which, unlike Watt's, had no condenser. There is reason to believe that he sent his drawings to England in 1787, that they were studied by Trevithick, one of Watt's rivals, and that thus the high-pressure engine became known in England. Evans' case again proves that the *Zeitgeist* of the eighteenth century simply demanded that some one should invent a steam prime mover. If Watt had not supplied the Newcomen mine-pump with a separate condenser there can be no doubt that the Evans high-pressure, non-condensing engine would have been introduced; for Evans knew nothing of Watt's condenser and developed his own invention, with nothing more to aid him than a book in which a description of Newcomen's engine appeared. By 1803 Evans was a regular builder of steam-engines. He was perhaps the first American to apply steam industrially.

The other outstanding figure in the annals of American steam-engineering is George H. Corliss, whose valve-gear, invented in 1849, made it possible for an engine to be automatically controlled with a nicety utterly beyond the simple ball-governor of Watt. What we have here is not a key invention, but a magnificent intensive improvement which is of technical and social importance because of the great saving in fuel that it brought about. The invention was of a type that naturally followed a scientific study of the steam-engine. By Corliss' time the science of thermodynamics was well advanced. The steam-engine was properly regarded as a heat-engine. Coal contained heat, which had to be made available as energy. Because the United States was not so far advanced as Europe in theoretical science in the nineteenth century it follows that the principles of thermodynamics were better understood abroad than here, and that the major improvements in steam-engines were of European origin.

It would serve no useful purpose to trace the development of the steam-engine in America and to note the gradual displacement of the reciprocating engine in favor of the turbine in great central stations. We are concerned primarily with social effects. At first the steam-engine could be used only near the mine. So we find American industrial centers huddling near coal-mines or springing up wherever there was water deep enough and still enough to float a boat. The spreading of the railway net over the country was accompanied by a spreading of population, but to this day, with a freight rate of less than 0.9 cents a ton mile on the average, the cost of transporting coal must still be reckoned with in manufacturing.

The distance to which power can be transmitted by a steam-engine directly is limited to a few feet by the length of a belt or a rope. As a result the engine had to be used individualistically. Every factory proprietor the world over erected his own boiler house and his own power plant. By 1930, 270,000 individualistically operated steam-engines, required by as many mills and factories, sprang up in the United States. To these must be added 58,000 individualistically operated locomotives which haul more coal by weight than they do wheat, corn, oats, hay, lumber, steel and iron combined—haul power in the form of solid, black lumps.

With the invention of the incandescent lamp by Edison in 1879 this picture was destined to change. Edison was con-

INVENTION AS A SOCIAL MANIFESTATION 29

cerned primarily with competing with gas as an illuminant. A lamp, however, was not enough. What made the filament glow? Heat—electric heat. A current had to be supplied to the lamp. There were dynamos even in the days of '79 and '80, but the most potent source of electric energy in New York was the Western Union Telegraph Company's battery of 2,000 cells. To illuminate a small section of a city with a gigantic battery was technically absurd. The unflagging Edison was constrained to modify existing generators and to design a distribution system, junction-boxes, suitable bases into which lamps could be fitted, meters, in a word, the whole paraphernalia of the modern central station—something that historians of invention are apt to overlook when they appraise him as an inventor. Power houses of today differ radically from that which Edison built in Pearl Street in 1881. Yet that historic brick offense to the eye was the nest in which the enormous electric light and power industry of today was hatched. To be sure it was Westinghouse who brought about the adoption of alternating current, the kind which is now generally supplied and which Edison opposed. Central station practice nevertheless began in Pearl Street, and for no other purpose than that of enabling the electric lamp to compete with gas. We owe to Edison far more than the incandescent lamp—the immensely richer achievement of enslaving electricity in the service of mankind. When the incandescent lamp was invented in 1879 the average American wage-earner had at his command 1.66 installed horsepower: fifty years later the figure had risen to 23. Engineers usually estimate that a horsepower is equal to ten manpower. On this basis each American wage-earner has the equivalent of more than 230 helpers.

The Pearl Street station in New York made the electrical age a visible, incandescent reality. Energy became free. It could be flashed along copper wires, at first only a mile or two, and finally, as alternating current came into use, over more than two hundred miles. What began as a means of supplying the energy for lamps developed into a means of driving factory wheels and street-cars, washing household linen, churning butter, milking cows automatically and sucking dirt out of carpets. Gradually the United States was enmeshed in copper.

Edison had unwittingly set an example. If one steam-engine could supply all the energy needed to light some hundreds of distant lamps why should it not also drive motors in factories and homes? Why, then, should not the thousands of individualistically operated steam-engines disappear and give place to a few steam-engines and waterwheels in enormous central stations?

About fifteen years ago electrical engineers began to make studies which showed that we were on the threshold of a second Industrial Revolution. Stations can be interconnected so that a few of colossal size can take the place of many. Thus regional power pools can be created which can be tapped wherever land is cheap, railways and waterways are at hand, and labor is reasonably abundant. One of the first experiments in interconnection was made by the Middle Western Utilities Company. Its interconnected system serves nearly 4,000 small communities with a population of 6,000,000, of whom about 2,400,000 are employed gainfully. The power distributed to these 4,000 communities is equivalent to the working capacity of over 11,000,000 able-bodied men working 365 days the year round.

As transmission engineering develops and better insulators are discovered there will be fewer and fewer central stations, possibly as few as five hundred serving the entire United States. Muscle Shoals will be linked with steam stations several hundred miles away. Already the power-dispatcher dominates our lives. When superpower on a nation-wide scale comes, he will be the absolute lord of the electrical energy dispensed to thousands of factories and homes in scores of cities.

"Mankind never before has grasped such a tool," Herbert Hoover has said, "and this because, until electricity came, energy was not distributable." After Watt, workers flocked to the steam-engine; after interconnection or superpower, energy was flashed to the workers wherever they might be. What we

INVENTION AS A SOCIAL MANIFESTATION

behold is a centralization of the production of energy and a decentralization of industry. The cities seem to be very slowly emptying themselves of their industries according to some statistics gathered by the Metropolitan Life Insurance Company and the Middle West Utilities Company. This does not mean that great cities will disappear or that industry will abandon the metropolis altogether. New York and Chicago must feed and clothe themselves, which necessity alone makes it impossible for them to dispense with factories entirely. A few special industries will also remain, partly because they need no great structures, partly because of the market at their doors. But the major industries seem destined to redistribute themselves in small towns, where energy can be drawn from the overhead copper mesh and where land is cheap.

The engineers of great public utility corporations planned this second industrial revolution of which we see the beginning. Hitherto the introduction of a great invention has been marked by much head-shaking and social maladjustment, so that students of civilization have sometimes questioned whether humanity was any better off because of the machine. This time there was deliberate, organized planning, inventing and designing of a highly intensive kind to achieve a definite social and economic end.

MACHINES THAT MAKE MACHINES

A steam-actuated orange-peel bucket at the end of a wire rope drops into a pile of ore. The steel fingers open, bury themselves into the mass, then rise, clutching five tons. What we have is clearly a huge mechanical fist. Viewed with a romantic eye, a gantry crane turns out to be a gigantic system of steel muscles, compared with which the human biceps and triceps seem puny. Even a locomotive suggests the iron horse to which it is so often likened. The thousands of machines that do the work of hauling, lifting, pressing, punching, boring, twisting, and bending are highly specialized, artificial organisms. Industrial processes are carried out, for the most part, by tireless, brainless automata—artificial extensions of the human body controlled by the human brain.

With the introduction of the steam-engine the invention of these substitutes for the human organs was stimulated. There never was a time when men did not invent. But after the steam-engine, invention acquired a new significance, a new impetus, especially in a country like the United States which, for generations after it came into being, suffered from a dearth of labor. Machines were devised to build other machines, and thus a process of creation and reproduction was inaugurated which seems appalling to some sociologists and romantically captivating to others.

The first modern machine that had to be built by another was the steam-engine itself. Before the industrial revolution most mechanism was of wood, and this despite the skill that had been acquired in casting and working iron and other metals. The great demand for steam-engines made it necessary to invent metal-working machines. Had it not been for testy John Wilkinson, Watt would never have been able to build an acceptable engine. No one could bore a cylinder accurately enough. Watt had to stop the space between piston and cylinder with "all the appliances of a hat, *papier-maché*, grease, black-lead powder, a bottle of oil to drain through the hat and lubricate the sides, and an iron weight above all to prevent the piston leaving the paper behind in its stroke." It was Wilkinson, one of whose specialties was cannon-boring, who made it possible to dispense with these precarious aids. He succeeded in boring a fifty-inch cylinder so well that Boulton, Watt's partner, could write in 1776, that "it does not err the thickness of an old shilling in any part." Today tolerances of but one ten-thousandth of an inch are the rule in many automobiles.

By the end of the eighteenth century some notable inventions had been introduced which made it possible for steam-driven machine-tools to build other machines. The greatest of these was the slide-rest. Before it was invented the tool had to be held against the spinning metal. Human skill alone con-

INVENTION AS A SOCIAL MANIFESTATION 33

trolled the depth of the cut. The slide-rest was an iron fist, which, in the form of a carriage, was screwed along the lathe-bed. Thus a tool could be held nervelessly and rigidly against the metal to be pared. It is another evidence of the spirit that animated society of the eighteenth century that the slide-rest was independently invented in England and the United States —in England by Bramah and Maudslay, and in the United States by David Wilkinson.

The United States lacked mechanics after it became a republic, but not mechanical skill or inventive ability. Out of its lack, perhaps, was born Eli Whitney's great invention of interchangeable parts—an invention which is as characteristic of the industrial revolution as the steam-engine itself, yet one which is too often ignored in histories of invention. Mass production is dependent not only on abundant and cheap mechanical energy, but on the ability to produce machine parts which shall be indistinguishable from one another. Wheels, shafts, levers, pins, plates must all be gauged with such precision that they do not vary in important measurements by more than a negligibly minute fraction of an inch, and that they will fit together without filing and scraping. The broad principle of interchangeable parts is found in Gutenberg's movable types, which had to be cast from molds, so that the "a's" were all alike and the other letters of the alphabet as well. Jefferson had written from Paris that the principle of interchangeability had been applied to mechanism in France, in other words to a definite arrangement of mechanical parts. Nothing came of this innovation immediately. Whitney reinvented it to fill a contract for 200,000 muskets to be used in the war of 1812. Never had two muskets been exactly alike. Before the astonished eyes of the Secretary of War and a few American officers Whitney took the parts out of a miscellaneous assortment in a box and assembled ten perfect guns. The day of mass production and standardization had dawned.

Whitney is the direct industrial ancestor of Henry Ford and all other quantity manufacturers of identical machines and articles of merchandise. It was not enough to devise a new system of manufacture. A factory had to be organized to fit the system. Whitney became a pioneer efficiency engineer. For months he was busy with building "machines for rolling, floating, boring, grinding, and polishing." He did simply what any Detroit automobile manufacturer does when a new car is to be produced, and must be damned or praised with Watt as the unwitting creator of the Machine Age and automaticity in manufacturing, and as the destroyer of old crafts and the transposer of old skills to the machine. For decades Europeans spoke and wrote of the "American system of manufacture," meaning Whitney's method of assembling machines from interchangeable parts.

By the time that Whitney applied himself to gun-manufacturing the more important machine tools had been invented. Not only was there the lathe in which the work is revolved against the tool, but the planer, in which the work is fastened to a table and slides against the tool, the boring-mill, and the shaper in which the work stands still while the tool moves back and forth. In 1818 Whitney himself invented the milling-machine in which the tool rotates while the work is fed against it. But the most human of these machines that make machines is the turret lathe which was invented by Stephen Fitch in 1845 and improved in 1854 by Stone and Howe, two ingenious mechanics of Hartford. In the turret lathe one tool after another was presented to the work, the turret being at first turned by hand and later mechanically. Out of the turret lathe came the modern automatic screw-machine which turns out threaded bolts, screws, indeed almost any part of any shape. Add to this the typically American grinding machine introduced by Joseph R. Brown, together with punches, presses, and drills, and the list of machines that make machines is fairly complete.

Given machine tools and the interchangeable-parts system and we have the essentials of American mass production. Without the system and without the tools it has been esti-

INVENTION AS A SOCIAL MANIFESTATION

mated that it would require one million men to build ten thousand Ford cars by hand at a cost of $10,000 each.

Dexter Kimball, Dean of Sibley College, Cornell University, has pointed out that what we behold here is something much more significant than the mere subdivision of labor—the transfer of skill and intelligence to the machine. It might be supposed that the net result would be a general reduction in wages. We find an actual average increase when the machines to be tended are especially large and complicated. It is what we call "unskilled labor" that suffers most—workers engaged in purely repetitive tasks which can be performed unthinkingly. The output of work by the machines that have been made by other machines is enormous. It is thus pictured by Thomas T. Read:

> Although we ordinarily think of China as a country having nearly four times as many people as there are in the United States, the United States has the equivalent of many times the number of effective workers that there are in China. In short, the United States may be thought of as a country in which the work done is equivalent to the work that could be done by ten times as many people as there are in China, or almost forty times as many people as there are in the United States. ... The output of work per person in the various countries of the world is as follows:

China 1	Poland 6	Germany 12
British Indian . 1¼	Holland 7	Belgium 16
Russia 2½	France 8¼	Great Britain .. 18
Italy 2¾	Australia 8½	Canada 20
Japan 3½	Czechoslovakia .. 9½	United States .. 30

> A blacksmith works eight hours a day at the rate of one-tenth horsepower and is paid ten dollars for it. Two pounds of coal costing less than one cent will do the mechanical equivalent of the blacksmith's work. Hence one blacksmith at ten dollars, aided by eighteen pounds of coal, at ten cents, will do as much work at half the cost as ten blacksmiths at two dollars per day each. ... The per capita output of work in this country is so much larger than the output of work in any other country that the consequent divisible wealth per capita is very much greater.

But there is a dark side to this effect of the machine. We are confronted with what economists call "technological unemployment." The machines that have been made by other machines have thrown men out of work, as the following table shows:

DECLINE IN PRODUCTIVE WORKERS (DEPARTMENT OF COMMERCE YEARBOOK, 1930)

	1919	1927	Decline
Agriculture	11,300,000	10,400,000	900,000
Mining	1,050,000	1,050,000	—
Manufacturing	10,686,000	9,868,000	818,000
Transportation	1,913,000	1,737,000	176,000
	24,949,000	23,055,000	1,894,000

Because of their number it is impossible even to list here the machines that have been made by other machines and that have brought us face to face with the problem of technological unemployment. They range from canning apparatus to ore-unloaders, from typewriters to cigarette rollers and packers. A century ago any of these mechanisms would have been hailed as an epoch-making invention, comparable with the greatest ever conceived by the human intellect. Today we accept them as a matter of course. Invention has become a profession, a branch of engineering.

By the side of Whitney in this pageant of the machines that make machines looms the impressive, studious figure of Frederick Winslow Taylor. He, too, was the inventor of a principle. Until he began his work in the Bethlehem Steel Works about 1880, machines were arranged only with an eye to their general relations to one another and were left more or less to tenders, each of whom was a law unto himself. Taylor saw that the mill should itself become a colossal machine in which furnaces, cranes, lathes, grinders, and men were but elements. The output of machines and men had never been scientifically measured. No one knew what was a fair day's work. To determine what output should be, Taylor studied machine tools, and found that even their makers had either nebulous or entirely wrong notions as to the manner in which their products could be most effectively used. With Maunsel White he began

INVENTION AS A SOCIAL MANIFESTATION 37

experimenting on his own account, and in the course of twenty-six years converted 800,000 pounds of steel into chips and shavings.

Out of this work came their invention of high-speed tool steels (1900)—steels which are alloys of chromium and tungsten or vanadium with steel, which proved to be several hundred per cent better than the old so-called self-hardening steels and which could be permitted to become red hot. The effect was electrical. Although the old principles of Maudslay, Whitney, and the rest were still good, machine tools of a faster type could be built, and this about 1906 just when the automobile industry had begun to think of sales running to tens of thousands a year.

Pushing to the extreme his study of machine tools, the machines that make machines, Taylor arrived at what we now call scientific management, or efficiency engineering, a major contribution to engineering and industry, one which has aroused as much ire and resentment as the transfer of skill and intelligence to the machine.

What is known as Taylorism in Europe, and is erroneously regarded both here and abroad as an inhuman speeding up of the worker with the aid of time and motion studies, has been considerably modified by Taylor's disciples and successors. Instead of the old military organization, by which discipline could easily be maintained, we now have Taylor's system of functional management by experts. The standardization of machines and processes inseparable from the application of Whitney's principle of interchangeable parts is pushed farther to include the machine-tenders and their superiors. Before Taylor's principles were published the foreman saw to it that discipline was maintained, that new men were instructed, that the work for the day was laid out, and that operations were supervised in a general way. Today these functions are assigned to trained individual task-masters whose accurate studies and records show precisely whether or not men and machines are producing a standard amount of work in standard time, and indicate what extra rewards are due to men whose efficiency is above the standard. Without scientific management of some kind it would hardly be possible effectively to administer and control the enormous organizations that mine, smelt, and roll iron and manufacture automobiles and the countless machines on which American industrial civilization depends. Taylor integrated machines and men and converted the living muscles and the steel wheels of a factory into a huge mechanism which could be managed by a super-machine-tender—the president of a corporation.

THE REVOLUTION IN AGRICULTURE

Even a mechanized society cannot live by power alone. It must clothe and feed itself. Man had done both for thousands of years. Looms go back to the prehistoric Swiss Lake Dwellers who built their huts on piles, and agriculture as far back as there are human records. Still it remained for the eighteenth and nineteenth centuries to mechanize the spinning and weaving of fibers and the harvesting of wheat and thus to solve in terms of a machine society the age-old problem of keeping body and soul together. Two Americans, Eli Whitney and Cyrus H. McCormick, were in the main responsible for the solution. Both were products of the industrial revolution.

Before Whitney appeared on the scene, spinning was highly developed and a beginning had been made in power weaving. Cotton was the most abundant of spinning fibers. Yet cotton fabrics were relatively expensive because of the cost of picking out the seeds from the bolls by hand. Indeed, fine cotton cloths were as much sought after as fine linen. Thanks to the flying shuttle of Kaye (1733), the spinning-jenny of Hargreaves (1764), the spinning-frame of Arkwright (1769), and the spinning-mule of Crompton (1779), all English inventions, the means for preparing large quantities of cotton for the loom were far in advance of the means for cleaning the bolls. The finest and most abundant spinning fiber on earth was handicapped by its seeds.

Before Eli Whitney invented his gin in 1792 a skillful slave could clean about four pounds of cotton in a day. The total

INVENTION AS A SOCIAL MANIFESTATION

American crop of 1791, the year before Whitney's invention appeared, was only two million pounds. Ten years later 40,000,000 pounds were raised. After that the crop nearly doubled every ten years. In 1811 it was 80,000,000 pounds; in 1821 it was 177,000,000; in 1826 it had risen to 320,000,000. Today it is over 7,000,000,000 pounds. This extraordinary increase in production is directly traceable to the gin. No labor-saving invention had a more profound influence on the formative period of the United States. If it is treated here as part of the record of the last century of invention it is because it was not the least important factor in precipitating the Civil War.

Before the invention of the gin the South was beginning to realize that slave labor, originally introduced to cultivate tobacco, was probably unprofitable. After Whitney demonstrated that cotton could be cleaned by machinery the demand for land and slaves increased enormously. The effect on American society, politics, and history was startling. A planter aristocracy arose which stood out in sharp contrast with the democratic North, which felt bound to uphold the traditions of the Cavaliers from whom it claimed descent, and which acquired political astuteness in maintaining its social and economic position. The effect on the rest of the world was no less remarkable. Cotton became the king of fibers, because it was the cheapest and most abundant of all. Flax and wool gave way to cotton in the manufacture of much cheap clothing. Patched trousers and rags were not so conspicuous as they had been in the sixteenth and seventeenth centuries. Even savages, once stark naked, wrapped themselves in cotton cloth.

This is not the place to trace the history of the Civil War, but it is the place to point out that the political struggle that centered around slavery and the tariff before the Civil War was distinctly the result of the cotton gin's invention. We find the South selling some cotton to New England factories, but far more to Great Britain. There is a clamor for a protective tariff from the North, and fierce opposition from the South, which sees in a high duty only dearer clothing for slaves and higher bagging and baling costs. Law after law is passed to increase the duty. Whether or not states shall be slave states or free states becomes a bitter political question. The very constitutionality of the procedure whereby one section of the country is protected at the expense of another is attacked. The doctrine of nullification gains ground—the doctrine that a state can declare void any statutes of the federal government opposed to its views. The Missouri Compromise is repealed in 1854, the deciding turn in the road which leads straight to the Civil War. Then follows political chaos. The Republican party is born of the cotton-gin. Organized in 1854 against slavery and the "revolting aristocracy of the South," it becomes the strongest party by 1858. It is still more triumphant after the war. Nowhere in the whole history of invention is a clearer case of the consequences that follow the introduction of a machine to be found.

Comparable in its social and economic effect to Whitney's cotton-gin was the crude reaper demonstrated by Cyrus H. McCormick before an astonished gathering of farmers in 1831 in Rockbridge County, Virginia. Here was a horse-drawn machine which cut ripe wheat and threw it on a platform from which it could be raked into piles. It was not the first reaper, by any means. The problem of harvesting grain mechanically had engaged even the Romans. From the crude Gallic header of A. D. 70 described by Pliny to McCormick's time is a stretch of nearly two thousand years. Fully a hundred mechanical reapers had been invented in that time, seventeen of them between 1786 and 1831. Even in the United States McCormick was not alone. He had in Obed Hussey a rival pioneer more brilliant as an inventor than himself, a sailor who knew very little of practical farming. Yet it is to McCormick that we owe the revolution in agriculture—a revolution that was as significant as the revolution brought about by the introduction of Watt's steam-engine. Indeed, it may be said that McCormick's invention supplemented Watt's. The factory drew labor from the farm, and to make good that loss some form of reaping-machine was clearly needed. If McCormick is

INVENTION AS A SOCIAL MANIFESTATION 41

ranked among the great inventors it is not only because of a certain mechanical ingenuity and doggedness, but because of extraordinary business ability. With him begins the mechanization of agriculture.

In business affairs inventors are notoriously incompetent. As manufacturers they have usually been conspicuous failures. McCormick ranks with Richard Arkright, George Stephenson, and Sir Henry Bessemer as an exception. He invented not only a reaper, but the way to manufacture and sell it. He became perhaps the first real American captain of industry, a figure that looms large even beside such eminences as Vanderbilt, Rockefeller, Carnegie, Hill, and Ford. The farmer from Virginia turned out to be an empire-builder in a very real sense. With Chicago as his capital he ended by ruling the entire agricultural world. He began with a blacksmith's shop, and in fifteen years achieved a world-wide reputation as a manufacturer.

McCormick's invention appears in its true light when we consider that it had always been difficult for the world to feed itself. From the days of Abraham to the introduction of the reaper, bread was the product of sweat and toil. It was far easier to cultivate the soil with the primitive plows that had dug their way through the ground for hundreds of years until metal began to be used for the shares than to reap, and this because harvesting is a matter of days, and cultivating of weeks. With only the sickle and reaping-hook a farmer could cut not more than two acres of grain in a day. Then came the slow work of raking, binding, hauling, and threshing. Such was the demand for labor that in 1831 agriculture kept four out of five Americans gainfully employed. McCormick's reaper at one stroke solved both the labor problem and the speed problem. Even the crude invention of 1831 was about ten times as efficient as a man. It reduced the cost of harvesting a third and enabled a farmer to cultivate more acres.

With the Civil War came the supreme test of the reaper. The Commissioner of Agriculture expressed the opinion that the wheat crop of 1862 could not have been harvested without the reapers of the west, each of which released five men for military service. Secretary Stanton even went so far as to say that "the reaper is to the North what slavery is to the South. By taking the place of regiments of young men in Western harvest-fields it releases them to do battle for the Union at the front, and at the same time keeps up the supply of bread for the nation and the nation's armies. Thus, without McCormick's invention I feel the North could not win."

It has been said that the invention of the McCormick reaper moved civilization westward at the rate of thirty miles a year. One hundred years ago the inhabited area of this country was largely confined to a few thinly settled states along the Atlantic seaboard. After the introduction of the reaper immigration began the drift westward to the vast open prairies. Their opening shifted the peak load of agriculture from reaping to plowing. If more crops could be reaped it clearly required a more efficient plow to till more acres.

The best wooden and cast-iron plows were useless in the sticky soil of the prairies. John Lane made a steel plow in 1833 with which it became possible to furrow the new soil. Hard upon him came John Deere and William Parlin with their steel-faced plows, and James Oliver with his chilled-steel plow (1853). Once more the peak load in agriculture shifted—this time to threshing. The flail was as antiquated as the sickle. Hiram Pitts' thresher spelled its doom in 1834.

We have now the reaper, the steel plow, and the thresher. After they were introduced widely we hear less of famine and more and more of overproduction, except in time of war. The half-century that followed the invention of the McCormick reaper was crowded with new labor-saving agricultural machines. The Marsh brothers invented (1858) a way of lifting the cut grain by a canvas conveyor from the reaper platform to a table, so that two men standing on a footboard could bind it into bundles. McCormick bought the patents and adopted the principle of this harvester. Charles Withington (1872) devised a method of binding the sheafs mechanically with wire. McCormick bought his rights, too. Hand labor had now

INVENTION AS A SOCIAL MANIFESTATION 43

been eliminated from cutting and binding. A boy old enough to drive horses could reap and bind a whole crop. Appleby (1878) successfully substituted twine for wire in binding by inventing a knotter. McCormick bought shop rights under the patents.

Out of the horse-drawn reaper of McCormick came the gasoline tractor "combine" that cuts, threshes, and bags grain for the market. Two men with a sixteen-foot combine will cut, thresh, and clean fifty acres a day. With the hand implements of 1831 it would take fifty men to cradle and hand-bind the crop in the same length of time, or one hundred men with sickles to cut it and twenty-five more to rake and bind it. An additional seventy-five men would be needed to thresh it with flails in a day.

Some frequently cited figures of the Department of Agriculture are here very much to the point. We are told that between 1855 and 1894 the amount of human labor required to produce one bushel of corn declined from four hours and thirty-four minutes to forty-one minutes. Similarly the human labor required to produce a bushel of wheat between 1831 and 1894 declined from three hours and three minutes to ten minutes. It is probably wrong to attribute such startling results entirely to labor-saving agricultural inventions. Fertilizers are also necessary in farming, and these have no small influence on the size of crops. Still, if we subdue the picture evoked by the Department of Agriculture's statistics it cannot be denied that the machine has made it possible to farm more acres with less labor than ever before and that with the general introduction of the machine there began, in America at least, that general drift of rural population to industrial centers which is one of the characteristics of this mechanical age. While about 80 per cent of the American population was required in 1831 to produce food for the 20 per cent who dwelt in cities, and for export, the proportion had fallen to 26 per cent by 1920. In the last decade alone some 3,500,000 have flocked to the cities. And still there is an overproduction of wheat.

The mechanization of agriculture followed a different course from that of industry. It was impossible to install a boiler and a steam-engine on every farm or to do away with the horse and the ox for plowing and reaping. Steam-tractors were invented in Napoleon's time—crude, lumbering machines utterly useless on the farm. Even in the United States farm plows and reapers could be hauled by steam with a semblance of profit only on the great "bonanza" farms of the Far West. To conjure up a picture of what is involved in capital investment in the hauling of wood or coal (the only possible fuels until gasoline was widely distributed) is to realize that the horse and the ox were destined to remain the farm's principal motors for some generations. Twenty years after the invention of the reaper agriculture in the United States had a primary power plant of six and one half million horsepower—all animal. Fifty years later this increased to 23,500,000—still largely animal. Now the total is about 65,000,000—mostly mechanical power. The significant fact is that with the introduction of the gasoline engine forty million horsepower were added in the last thirty years against only 17,000,000 between 1850 and 1900.

TRANSPORTATION

If ever a country needed the steamship and the railroad it was the United States. As late as 1784 the freight rate from Philadelphia to Erie, a distance of roughly 400 miles, was equivalent to about $250 a ton. Before that merchandise and produce which could not stand a freight charge of $15 a ton could not be carried overland to a consumer only 150 miles away. Farms and villages were of necessity self-supporting. Communities tended to establish themselves near waterways. Before the Revolution a man traveled usually on horseback or by boat—preferably by boat.

With fine navigable streams it was to be expected that even Colonial Americans thought of adapting the steam-engine to boat propulsion. Europeans had the same notion. But such adaptations are more obvious than easy. Before the end of the

INVENTION AS A SOCIAL MANIFESTATION 45

eighteenth century Rumsey had built steamboats that ran on the Potomac and the Thames, and John Fitch had been notably successful on the Delaware (1790). In 1802 Symington's *Charlotte Dundas* was a conspicuous success. A year later Oliver Evans built a steam-dredge for the city of Philadelphia and ran it under its own power as an automobile on land and as a steamboat on the Schuylkill. Inspired by Fitch, Colonel Stevens of Hoboken in 1804 built a remarkable twin-screw boat with a tubular boiler. In the light of these early signal successes it is preposterous to hail Robert Fulton as an original inventor—a claim which he never put forth himself. Indeed, he knew all that was to be known about the experiments of his predecessors by the time the *Clermont* was ready to make her historic voyage up the Hudson in 1807. A much more daring venture was the *Savannah*, which crossed the Atlantic to England in 1819 and was the forerunner of the floating trans-oceanic palaces of today.

By 1831, when the first American train was hauled by a locomotive (the *De Witt Clinton*) the steamboat was well developed in America. The Hudson, the Ohio, the Mississippi, the Missouri all had their steamers—vessels which would compare favorably in size with river boats of today, even though they lacked the luxuries that we demand. When the McCormick reaper converted the Middle West into a vast granary the Great Lakes became a highway for wheat, which was carried by both steamers and sailboats. In our own time necessity for transporting commodities in bulk—chiefly iron ore from the Mesaba district—has led to the invention and development of such characteristic American craft as the Lake ore-carriers. Twelve thousand tons of ore pour into the hold of such a vessel like water, to be unloaded in a few hours at South Chicago, Ashtabula, or Conneaut by electric machines under the control of a single man. Nowhere in the world is there a parallel for such efficient low-cost handling of raw material.

Even a country so richly endowed with waterways as the United States could not dispense with the railroad. Tracks can be laid anywhere. Spurs can be run to the very factory. A hundred years ago it was easier to lay railway tracks than to build adequate highways.

The development of the American locomotive follows much the same course as that which we have recorded in the case of the steamboat. We find little originative invention, but an ingenious adaptation of European practice to American conditions. The very first locomotive that puffed in the United States was imported from England. It ran on August 9, 1829, at Honesdale, Pennsylvania. The experiment was enough to prove that English locomotive practice could not be followed in America.

English locomotives were not designed to run on the flimsy American tracks of the time; moreover, they burned coke and not wood. Pioneer railroad-builders in England were justified in building expensively and elaborately. Great Britain was densely populated and comparatively small in area. The need of cheap, mechanical transportation, in view of the inadequacy of turnpikes and canals, was passing. So the roads were built straight and level, with fine masonry and iron bridges and tunnels. In the United States freight and passengers were scarce. Thus it happened that in England the country developed the railroads; in the United States the railroads developed the country.

It would serve no useful purpose to trace here the step-by-step adaptation of the locomotive to American conditions. Straight tracks were expensive. Hence curves abounded. Jervis introduced the swiveling truck to round them easily. Tracks were light rails on stringers, the whole apt to sink under a weight. Jervis applied the snowshoe principle; that is, he spread the weight over eight coupled wheels. Harrison's equalizing lever distributed the shock or hammer-blow to which a locomotive was subjected when it bumped over a wavy track. These are not great inventions, merely the ingenious solutions of first-rate mechanics of problems that were not insuperably difficult. By 1845 the eight-wheel American passen-

INVENTION AS A SOCIAL MANIFESTATION 47

ger locomotive was so thoroughly standardized that it remained virtually unchanged for fifty years.

The discovery of gold in California had as much to do as any circumstance with developing the huge, long-haul American freight and passenger locomotive. Before the first transcontinental railroad was built, it took from May until November to cross the continent in a prairie schooner drawn by six or twelve animals, the whole costing from $3,600 to $7,000. The voyage around the Horn was rough and precarious. By 1858, when gold was discovered near Pikes Peak, the need of a railroad across the continent was acute. In the 'sixties as many as 500 westward-bound prairie schooners passed Fort Kearney, Kansas, in a single day. At one time a single firm employed as many as 6,000 wagons and owned over 75,000 oxen. The fast passenger coaches which began to run by 1858 traveled day and night and covered the distance from St. Louis to San Francisco in somewhat more than three weeks. The fare varied from $100 to $600, including meals. The pony express, inaugurated solely to carry mail, took ten days to reach the Pacific coast from Missouri. The first postage rate was five dollars for letters weighing less than half an ounce, and ten dollars for those weighing from one half to an ounce.

Here we have the economic background of American long-haul transportation. Enormous distances and natural resources of a richness that could only be surmised were enough to spur inventors and engineers to develop the railway. This is not the place to describe the building of the first railway across the continent, a purely engineering undertaking, comparable in its way with the building of the Panama Canal and carried on only by fighting off Indians and at a cost of $830,000,000 to the government for 1,800 miles of track stretching from the Mississippi to California. The point is that only powerful locomotives hauling long trains through sparsely populated regions could possibly pay a profit. So it came about that American passenger and freight locomotives are enormous compared with those of Europe. They are the products of able engineers rather than of inventors of the Watt and Edison type.

In 1924 there were 65,358 locomotives in operation. Today there are about 58,000. Indeed, there are fewer locomotives in America today than there were in 1911, but these have 50 per cent more tractive power. The average freight-train now comprises forty-nine cars—eleven more than were hauled in 1921, and much larger cars, besides.

Until the railroad came, life on an American farm or in an American community was not technically different from what it was in Julius Cæsar's time. A city depended on the surrounding country for its food; a farmstead was self-sufficient. The railroad became a tentacle that reached farther and farther out over the continent to drag back foods that were ducal luxuries a century ago. Consider the average American breakfast of today in a town on the Atlantic seaboard. Coffee arrives from Brazil, 5,500 miles from New York; it is shipped hundreds, even thousands, of miles to millions of homes. The fruit from which the orange juice was squeezed was grown in California, 2,500 miles away. The lamb chop traveled 1,800 miles from Wyoming or Montana. The piece of toast that is considered almost a necessity is made from wheat grown in Kansas, 1,500 miles away. The oatmeal was sown and gathered in Iowa, 1,200 miles distant. The butter may have come from Wisconsin, 900 miles. All told a distance of possibly 15,000 miles had to be covered by motor-trucks, steamers, and railway trains to make the typical American breakfast possible. As a result the American bill of fare has changed. There is a greater variety of food. Tropical fruits are no longer luxuries. It is worth noting, too, that over 50 per cent of the commercial fruits and vegetables consumed in the eastern part of the United States are grown west of the Mississippi.

It is impossible to write the early history of the locomotive without writing that of the automobile. The French Captain Cugnot had built three steam-tractors by 1769 to haul field-pieces—all unsuccessful. Richard Trevithick experimented with a steam-coach in 1802. Oliver Evans' steam-dredge, al-

INVENTION AS A SOCIAL MANIFESTATION 49

ready mentioned, found no difficulty, apparently, in running on land. Sir Goldsworthy Gurney ran steam-coaches successfully as early as 1823. By 1833 as many as twenty steam-vehicles were traveling in and around London, and a dozen companies had been formed to operate what were called "road locomotives." Vested interests succeeded in choking these enterprises. The Road Locomotive Act was passed by Parliament, an Act which imposed a prohibitive tax and which stipulated that a man carrying a red flag should precede a road locomotive and warn people off the road. England never regained the chance to become the leading producer of automobiles.

A successful automobile is dependent on two factors—a good engine which burns a cheap fuel, and good roads. The early steam-coaches burned coal or coke under boilers. When the first oil-well was successfully drilled in Pennsylvania by Colonel Drake in 1859, and Doctor Kier discovered in 1850 that it was possible to distill from "coal oil" light constituents which came to be known as kerosene and gasoline, the cheap, convenient fuel at least was at hand. Had the internal-combustion engine not been developed in successive stages by Le Bon (1799), Lénoir (1860), and Otto (1876), there can be little doubt that we would now be rolling over the road in steam-carriages. Indeed, by 1900 there was actually something like quantity production of steam-cars in the United States. An automobile of some kind was inevitable. It was also inevitable that in view of the discovery of enormous deposits of petroleum the fuel of that automobile would be a derivative of petroleum.

The gasoline automobile of today is an evolution. It assumed its present technical form in the 'eighties of the last century at the hands of Daimler and Benz, two German engineers. Its subsequent history in the United States parallels that of the steamboat and the locomotive. In other words, we find little originative invention, but ingenious adaptations to American needs. The country was huge, but its roads were poor. Hence the first automobile-makers, especially Henry Ford, were more concerned with producing cars that would not break down as they bumped over rocks or churned mud than with the niceties of technical construction. The country was rich. To develop the market, quantity production was essential. Here America displayed characteristic inventiveness. Eli Whitney's principles of standardization were pushed to the extreme. Machines were invented to slice off excess metal from a score of cylinder blocks at once, like so much butter; to drill eighty-one holes in a casting at a single operation; to transfer human skill to mechanism and still secure fits so close that errors involving more than one ten-thousandth of an inch were not tolerated. From the Chicago packers Ford learned the principle of what came to be known as "progressive assembling," which means that moving chains and conveyors carried frames to crews of workmen, who fitted the necessary parts. By 1912 it took only fourteen man-hours to assemble a Ford car at a cost of $8.75; two years later improvements in progressive assembling had reduced the time to two man-hours and the cost to $1.25. Such methods could hardly be introduced without adapting the car to them. Whatever differences there may once have been between American and European cars (the differences have almost disappeared) are to be traced to the exigencies of mass production.

In 1896 Barnum and Bailey advertised a "horseless carriage" as an attraction; the first man who tried to drive an automobile through Central Park was arrested for disorderly conduct; Elwood Haynes, a pioneer inventor, was ordered by the police to remove himself and his vehicle from Chicago's streets. Today there are 25,000,000 registered automobiles in the United States—one to every five inhabitants.

The social and economic effect of these millions of cars has been profound. Farmers deliver their produce to cities a hundred miles away. Real-estate values within twenty miles of a railway station have soared. London, New York, Chicago, and Philadelphia have evolved from cities into regions. Suburbs are scarcely distinguishable from the cities into which they merge. Were it not for such signs as "Welcome to Evanston" or "This is Yonkers" we would hardly be aware of the

INVENTION AS A SOCIAL MANIFESTATION

fact that we have passed from one self-governing community to another. Transportation of pupils to distant schools is now a recognized function of educational boards. The growth of consolidated schools is a direct consequence of the motor-bus.

The rise of the automobile has been accompanied by a partial shifting of transportation from the railway track to the highway. In 1921 day-coach passenger revenues were $795,-402,216. Today they are less than $430,000,000. Short-haul railway freight traffic has suffered similarly. It is clear that the railways face an economic crisis and that they must become a link in a national system of transportation of which both engine-driven road vehicles and aircraft form parts. The signs of this transformation are here. Already the railways have begun to feed themselves with traffic brought to them by road and air vehicles.

That neither airship nor airplane is likely to compete with the railroad or the automobile as a carrier of goods became apparent when the first men succeeded in transporting themselves through the air. Speed is an aircraft's great advantage over land vehicles—speed and directness of flight. Speed is a matter of energy. It must be paid for in fuel. The future of air travel seems especially reserved for the transportation of goods and persons under circumstances where speed is more important than cost. This is also true, in a measure, of both land and water transportation. If we must reach Europe in less than five days or Chicago in less than twenty-four hours from New York, we must pay the cost of the extra energy involved.

Speed is of more importance in an airplane than in any other vehicle. Before it can rise an airplane must be in motion. When it is aloft it must continue to move at not less than a certain critical rate if it is not to sink to the ground. Motion—speed—is one secret of an airplane's ability to remain in the air. The other is proper control in a medium which, if we could but see it, would appear much like the Whirlpool Rapids of Niagara—all swirls, billows, and currents that move up and down.

The invention of a machine which would soar with outstretched wings, like a bird of prey, was a far more difficult process than the construction of a locomotive or automobile, once the steam-engine was in existence. Moreover, the principles of aërodynamics had first to be discovered before it was possible not only to build but to fly an airplane. Dozens of mechanics, who never pretended to be scientists, developed the locomotive and automobile. Indeed, they rather prided themselves on being "practical" men.

The "practical" men who wanted to fly naturally turned to birds and tried to imitate their wing-flapping. All of them failed, and some of them were killed. What was needed was a knowledge of the air and how surfaces in motion would behave in it. Knowledge of this kind was slowly accumulated in the early nineteenth century by such pioneers as Sir George Cayley, who made successful glides with motorless wings and showed that stiff surfaces in motion can support a load. His experiments were supplemented by those of Otto Lilienthal, who lost his life after having made hundreds of successful glides; of Samuel Pierpont Langley, who succeeded in flying a small, unmanned steam-driven airplane in 1896; of Octave Chanute, who adopted Lilienthal's gliding methods in the United States.

By the time the Wright Brothers of Dayton, Ohio, were ready to begin their studies the general principles on which a flying-machine must be built were fairly well known. They might have been established sooner had there been an engine sufficiently light and strong. In an airplane ounces must be saved. Unable to procure engines sufficiently light, Langley and Maxim had to build their own. By 1900 the automobile engine had been well developed. It was heavy, as airplane engines now go, but it could be made light enough to drive a machine in the air. Without such a light engine no amount of scientific research on wings and rudders could achieve a practical result. Hence the Wrights were fortunate in experimenting at a time when the problem of generating energy was not nearly so baffling as it was before the automobile appeared.

INVENTION AS A SOCIAL MANIFESTATION 53

A mistake is usually made in regarding the Wright Brothers as inspired tinkers. They deserve to be regarded as experimental physicists of the first rank. So scientific were they in their methods that they tested surfaces in a wind tunnel of their own construction. There was no other road to success. The cut-and-try method leads to nothing in the air. Surfaces must be correctly proportioned and shaped, whether they be wings or propellers.

Nearly all the elements of the first successful Wright mancarrying airplane were old. The same may be said of the telegraph, of the harvester, of nearly every successful invention. Yet the combination in each case turned out to be new. The successful synthesizer must know what to select from past successes and failures. Because they had this ability to select, the Wrights deserve a place beside the great in the annals of invention. Their great contribution was a means of maintaining side-to-side balance. Before their time, gliders, such as Lilienthal, threw their weight about to prevent the machine from being overturned by a gust of wind. Some of them were killed because they were not quick enough. The Wrights saw that it was necessary to shift not the center of gravity, as their predecessors did, but the center of air pressure. In other words, if one wing dipped they warped it to increase the air pressure or resistance beneath it to raise it again. The same result is secured by the *ailerons* which have taken the place of wing-warping devices.

The development of the flying-machine after the Wrights made their first successful flight on the sands of North Carolina in 1903 was the result of intensive scientific and engineering research. Eiffel and Prandtl at Paris and Göttingen proved that it is easier to part the air (or any other medium) with a smooth, correctly designed bulk than to rake it with a multitude of projections. Sharp prows gave way to blunt, curved fuselages. The pilot no longer sat on the lower wing of a biplane and saw the earth swim past between his legs; he sat in a deep cockpit or a completely inclosed cabin. Engines were developed which weighed no more than three pounds to the horsepower.

The World War proved to be a laboratory for the perfection of the airplane. In three years of fighting above the clouds and of dropping bombs more was done for aviation than was possible in ten years of peace. The airplane emerged from the war far swifter, far better able to carry heavy loads, than it was in 1914. After the treaty of peace, the era of commercial passenger-carrying and of the first transatlantic flights began. Today the commercial airplanes of the United States fly nearly three million miles annually and carry about 200,000 passengers.

THE TRANSMISSION AND RECEPTION OF IDEAS

Language is the greatest of human inventions, and next to language comes writing or a method of recording language. It is the function of language to communicate ideas. Whether we speak or write we transmit and receive ideas by a code of signals—sounds in the one case and visible symbols in the other. When mass appeal becomes important, ways are devised to multiply written signals. Block printing is invented. Then comes a tremendous impetus with Gutenberg's movable type, making cheap books and newspapers possible.

A free press is inseparable from a free people. It is therefore natural that the United States should have contributed so notably toward the spread of ideas and knowledge through the printing-press. Before Mergenthaler appeared with his linotype (commercially introduced in 1886) and Lanston with his monotype (completed in 1897) type was always set by hand, despite the efforts of British inventors to substitute machines. Mergenthaler and Lanston gave us machines that cast their own type—the one in whole lines and the other in single letters afterwards assembled into lines. Ninety per cent of all American type is probably now set by the linotype and monotype.

Leadership in printing-press invention swung to the United States between 1830 and 1840 and has remained there ever

INVENTION AS A SOCIAL MANIFESTATION

since. Robert Hoe gave the first impetus with presses that were frankly based on English designs. His son Richard made the family name known all over the world. His "type-revolving" press, invented in 1846, printed 2,000 newspapers in an hour —twice as fast as the best cylinder presses of the day. The Civil War came and with it a thirst for news that could not be too rapidly satisfied. Faster and faster presses were built by Hoe. With William Bullock of Philadelphia came the first machine (1865) to print on a continuous web of paper instead of on single sheets. Thereafter newspapers were printed by the mile rather than by the page. The Hoes improved on Bullock with their rotary presses. Newspapers increased in size from four pages to eight, twelve, sixteen, and more. Rotary presses were combined so that quadruple, sextuple, and octuple, presses appeared. With the last, thirty-two-page papers were printed. Then came double-sextuple and double-octuple machines.

In these presses paper is pulled through. Speed is therefore limited by the tensile strength of the paper. About 1914 Henry Wise Wood applied a new principle. He *carried* the paper through and saw to it that vibrations which tended to tear the paper at high speed were suppressed. His first press, set up in Philadelphia in 1917, printed 120,000 sixteen-page papers in an hour—more than twenty-five times as many daily newspapers as were read in all the United States a hundred years ago.

It would be rash to say that the art of printing has gone as far as it can go. On the other hand, it is difficult to imagine the printed word exerting any greater influence than it does today. What we have in our books and newspapers is something that still harks back to the Middle Ages—to a group of monks writing at the dictation of a master scrivener in order to reproduce a missal over and over again, and to Gutenberg who speeds up this process mechanically. When Volta invented the primary electric cell, Faraday discovered induction, and Gramme invented the first dynamo, a new era of communication dawned—an era in which electrical directness of appeal became the goal.

We commonly regard Samuel F. B. Morse as the inventor of the telegraph. Certainly the instrument which he patented in 1837 and successfully tested in 1844 had the essentials of the telegraphs we know. But Cooke and Wheatstone in England and Steinheil in Germany were also successful telegraphers. Morse, like so many inventors, was a marvelous, indefatigable organizer of available information rather than a great designer. Without Joseph Henry's researches he would have been helpless. His partner, Vail, invented most of the mechanical improvements. The code originally was an adaptation of that used by navies for semaphore signaling, although in its final form it was new. The application of telegraphy to transoceanic submarine cabling (1857-1866) through the energy of Cyrus Field was an engineering feat of the first magnitude. Since only feeble currents could be transmitted, a receiver far more delicate than any ever invented was required. Sir William Thomson (afterwards Lord Kelvin) devised it in the siphon recorder.

When J. B. Stearns showed how two, and Thomas A. Edison how four, messages could be sent over a single wire, telegraphy became an important social force. Even these systems made it necessary to employ skillful human senders and receivers. A method which would make it as easy to telegraph as to play an automatic piano by compressed air was wanted. It came early (1855) from David Hughes of Kentucky. The chief difficulty was to synchronize sending and receiving apparatus. Nearly all high-speed printing telegraphs owe their merit to the success with which they have overcome this difficulty. So we now have the Baudot, Rowland, Buckingham, Murray, and about a dozen other systems.

There is no essential difference between radio and wire telegraphy. Historically, radio goes back only to 1896; theoretically it should have been invented in the middle of the last century. The reason for the delay is to be found in the fact that inventors did not realize the nature of the forces with which they were dealing. It was not until Clerk Maxwell mathematically studied the transmission of radiation by the ether

INVENTION AS A SOCIAL MANIFESTATION

that radio became possible. He proved that light is an electromagnetic wave motion and indicated that there must be invisible waves of light. It remained for Hertz, the German physicist, to invent the necessary wave generator and the necessary eye to "see" the waves sent forth (1887). Still, radio telegraphy was not yet born. Not one of the great physicists who were fired by Hertz' work realized that they had in their grasp a new means of communication. It remained for the twenty-two-year-old Marconi to improve on Hertz' apparatus and send the first telegraph signals through space (1896).

The telephone is likewise theoretically based on the same principle as the telegraph. Again we deal with signals—what engineers call modulation and demodulation. But Alexander Graham Bell (1876) and his immediate successors were aware of it only vaguely, if at all. Now it is so perfectly realized that a message may be transmitted partly by wire and partly by the ether, as in transoceanic telephony. In telephoning we convert sound vibrations into electric vibrations and the electric vibrations into sound vibrations again. What we hear, then, is an electrical duplicate of the voice at the transmitter. Whether these vibrations are transmitted by a wire or by the ether of space is immaterial. Like Marconi's wireless telegraph, the radio telephone should have been invented a generation sooner than it was. Pioneers in radio telephoning were Valdemar Poulsen (1903) and Reginald Fessenden (1903). Organized research on the part of the great electrical manufacturing companies here and abroad have brought radio telephoning to its present pitch of perfection.

It struck inventors soon enough that if it were possible for a Morse or a high-speed printing telegraph to send signals corresponding to the letters P-R-E-S-I-D-E-N-T over a wire, it might be possible to send pictures too, by splitting them into pieces electrically and reassembling the pieces at the receiver. The early instruments were simple modifications of telegraphs. It was merely necessary to fill the hollows of an engraving on metal with a non-conductor, to wrap the metal around a cylinder, and to let a stylus in a telegraph circuit travel over the cylinder like the needle over an old-fashioned phonograph cylinder. At the receiving end was another stylus which was either an ink-filled pen traveling over a paper wrapped around a drum or a point moistened with a chemical solution and in contact with a suitable piece of paper. As the transmitting pen completed the circuit, which it did whenever it touched metal, the stylus at the receiver traced a line of corresponding length on the paper. So a picture in parallel lines was traced by the sender and receiver. Here we have a form of what is called facsimile telegraphy. With a beam of light to take the place of the stylus at either end and photo-electric cells to convert, as it were, light into electricity or electricity into light, it is now possible to transmit and receive photographs over a wire or through space. The beam scans the original line by line. At the receiving end a similar beam travels line by line over sensitized paper. Again we have a series of parallel lines, but lines so closely drawn that they cannot be picked apart by the eye. The photo-electric cell is the signal device. It dictates how long the beam at the receiving end shall play upon the sensitized paper; also with what intensity. Without the photo-electric cell the process would be an impossibility. Such a cell has the property of fluctuating in conductivity with the amount of light that illuminates it at any given moment.

Suppose we scan not a photograph of a face line by line, but the face itself with a beam of light. Here we have essentially the principle of television. In the motion-picture theater we see from twelve to sixteen pictures a second. So rapid is the projection that the eye is presented with a new picture before the impression of the old has quite disappeared. We have the phenomenon of retinal persistence. Similarly, the scanning process must occur so rapidly in television that we are unaware of the line-by-line dissection at the transmitting station and the line-by-line reassembling at the receiving station. The problem is far more difficult than that of the motion-picture. To present sixteen pictures to the eye in a second is one thing, but to present sixteen images in a second each of which is in

INVENTION AS A SOCIAL MANIFESTATION 59

turn composed of lines is another. A crude television apparatus must deal with at least 25,000 changes of light and shade in a second. Each of these changes constitutes a signal as truly as if it were a dot or a dash in the Morse code or a spoken word. We have only to think of a telegraph which is capable of transmitting 25,000 dots and dashes in a second to realize the nature of the task that confronted the pioneers in television. The resultant images were about as coarse as halftones printed on newspaper stock. A really acceptable television image should consist of about 250,000 changes of light and shade in a second. To whom belongs the credit for having transmitted and received the first image by wire it is difficult to establish. C. Francis Jenkins in this country, John L. Baird in Scotland, and Denis Mihály in Hungary were working simultaneously at the problem and all of them had achieved some measure of success by 1925.

Once the interrelations of motion-picture principles and television are recognized, it is easy to understand why television was so late in coming. But it is not so easy to understand why motion-picture photography came as late as it did. The reason is probably the need of the photographic film which was invented by the Reverend Hannibal Goodwin in 1887. Coleman Sellers of Philadelphia had patented his "kinematoscope" in 1861, the first modern motion-picture machine. Before that, various toys had been invented to display a score of photographs of a body in motion so rapidly that the impression of continuous motion was conveyed by the phenomenon of retinal persistence. So many inventors brought the modern motion-picture camera and projector to perfection that it would be unfair to give any one of them the sole credit. The Lumières in France and Edison in this country were undoubtedly important factors. Edison's kinetophone seems to have been one of the earliest attempts to synchronize sound and motion-picture photography. In this case the sound was recorded on a phonograph. Today the sound is recorded on the film itself, so that by means of photo-electric cells the very light that projects the picture also translates bands of light and shade into spoken words.

Let us now try to evaluate the progress we have made from printing through telegraphy, telephony, telephotography, motion-picture photography, and television. We cannot but be struck with the importance of the new means of communicating ideas. Motion-pictures are seen not only by 120,000,000 Americans every week, but by hundreds of thousands of Europeans and Asiatics. In 1926 there were 27,783,963 telephone stations in the world sending messages over nearly 90,000,000 miles of wire. In 1931 there were 1,255 licensed broadcasting stations, of which 605 were located in the United States. Fully 15,000,000 receiving sets are in use by the American radio audience alone.

When we study statistics such as these it is evident that the new ways of communicating ideas must wear down national individualities and standardize the whole civilized world. Electricity has shaken the very foundations of economics. It would probably be impossible to hold together so far-flung an organization as the United States Steel Corporation or the great German chemical trust without telephone, cable, and telegraph.

There are disadvantages in the immediacy with which an idea is now communicated and made known to millions simultaneously. The late James Bryce was even of the opinion that the World War might have been averted had there been no swift means of communication at the disposal of mankind. Statesmen might have had more time to compose their differences. As it was, the whole world was electrically ignited at once.

We used to talk rather fantastically about "annihilating space" when the telegraph and telephone were introduced. Space annihilation indeed! Not until radio broadcasting came, about 1921, did we realize what the term meant. Millions now hear the living voice of the President. When the Philharmonic Orchestra plays in New York lonely ranchmen on the prairies

INVENTION AS A SOCIAL MANIFESTATION

and sailors at sea hear. Dr. Foster of Newark felt constrained to open a radio sermon with this apology:

> I cannot address you as citizens of Newark because my voice is being heard beyond the limits of the city. I cannot address you as fellow Americans because my voice is heard perhaps in Cuba, in Canada, and in Central America. I cannot address you as brothers of my faith, because only a very insignificant part of the great number who are listening are of my own faith. And, therefore, I must address you as fellow human beings.

Here we catch a glimpse of the social destiny of electrical mass appeal. The world seems to shrivel up until it becomes a ball that can be held in the hollow of the hand.

Such considerations as these lead Professor Shotwell to remark in his Introduction to the *History of History*:

> If the test for the distinction between pre-history and history is the use of writing, we may be at another boundary-mark today. Writing, after all, is but a poor makeshift. When one compares the best of writings with what they attempt to record, one sees that this instrument of ours for the reproduction of reality is almost paleolithic in its crudity. It loses even the color and tone of living speech, as speech, in turn, reproduces but part of the psychic and physical complex with which it deals. We can at best sort out a few facts from the moving mass of events and dress them up in the imperfections of our rhetoric, to survive in the fading simulacra . . . of the world. Some day the media in which we work today to preserve the past will be seen in all their inaccuracy and crudity when new implements for mirroring thought, expression, and movement will have been acquired.

From the sketch that has been presented of the evolution of our methods of communication within the last century it is clear that the tendency is toward creating the illusion of reality, of hearing, seeing, and feeling of experiencing events. In telephoning we deal with one sense. We transmit our ears through space. With the motion-picture film and television we have solved the problem of transmitting our eyes. Conceivably we may deal likewise with other senses. We are literally disembodying ourselves electrically and sending our personalities over the earth. What the social effect will be is enough to stagger the imagination.

THE END OF "HEROIC" INVENTIONS

No historical sketch of invention would be complete without contrasting the methods of invention of a century ago with those of today. As we come down to our age we notice how hard it is to attribute some marvelous machine or process to a single man. We have insisted in the very beginning of this sketch that inventions are not attributable to single men—that they are composites. Even as early as 1857 Hodge, a civil engineer, testified before the House of Lords that "the present spinning machinery . . . is supposed to be a compound of about 800 inventions." This is the usual story. In the course of time inventions are consolidated.

Invention always implies research. The "heroic" theory of invention, the notion that an idea flashes from a brain and gives the world a sudden, fresh impulse, must be dismissed. Watt's engine did not leap from his brain to the drafting-board. New machines had to be devised to bore cylinders with the requisite accuracy; studies had to be made of heat losses; model after model had to be built in order to proportion moving parts correctly and to determine the precise point at which steam might be cut off and used expansively. So it was with the linotype, the typewriter, the phonograph, the carbon-filament incandescent lamp, with every one of the machines and processes that seemed to us so perfect when they took their places in the factory or the home. For the most part it was crude, rule-of-thumb, cut-and-try experimenting, but nevertheless it deserved to be dignified by the name research. It was research akin to the tedious, experimental quest of the apt word that distinguished the compositions of the lyric poet, and to the preliminary sketching of an artist before he ventured to paint a mural decoration. Moreover, it was so costly and so uncertain as to its outcome, when empirically conducted, that a penny-counting manufacturer might well regard it as too hazardous for financial support.

A dazzling example of what organized research could ac-

INVENTION AS A SOCIAL MANIFESTATION 63

complish was that presented by the German chemical industry. Mauve, the first aniline dye, had been discovered by Perkin, an English chemist, in 1856, but the coal-tar chemical industry was German before 1900. Hundreds of coal-tar dyes, drugs, explosives, photographic developers, perfumes, and flavors flowed from Germany—all products of organized industrial research. German chemical companies, soon merged into a powerful trust, exacted tribute from every textile dyer, hospital, physician, invalid, food manufacturer, baker, confectioner, and printer in the world. If organized research could succeed in devising new chemical products it could also succeed in developing machines, telegraphs, illuminants, automobiles, and other devices.

At present there are more than a thousand American industrial laboratories in which some 30,000 professional scientists, engineers, and inventors are kept at work at a cost of nearly half a million dollars a day. So all-pervading is the spirit of organized research and invention that the small company, which is not rich enough to maintain a laboratory of its own, is almost sure to belong to a trade association that conducts scientific investigations for the benefit of its members. About ninety such associations support coöperative research and experimental laboratories at a total annual cost of $25,000,000.

Although such distinguished names as Coolidge and Langmuir are associated with the modern incandescent lamp and Carty and Scribner with the telephone, group invention and research are apt to be anonymous. How can it be otherwise when a single organization, such as the Bell Telephone laboratories, employs at least a thousand trained scientists and engineers? Group inventions reflect an institution rather than a personage. They have character, but it is the character of a railroad, a community, a disciplined society. There is an utter absence of the old picturesque and pathetic struggle before success is achieved against great odds. Morse, grateful for a loan of ten dollars, which warded off starvation, Howe glad to earn his passage across the ocean by cooking for steerage passengers, Goodyear cast into prison for debt—such figures will be increasingly rare. Group inventors are well paid; they eat three meals a day; they go about their work of creating new mechanism as methodically as if they were clerks; they belong to a professional class which has nothing to do with such unstandardized mortals as cranks and other eccentrics; they live and die like their fellows in any prosperous community. The Samuel Smiles of the future will find it difficult to write appealing human biographies of the men who worked together in laboratory crews.

In these laboratories of industry every possible phase of a subject is considered. Thus we find that the Bell Telephone laboratories study acoustics and the relation of the ear to sound. In the General Electric laboratories not only are lamps improved, but chemical reactions in exhausted vessels are studied. The chemists and physicists of the Eastman Kodak laboratories range over the whole field of organic and inorganic chemistry and are probably the best informed men in the world on the chemical effect of light. Similarly, in the Du Pont laboratories the subject of cellulose is mastered, and with it paints, varnishes, artificial silk, and hundreds of substitutes for bone, ivory, wood, and leather. The hired inventors and research engineers have at their command resources of which the outsider is hardly cognizant—splendidly equipped libraries, experimental apparatus, patent lawyers to guide them, money, and time. Relations are established with foreign companies which also engage in systematic invention and research, and arrangements are made to exchange patent rights. The world is thus enmeshed in the dragnet of a well-managed, far-seeing research organization; nothing of the remotest value in technical literature, nothing that has been discovered in foreign laboratories can escape.

Compelled to compete with organized research as it is conducted in the well-organized industrial research laboratory, the outside "heroic" inventor who worked picturesquely and alone in a garret is disappearing. Edison may prove to have been the last and greatest of the type. If the heroic inventor or discoverer survives at all he is sure to be a trained scientist rather than an empiricist who, like Goodyear, patiently and inspirationally performs ten thousand experiments before at last he achieves the vulcanization of rubber.

CHAPTER THREE

INDUSTRY

In the early stages of its development the prime objective of machine industry was conceived to be the enrichment of the capitalists. Under that conception of things, factories were badly built, wretchedly ventilated, if at all, and crowded with revolving wheels and rolling belts unsafeguarded. Around these factories spread the dark and hideous cities of modern times. Men, women, and children were employed long hours and received, as a rule, a mere "subsistence wage" based upon a minimum standard of life. But in more recent times machine economy has been undergoing radical changes. For the single goal of private greed is being substituted a social objective—mass production for the people under favorable conditions of labor, a wide distribution of wealth through increased wages, the general diffusion of the benefits of civilization with a view to its stability. In the following pages the outstanding phases of this transformation, still in its infancy, no doubt, are summarily described by their promoters.

Industry

By Henry Ford *in collaboration with* Samuel Crowther

TRACING the growth of industry through a century is not primarily an inquiry into the development of tools, machines, sources and methods of power, products, and methods of production. For all these things are only the facilities of industry and not industry itself. They are, without management, only more or less interesting examples of ingenuity. And management itself is only a trivial affair unless it has definite social objectives.

THE SOCIAL OBJECTIVE OF INDUSTRY

The great progress of the century has been in the enlarging of the conceptions of management to a point where it is now generally accepted in this country—although the principles are by no means always followed—that industry does not exist merely to make things to be sold, but has to do with both the creation of wealth and its distribution.

That is to say, industry now exists—in theory and to some extent in practice—as a large public service and not as a private perquisite. The older forms of industry have remained, but the objectives are new. And these new objectives have taught us to develop and to use in a big way the inventions that from time to time have come forward. To me, all the mechanical developments of the century are not very important as compared with the knowledge we are gaining of how to use everything in the public interest while preserving the full advantages of individual initiative. For otherwise the facilities of industry might be only a curse instead of a blessing.

THE TRUE NATURE OF THE MACHINE PROCESS

We have during this century definitely moved away from hand craftsmanship to a machine craftsmanship, and this has caused a confusion of thought because of the conception that improvements in machinery are only labor-saving—that is, man-displacing. There is no point in saving labor unless through the saving at one point we provide at some other point for an increased use of labor through an increase in the volume of wealth. The proper and profitable use of machinery for production demands a very different plan of industrial organization from that which held a century ago in the days when machines were only aids to man power. The phrase *labor-saving* has held over, but now machinery is *labor-serving*, and whether a machine be regarded as labor-saving or as labor-serving depends on the point of view of the management.

There is no conflict between providing goods and providing work, and there is no conflict between low costs and high wages. Only through the machine, however, can these conflicts be avoided. A man working by hand cannot ordinarily produce enough wealth to enable him to live on other than a very low scale, and what he produces is necessarily so expensive as to be beyond the buying power of all but a few people. The simple hand tools added enormously to a man's power of production, but the further addition to productive capacity had to come through the use of manufactured power. This must be used through the medium of machines and, as we learned to make better and more comprehensive machines, we also learned that we were doing far more than merely extending the power of the hand. We found that, with power, we could do things which could not otherwise be done. Most of our important machines today do not save labor except in the larger sense—they either do things which could not be done by hand or do things which would not be done at all had not the machines come in to make the product cheap enough to

INDUSTRY

meet a universal need. The automobile is an instance of a commodity that has been made possible by the machine, and it is much more than the extension of the power of man. It is a thing of itself. A man can go faster on roller skates than he can on foot, and on a bicycle he can make his strength still more efficient, but the personal strength of the driver of an automobile is not of moment and the automobile is not a labor-saving device. Up to a point, a drop-forging machine can be looked upon as performing operations within the scope of a blacksmith, but very quickly it reaches areas where hand work cannot penetrate, and an upsetting machine, which presses metal into a shape, is entirely beyond anything which a mobilization of human beings could effect.

It is not only fantastic, but most misleading, to translate horsepower into terms of human energy. A factory which uses thirty or forty million horsepower a month is not saving labor—it is not substituting for the services of some millions of people. It is doing what those millions could not do. It is not choosing between men and machines. It is doing work that could not be done by men.

This has forced on us a new theory of industry, and this theory constitutes the really great contribution of America during the century. The older economists and industrialists did not understand the machine. They thought of it only as labor-saving. They could not comprehend that it had to create not only products, but also consumers for those products. They could see in the machine only a ruthless means of making the poor poorer and the rich richer. They could not see that a large factory—which is only a large machine—could not exist if the consumption of its production had to depend on the rich. The great consuming power comes from the wage-earners and it is the wage payments which directly or indirectly must finance the purchase of the products of industry. If power and machinery be used solely to decrease costs to their proprietor—the proprietor paying low wages to the workers and charging high prices to the consumers—then very quickly the markets will become stagnated and the machines will not only be an evil, but will be idle and of no avail. But if the aim is ever to increase wages, while at the same time decreasing the cost of the wages per unit of production, so that the higher wages give lower-cost goods and the lower costs are passed on in the price of the product to the consumer, inevitably the markets must broaden and the standards of living rise.

THE CHIEF CHARACTERISTIC OF AMERICAN INDUSTRIAL PIONEERING

It is just this that American industry has been accomplishing. It has steadily used power and machinery to lower prices and raise wages and has so immeasurably broadened its markets or, to put it another way, raised the standards of living, that, year by year, vast categories of goods and services which were once in the luxury class have come over into the necessity class.

The progression has not been easy. The land over which industry is traveling has never been charted and it now and again gets mired, but steadily through any considerable period, notwithstanding the accidents, the production of industry has been consumed at a rate greater than the growth of the population. We are beginning to understand that industry, under enlightened management, is able to achieve a larger and more satisfying distribution of wealth than was thought possible or even desirable a century ago.

How, then, has all of this come about? Our country has large natural resources, but the explanation will not be found in these resources. For other countries, too, have large resources—the Russian stores of wealth are probably greater than ours. We have a large home market, but this home market has been created. It was not standing at attention waiting for us. China and India have more people than we have. We have splendid machinery—but there is plenty of first-class machinery in other countries. The explanation will

INDUSTRY

rather be found in our effort to pioneer through industry as our forefathers pioneered through the great country. We grew into industry as a necessity and with few traditions. The pressure of necessity never permitted the industrial mind to close and the opportunity never arose to create industrial classes with a sharp cleavage of living standards between the employer and the employed. The wage-earner of one day has been the wage-payer of the next. And thus through our whole industrial body has moved a continuous life which has kept the minds of men open to receive ideas.

The pioneering spirit—the spirit of intense individualism—which has distinguished every step of America's progress was probably due in part to the nature of the country and in part to the character of the early settlers. This spirit carried over into industry and it was a child of necessity. The necessity was born of law; although, as usual, when law steps into the field of economics, the results were exactly opposite to those which had been contemplated. Great Britain, with the intention of preventing the Colonies from gaining an economic independence, strictly prohibited the entry into this country of machinery or the models or specifications of machinery. This prohibition extended to the setting up of many kinds of manufacturing, particularly the making of iron. The laws threw our forefathers on their own resources. Almost every farm and village had its own contrivances and devices, and back to them may be traced many of our best modern industrial practices. For many years I have been collecting examples of ingenuity in the home and village industries of the early days, and some of them are very remarkable. "Yankee ingenuity" is not just a phrase—it is a fact. Developing out of this ingenuity, forced by necessity, came a remarkable line of American inventors. Other countries have produced research students and scientists, but we have produced a unique list of inventors—Franklin, Whitney, Morse, Fulton, Bell, Goodyear, McCormick, Howe, Westinghouse, Edison, and many other men whose contributions have been of such importance that simply as individuals they must be ranked among our great national resources.

DEVELOPMENT OF MECHANICAL EQUIPMENT

There are two great divisions of industry. The first is the division of mechanical equipment—that is, the power and machinery. The second is the division which coördinates the mechanical equipment to produce goods that will benefit the public—both in use and in the producing.

During the early days and in some measure until after the Great War, our industrial leaders were handicapped by financial dependence, for when we acquired political independence we acquired only the right to economic independence and not the thing itself. We were dependent on the investors overseas for money. Slowly we forged a financial independence, and so today the industrial structure is our own as well as of our own making. Our industry today depends upon the manufacture of power.

Cheap electricity is an essential of modern American industry. Without electrical power we could not have modern factory organization. The direct use of either water or steam power is very limited. Take steam power. No large modern mass-producing establishment could be operated by steam power, because no single steam plant could be constructed large enough—too much power would be consumed in the transmission. If steam power is relayed by shafts and belting for any distance, too great a proportion of the power is used up before it reaches the machine. And thus industry had to await more mobile power before it could enter into the organizing of its machinery for efficient production.

It is in this field that Thomas A. Edison performed one of his many great services to industry. Before his time the dynamo was hardly more than ornamental—just an added cost, for it was only 50 per cent efficient. Edison produced a dynamo 90 per cent efficient and thereby inaugurated the Electrical Age, which is the basis of the present industrial era. He took mechanical power, no matter what its origin, and turned it into electrical power; he carried the electrical power with comparatively little loss to where it was wanted and turned it back into mechanical power.

Here was what industry had always lacked—mobile, economical transmission. Edison provided power at the end of a wire and there was no longer any need to establish a steam plant wherever a machine was set up. This gave controllable power connection to each individual machine in the shop and soon each machine had its own motor. The motor-in-head machine is one of the achievements of the century. Take this single element away from modern industry and it would cease to exist in its present form. There could be little of the efficiency, cleanliness, precision, and safety which mark it today. The great change has been that today the machines may be organized around the work, whereas before the work had to be organized around the machines.

The elementary tools of industry—the machines which are used to make machines that produce goods for consumers— were in the early days of American industry brought from abroad, chiefly from England, which led the world in metal-working tools. But the labor-saving machinery was invented by Americans to make up for a lack of skilled labor. With the aid of electricity and the coming of high-speed tool steel, the principles of these tools have been developed in hundreds of directions, so that today, as later will be described, a modern factory is not just a collection of machinery, but is in itself a master tool for the performing of certain operations. All of our machinery has been gradually refined to meet the necessity of precision manufacturing and the necessities of continuous manufacturing. Enormous progress has come in the making of stronger and lighter steels and there is now a group of alloys and special metals, most of which are rustless and which will operate still further to change the face of industry. Electricity has made possible another long line of developments through welding, which is now making possible operations that were heretofore impossible and is supplanting a whole field where casting had formerly to be depended upon.

MASS PRODUCTION FOR THE MASSES

The automobile offers a convenient and adequate example of all this. The automobile, whether it is considered as a mechanical product, a method of transportation or a social force, is also an outstanding example of converting a high-priced luxury into a low-priced necessity through the peculiarly American method of coördinating machinery which has most inaccurately been termed "mass production."

The methods of this production and the consequences flowing out of them offer the clearest illustration of industrial developments of the century. Mass production is often confused with quantity production, and it is taken for granted that it is synonymous with machine production. It is true that in mass production great quantities of goods have to be produced, else the methods would not justify themselves, but one can have quantity production without the real essentials of mass production, and of course one can have machine production without either the quantity or the mass elements.

Mass production may be defined as the concentrating upon a single standardized commodity of all available and pertinent methods of manufacturing, so that the factory in a sense becomes a single purpose tool. Mass production has not needed to call to its service any very new machine ideas, but it has used old machine principles in a new way, and more particularly it has combined these machines in such sequences and has so correlated designs and making as to give a wholly new manufacturing technique which has very little in common with the old technique.

The methods of mass production are not of universal application. They are restricted to articles for which a mass consumption can be created. Obviously there cannot be a mass consumption for one hundred thousand kilowatt turbo-generators, but undoubtedly the principles can be applied and

INDUSTRY

are very largely being applied to a very wide range of the necessities of life and in addition are constantly making necessities out of articles which once were luxuries. This process is so new that the limits of its application are not yet discernible. The experiences of the Ford Motor Company on this point are relevant, for it is generally conceded that this company has been the pioneer in developing the methods. The company has found that mass production must precede mass consumption, for not otherwise can the use and price objectives be achieved.

Mass consumption depends both on utility and on price. The production creates the demand instead of the demand creating the production. The demand ever widens. If production be increased 500 per cent, costs may be cut at least 50 per cent, and if the decrease in cost be used both to raise wages and to lower prices, the demand for the article will be increased tenfold. Mass production begins with the design. The design may be simple or complex, according to the state of the art of manufacturing, but in any event it must be one to which machine processes can be made applicable. At the same time its appeal must be so universal that at a price it will be in large demand. It is unlikely that at the time of making the design any actual demand will exist; so the design has to fill a need rather than meet a demand. No article can go into mass consumption unless it be finely designed and finely made. A continuing mass consumption cannot be created simply on price.

THE SCIENCE OF INDUSTRIAL MANAGEMENT

In the manufacturing of the article, three fundamental principles are applied. These are:

(1) Planned, orderly progression of the commodity through the shop.

(2) Delivering the work to the workman instead of leaving it to his initiative to find the work.

(3) Analysis of operations into their constituent parts.

To plan the progress of material from the initial manufacturing operation until it emerges as a finished product involves the delivery of material, tools, and parts at various points along the line, and to do this successfully with a progressing piece of work means a careful breaking up of the work into its most minute divisions.

This system is practiced not only on the final assembly line, but throughout all the various arts and trades involved in the completed product. The final automobile assembly line shows hundreds of parts being quickly put together into a going vehicle, but flowing into that are other assembly lines on which each of the hundreds of parts has been fashioned. It may be far down the final assembly line that the springs, for example, appear, and they may seem to be a negligible part of the whole operation. Twenty years ago the same artisan would cut, harden, bend, and build a spring. Today the making of one leaf of a spring is an operation of apparent complexity, yet it is really the ultimate reduction to simplicity of operation. Trace the course of a spring leaf.

(1) A strip of steel prepared by the steel mill is placed in a punch press for cutting and piercing. The workman puts the strip into press until it hits a stop, then trips the press. The cut-off and pierced piece falls on a belt conveyor which runs along the loading end of a series of heat-treating ovens.

(2) A second workman takes the piece from the conveyor and places it on a conveyor which passes through a furnace in which temperature is automatically controlled.

(3) At the unloading end of the furnace the heated piece is lifted with tongs by a third operator and placed in a bending machine which gives the leaf its proper curve and plunges it in oil, the temperature of which is maintained at a definite degree by apparatus beyond the operator's control.

(4) As the leaf emerges from the oil bath, the operator takes it out and sets it aside to air-cool.

(5) The leaf is drawn by a fourth operator through molten nitrate kept at a regulated temperature.

(6) A fifth workman inspects it.

(7) A workman removes the leaf from the conveyor which carries it from the molten nitrate, and inserts a bolt.

(8) A workman puts a nut on the bolt and tightens it.

(9) A workman puts on right and left hand clips and grinds off burrs.

(10) A workman inspects it.

(11) He hangs the spring on a conveyor.

(12) The spring passes to a workman who sprays it with paint, and a conveyor carries the spring above the ovens where it was originally heated to dry the paint.

(13) The conveyor continues the spring to a loading-dock, where a workman removes it and loads it into a car for shipment to branches.

The spring leaf is given not as an illustration of a fixed practice, but because it presents the minute breaking up of operations and the synchronizing of work and material. The breaking down of any job into its elemental operations permits the use of single-purpose machinery and also minimizes the human skill required. If the work be carried forward on a moving platform or conveyor at a proper height and pace, the utmost in man production can be had without any of the objectionable features of speeding up—too fast a pace quickly results in spoiled work. For both humanitarian and business reasons we do not believe in hard work—in a man doing a heavy job which a machine can do as well or better. For machines do not tire and men do. That took us a step forward.

We discovered that machines could be devised to take the place of many hand operations or operations where the machine had wholly to be controlled by hand. This quickly led to machines that did more than one operation, as, for instance, boring a number of holes at once, or performing several operations on a part at the same time. The minute subdivision of labor was too frequently based on man power and neglected the possibilities of machine power. Hence the movement toward machines that would do many operations at the same time, and would group operations instead of separating them as before.

THE CRAFTSMAN AND ACCURACY

The actual development to this point has run counter to tradition and it required some years to train men to view themselves as general mechanics and artisans capable of managing almost any kind of machine. There always will be places for expert hand craftsmen. But they will not be used in production, for there is no use in wasting the time of an artist doing tasks which a machine can do quite as well and usually better. The spirit of craftsmanship may be in everyone, but the ability to become a real craftsman is reserved for the very few. This was equally true in the days when handicrafts had no competition from machines: the number of real craftsmen developed from the multitude of hand workers was really very small. The average man will learn just so much and no more about any job that he is put at, and there is no point in pretending that he is a mentally hungry human being with a gnawing appetite for knowledge or an artist rudely suppressed by a machine. It is impossible to suppress strong natural gifts or ability.

Large-scale production is utterly dependent on the ability to get accurate work. Contrary to a general notion, speed is impossible without accuracy; mass production cannot be had in roughly finished or inaccurate work. Mistakes cannot pass through a mass-production operation. Machinery has therefore been brought to a point where it will turn out accurate work. That fact controls the whole technique of mass production. It requires an accuracy unknown to hand production. For instance, automobile manufacturers today turn out as a matter of course an engine which is more precisely made than the finest watch. This is literally true not only for measurements, but also for balance. Very few watches are made to a thousandth of an inch, but that accuracy is quite commonplace in automobile parts of considerable weight. The revolving parts of an engine, such as the crankshaft and the flywheel, are brought to a balance, both dynamic and static,

INDUSTRY

approximating that of the balance wheel of a fine watch. This does not mean that accuracy has just been discovered, or anything of the sort. The real point is that the accuracy which was once reserved for the most expensive instruments is now available for the affairs of everyday life.

The accurate making of common utilities means far more than is at first apparent. The machinery of today, especially that which is used in general life away from the machine shop, has to have its parts absolutely interchangeable so that it can be repaired by a non-skilled man and also in order that freight and carriage may be saved by assembling the machine at the point where it is to be used. A practical interchangeability can be obtained without extreme accuracy, and the requirements of interchangeability have long since been met. The trouble has been to devise ways and means of attaining accuracy without at the same time so adding to the cost of production as to make the price of the machine beyond the purses of more than a few of the people. It has always been possible to attain accuracy, but not to attain both accuracy and complete interchangeability at a low cost. The economic effects of accuracy are a study all in themselves.

The attaining of accuracy at a low cost is a very gradual process that does not come to industry in a day. One of the very great advantages of a factory turning out only a single product is the opportunity that is given for the detailed study of the manufacturing of each part. We have discovered that, no matter how carefully we plan a method of making, experience always shows us a better way. That is the reason why we are now going away from the extreme subdividing of the earlier mass production. We are heading back to the old days except that whereas then one man did the whole of a job, now a machine as far as possible does all of a job. The ideal would be a completely automatic machine operating to whatever degree of accuracy might be required. That ideal, however, is very far from being realized except with rather simple parts. The best that we can do is to combine as many operations as possible. Some parts are reduced in cost more than one-half and the degree of accuracy more than doubled. Manufacturers are gradually learning that there is no one best method of universal application, but that the work to be done must determine the method. Even the principle of the conveyor moving the work forward may be by no means always the most economical.

THE DECENTRALIZING TENDENCY

The improvements in the methods of mass manufacturing and in transportation have brought another great change to American industry that only a few years ago could not have been foreseen. Our industries were formerly grouped in rather a small section not far from the Atlantic seaboard in order to be near the coal sources, just as in a previous period they had to be grouped with relation to water power. And so a large part of the cost of most articles was transportation. The tradition of hand manufacturing held over, in that it was taken for granted that an article had to be completed in a factory—even though it might be taken apart and knocked down for shipment. Then a very self-evident fact began to be realized—the parts under accurate manufacturing did not have to be first put together and then taken apart for shipment, but might be assembled at the point of use—for if they were made properly it was not necessary to have any fitting of parts.

From assembling at the point of use it was only a step to making parts where they could be made best—considering availability of power, raw materials, and point of use. And now the trend is away from the great factory and the factory town and toward the establishing of smaller factories in various parts of the country to save the wastes attendant upon unnecessary shipping of materials and finished parts and also the wastes inevitable in bringing together very large bodies of men in one place. It is true that the country still has very large factories, but they are in the more advanced companies comparatively small as compared with what they would be were the old principles of concentration adhered to.

The tendency is thus to decentralize and, although many mergers of manufacturing units have been effected, the idea of the vertical trust which once obtained is dying out, for it has been discovered that it is not economical for any one institution, no matter how large, to attempt to make everything that it uses from raw to finished materials. It has been found that specialists working under their own management can do better than departments of a larger organization, and it is within the present probabilities that at some time in the future the great corporations will not be manufacturers at all, but central points for design, finance, and sales.

This decentralizing has still another phase. It has been taken for granted that industry and agriculture had to be separated. Now it is becoming apparent that this was quite wrong and that they are not only supplementary, but perhaps can be carried on by the same workers, for agriculture is a seasonal, part-time job and many industries are seasonal and part-time. If these industries can be located in the agricultural sections, then the same workers can be used in both branches and one of the large difficulties of industry, as well as of agriculture, can be overcome.

THE WORKMAN AND THE MACHINE

Machinery and power organized to a purpose have brought a freedom to the worker that the old factory system never contemplated and which was impossible during the ancient craft days. This freedom is being gradually achieved and not all of it is realized. It was presumed that semi-automatic machinery which did not need much intelligence would take away the need for craftsmanship. It has done quite the opposite. The provision of jobs which can be well paid but which do not require much skill has opened the way for the profitable employment of the aged, the halt, and even the blind. In our factories we have certain jobs which we always fill with men who otherwise might, because of their condition, not be able to find employment. And yet even in these jobs some skill is needed.

Possibly the requirement is not for skill, but for a higher alertness than was necessary in the routine employments of a few years ago. The man has to know what he is doing—no matter how mechanical the task—or he will continue to operate when something goes wrong. The call is for something in the nature of craftsmanship, but of a kind different from what is usually comprehended in that word. At least some grade of intuition is demanded for even the lowest job. The lowest job in our factory today requires a higher grade of intelligence than the lowest job in the best hand factory of the older days. This is a fact of which those who praise the good old days are not aware. Craftsmen one hundred years ago were not comparatively so numerous as they are now. Only a few names have come down to us. Today, high-grade craftsmanship is the commonplace of industry. It has to be.

It is usually assumed that in the old machine shop all the men were highly skilled journeymen or apprentices on the way to becoming skilled. That was never the case. Even in the best shops the proportion of first-class machinists was very small, and a fair part of the work was always done by semi-skilled or unskilled laborers who were mostly hired for their strength. Today not only is the lowest job in our industries much higher than the lowest job in the old industrial set-up, but also the highest jobs call for more skill and craftsmanship than few, if any, of the older craftsmen possessed. We have in our industries no absolutely unskilled men, but, taking our lowest grade as comparable with the old unskilled, then the proportion of highly skilled men to unskilled is higher today than ever it was. And judging from our own experience, this stepping-up process will continue. We today employ a higher proportion of skilled machinists than ever we employed, and every development that we make calls for more such men. This will not be so extraordinary as at first it seems if only one looks over what has happened in the last twenty-five years of machine production.

INDUSTRY

THE DISTRIBUTION OF WEALTH THROUGH WAGES

But industry would be a futile thing if it did not distribute income and through this income distribute goods. Every improvement in the methods of American industry has found eventual expression in wages. On this point I cannot do better than to repeat what I have already stated in connection with the Ford industries—it illustrates the tendency of American wages in general.

During 1910 we paid an average hourly rate of 25 cents. This dropped 2 cents in 1911, but in 1913 stood at 26½ cents. On January 12, 1914, we made effective the $5 per day minimum rate. This raised the average hourly rate for the year to slightly under 60 cents. In 1915 and 1916 the average rate dropped several cents, but in 1918 it had risen to 67 cents. On January 1, 1919, we raised the minimum daily rate to $6. This made the hourly rate for that year 77 cents. It reached nearly 86 cents during the inflation of 1920, and in 1921, adding in the bonus then paid, reached 87½ cents. During 1922 the rate was slightly under 80 cents, but this was actually a much larger wage than in the two previous years, because its purchasing power was greater. During 1923 the average was above 82½ cents per hour, and in the following year this increased by a cent and in 1925 decreased a cent. In 1926 the average was over 85½ cents, equaling within a fraction of a cent the wages paid in the inflated currency of 1920. This increased by 10 cents to 95 cents in 1927, and decreased 5 cents during 1928. For 1929 the average was a little less than 92½ cents, and then on December 1, 1929, we increased the minimum to $7 a day. This brought the average hourly rate for 1930 to a fraction over $1.

That is, during twenty years our average hourly rates have quadrupled. Our present average hourly rate about equals the daily rate for unskilled labor in 1910—for at that time a dollar for a ten-hour day was considered a fair wage. During this period our costs of manufacturing have steadily decreased, so that today it costs us less than half as much to make a finely and accurately machined piece of highly specialized material as it did in 1910 to make a comparatively crude piece out of comparatively crude material. These are facts and therefore it is not necessary to speculate on whether or not high wages can be paid.

If wages have been multiplied by four in twenty years, then they can be multiplied by more than that during the next twenty years. It was generally accepted in 1910 that industry had reached a high state of perfection and there were those who were inclined to sit back and view it with extreme satisfaction. There are those who think that the industry of 1930 has reached a high state of perfection and who also are inclined to view it with satisfaction. These wages have been made possible not by forcing the worker—for he does less hard work today than he did twenty years ago—but by new processes, machines, and materials. With this twenty-year accumulation of knowledge American industry should be able to progress much faster in the next twenty years. Therefore wages can be expected to increase in the future at an even more rapid rate—provided the leaders of industry actually lead. If wages do not continue to increase, the fault will be a human one—it will be due to lack of intelligence.

We have, it would seem, during a hundred years made some very real progress.

CHAPTER FOUR

TRANSPORTATION AND COMMUNICATION

In the late autumn of 1843, John Quincy Adams made a long journey to Cincinnati to deliver an address at the laying of the corner stone of an observatory. After arriving at Cleveland by rail, barge, and stage-coach, he found that he had the choice between traveling over two hundred and thirty miles of "bad and dangerous roads to Columbus or four days by canal boat, on the Ohio canal." He chose the latter. "The boat was eighty feet long, and fifteen feet wide, and besides his own party was packed with the crew, four horses, and twenty other passengers." This experience in the good old days Adams describes as follows: "My heart sunk within me when, squeezing into this pillory, I reflected that I am to pass three nights and four days in it. . . . We were obliged to keep the windows of the cabins closed against the driving snow, and the stoves, heated with billets of wood, made the rooms uncomfortably warm. . . . About eleven o'clock I took to my settee bed, with a head ache, feverish chills, hoarseness, and a sore throat"—this in a compartment "with an iron stove in the center, and a side settee on which four of us slept, feet to feet" next to "a bulging stable" for the horses.—HENRY ADAMS, *The Degradation of the Democratic Dogma*, pp. 68 ff.

Transportation and Communication

By Edward Hungerford

THE beginning of the fourth decade of the twentieth century marks, to a curious degree, the completion of the first one hundred years of modern transportation in the United States. To appreciate this rounding out of a century of progress in American transport, picture the United States in the years from 1830 to 1833. The nation already was more than half a century old. It was quite firmly unified, politically; but industrially and even socially, comparatively little had been accomplished within fifty-five long years. The reason for this was not hard to discover. The states of the Union lay scattered along the Atlantic seaboard for more than a thousand miles. Frequently they were intersected by broad rivers—in many cases, estuaries of the sea. These streams, flowing to the ocean and sometimes valuable for inland navigation, were, nevertheless, barriers to proper intercommunication between the states. Prior to about 1825 the only comfortable—ofttimes the only practical—way to get about North America was by coasting-vessels; or in certain individual cases, such as upon the broad rivers of Maine, the Hudson from Albany to New York, and Chesapeake Bay from Baltimore to the sea, by smaller craft which essayed inland voyages—and consumed much time in these efforts.

INLAND WATERWAYS

Yet it was considerably prior to 1825 that two very significant early experiments in the propulsion of a boat by steam as a motive power had been undertaken—with varying degrees of success. As far back as August, 1787, John Fitch, Connecti-

TRANSPORTATION AND COMMUNICATION 87

cut-born and already come to the forty-fourth year of his life, had actually operated a forty-five-foot boat in this fashion in the Delaware River, just above Philadelphia. An early Newcomen engine and a boiler had been installed, and these worked twelve large wooden paddles, six on each side of the craft, rigged tandem-style, so that the boat actually moved up against the current of the river, even though in a rather stiff and cumbersome fashion. Fitch followed this experiment—in a large measure futile, yet probably the first steamboat in the world actually to transport a man—with two or three others, in which he discarded the tandem paddles for paddle wheels. Yet, because of the failure of his invention, he died a thoroughly disappointed and embittered man.

While John Fitch tinkered with this new idea upon the Delaware, James Rumsey made similar experiments upon the waters of the Potomac, near Shepherdstown, Virginia, with little result. The Rumsey craft had followed a still earlier one, his creation, operated by poles mechanically—but not by steam—which George Washington had once stopped to see (in model form) and approve. Even in that early day and across a crude and undeveloped country, rumors came to Rumsey of Fitch's doing a task of the same kind farther north. He then hurried to apply steam power to his mechanical boat. A boiler and engine were quickly fabricated at Baltimore and at Frederick and installed in an actual boat which, after several unsuccessful experiments, finally moved quite well under its own power, December 3, 1787, about four months after Fitch had succeeded in transporting passengers from Philadelphia to Burlington (twenty miles). The Rumsey steamboat scorned paddles or paddle wheels. It attempted to propel itself by forcing water by a steam-pump through a pipe running lengthwise through it and out at the stern, a method of propulsion which quickly proved itself quite impractical.

To make the inland and coastal water routes of North America more comfortable and dependable the genius of Robert Fulton was needed. He was a resident of Pennsylvania, who possessed the rare combination of great imaginative ability and shrewd, practical business sense.

Robert Fulton's *Clermont* was a crude enough little vessel, but an eminently practical one. In this last way it differed from the experiments of Fitch and of Rumsey; of John Stevens of Hoboken, also. When, in 1807, the echoes of the *Clermont's* paddle wheels first reverberated against the Highlands of the historic Hudson, steamboat transportation in America could truly have been said to have been inaugurated. Fulton had had, in addition to his other talents, a real skill as a painter, particularly of portraits. Yet, when his keen and imaginative mind began to lead him into fields of engineering, he quickly dismissed portraiture. During some long years that followed the American Revolution and the establishment of the United States he lived in Paris. There it was that he began studying the French patents that John Fitch had obtained long before, also the successful experimental steamboat, the *Charlotte Dundas*, that William Symington had placed in service on the Scottish canals at the beginning of the century. With all this material at his command, Fulton actually built a small steamboat (in the spring of 1803) and tried it out upon the Seine. The engine was too heavy for the boat and the small craft broke in two. For a time, Fulton abandoned his steamboat experiments. But three years later he was at it again, in the land of his birth and upon the waters of the Hudson River upon the banks of which he was to pass most of the remainder of his busy life.

So came into existence the *Clermont*, a stout little craft, built in the yard of Charles Brown of New York, and at once equipped with an efficient Boulton & Watt engine, especially imported from England. Came and conquered. On the seventh day of August, 1807, the *Clermont*, watched by a jeering and incredulous crowd, started on its first trip from the city of New York to the city of Albany. Started and completed its maiden trip, at once an acknowledged and complete triumph. The practical steamboat in the United States had at last been devised.

TRANSPORTATION AND COMMUNICATION 89

With such a fair start, steamboating now went swiftly ahead. By 1830 or 1833 it had grown into a considerable industry. By that time—a hundred years ago—a considerable fleet of steamboats already had made its appearance, not only upon the Hudson and the Atlantic coastal waters, but also upon the Ohio, the Mississippi, and other navigable rivers of the interior of the land. Upon the Great Lakes also.

The coming of the steamboat had done much to make these water trades both more comfortable and more dependable. The early experiments of Rumsey and Fitch, followed by the successful *Clermont* of Robert Fulton, brought steamboating to a sizable business. Yet it was not until 1836 that the brilliant Eliphalet Nott, president of Union College at Schenectady, created his huge Hudson River boat, the *Novelty*, which by the simple device of burning coal instead of wood (dangerous, costly, and increasingly difficult to obtain), opened a vast new field of opportunity for the steamboat in America. Anthracite had been discovered up in the hills of northeastern Pennsylvania, and was being mined. The completion of the very early Delaware and Hudson Canal, in 1828, from Rondout upon the navigable Hudson inland to Honesdale, Pennsylvania, 108 miles distant, had made it possible to put thousands of tons of this new fuel upon the Hudson and at the piers of the swiftly growing city of New York. . . . In a similar way the Chesapeake and Ohio Canal—an enterprise originally forecast by George Washington—was endeavoring to find its way up along the Potomac from Washington to the bituminous fields back of Cumberland.

Because the early canals played their own important part at the beginning of this century of progress in American transport, they are worthy of at least a paragraph or two in this telling. They began, geographically, in New England, with a short ditch from the young mill town of Lowell to the Charles River opposite Boston; there was another short waterway (the Blackstone Canal) back of Providence, and a much longer one from New Haven up to Northampton. The Delaware and Hudson was not the only water pathway to reach into the anthracite hills of Pennsylvania. By 1833, a canal (the Morris and Essex) had been dug from Communipaw (directly opposite the tip of Manhattan Island) to Phillipsburg across from Easton upon the Delaware. Here it connected with another canal, which ran—and still runs—from Philadelphia up along the west bank of the Delaware to Easton. Both canals shared the use of the Lehigh River and canal system, which was made passable for barges up to the very base of the coal-fields. A similar canal ran to the anthracite mines up the valley of the Schuylkill from Philadelphia.

All of these enterprises were overshadowed completely by the great Erie Canal, from Albany to Buffalo, 364 miles, which, after great effort and expense, had been completed by 1825 and at once became a most important traffic link between the navigable Hudson and the navigable Great Lakes—reaching with their tributaries far into the interior of America. Of course, vessels that were fit and safe for navigation upon the Lakes were entirely unfit for the locks and prisms of the Erie Canal—only designed then for small horse-drawn barges—but it was not particularly difficult to transfer loads at Buffalo, and this was done for many years. And so the Erie Canal came, a little more than a century ago, to be a most important factor in American transportation. It gave impetus to such important New York State cities as Syracuse, Rochester, and Buffalo, and, repeatedly enlarged, it did its own great part in the early upbuilding of inland America. Under another name (the New York State Barge Canal) and with a route very slightly altered from its original one, it still is one of the notable inland waterways of the world.

Most of its early fellows have disappeared. Some came after the 'thirties—the Miami Canal (from Toledo to Cincinnati) and others in Ohio, the canal from Lake Michigan at Chicago to the Illinois River and thence to the Mississippi and all of its tributaries. A few others were dug before—such as the Raritan Canal across New Jersey from Amboy to Bordentown, the canal from the Delaware to Chesapeake Bay, the James River Canal, reaching inland from Richmond, Virginia, and the

TRANSPORTATION AND COMMUNICATION

short link from Norfolk through the Dismal Swamp to the many navigable waters of eastern North Carolina. This last still is in use by very small craft who wish to avoid the arduous outside route around Cape Hatteras.

It was such exposed points upon our seaboard as Hatteras and Point Judith and Cape Cod and Cape Ann that made early intercoastal communication between our seaboard states frequently so hazardous and difficult. It is only in very recent years that the long-projected canal across the base of the Cape Cod peninsula has actually been achieved, as a really tremendous benefit to navigation.

HIGHWAYS

The answer to the successful intercommunication between the states lay elsewhere than upon the reaches of the sea.

Upon the highways?

It might have been if in 1833 we had had a little better highways—and a far better method of locomotion for them. True it was that the nation was making a real beginning toward a system of improved roads—modeled to no small extent upon those that the Romans built nearly two thousand years before. One had only to go to Baltimore or to Washington to see the beginnings of that remarkable project, built by federal aid, the National Road, whose ultimate western goal was St. Louis upon the distant Mississippi. And at St. Louis the National Road some day would connect with the Boonville Pike on to Independence, where the highway already forked—into the historic Santa Fé Trail and the almost equally historic blazoned way to Oregon.

The National Road, whose actual legal eastern terminal was at Cumberland, Maryland, where it connected with the excellent road already extending eastward to Baltimore and to Washington, was authorized in 1808 by Congress, which appropriated a liberal sum for its construction. Because of the inherent difficulties in handling supplies and the like for the builders of this important trunk highway, work proceeded slowly upon it for a long time, and it was not until 1818 that it reached the east bank of the Ohio, at Wheeling. Some years later that stream was spanned at that city by one of the first of the famous Roebling suspension bridges, a stout structure which, although paralleled by a more modern bridge, still stands faithfully to its appointed task.

Wheeling reached—and at Wheeling there was quick connection with the river boats not only for the Ohio and its many navigable tributaries, but for the distant Mississippi, the Missouri, and the hundreds of miles of waterway that in turn poured themselves into them—traffic began to pour itself into the National Road, in swiftly increasing volume. Its tollgates were busy collecting their revenues, taverns and brisk towns sprang up along its route, all fattening themselves upon its traffic. The road-builders resumed their activities west of the Ohio. They kept on from Wheeling to Zanesville and Columbus and Indianapolis. St. Louis ceased to be so distant a goal. For years they labored, and traffic followed upon their very heels, until finally they approached Vandalia, the ancient capital of Illinois. There they were halted—forever. That was in 1851. The National Road had been superseded by the highway of wood and iron that men called the railroad. All of which anticipates, however.

Other and important highways were planned and built—across New York, across Pennsylvania, Ohio, and other Eastern states. Some of these were important as engineering projects and as real lanes of traffic. A typical one was the Great Western Turnpike, leading out of Albany, over the New York hills to Richfield and Syracuse and Canandaigua and Batavia to Buffalo and the Great Lakes. It ran roughly parallel to the Erie Canal, although many miles to the south of it. At Albany it made connection (by ferryboat) with the much earlier Albany Post Road, which shared with the Boston Post Road the advantages of direct entrance into the waterbound island of Manhattan.

Some of these post roads were good, and others were very bad. Over most of them there flowed a pretentious traffic—stage-coaches and wagons, post-chaises and private carriages,

TRANSPORTATION AND COMMUNICATION 93

men and women on horseback and men and animals afoot. Traveling was done under peculiarly difficult conditions much of the time. But its volume was heavy. It followed the highways and cursed at them and at many of the vile inns along the way, because nothing better overland had offered itself—prior to 1830.

RAILWAY BEGINNINGS

In that memorable year something very important in the annals of American transport came to pass.

In the suburbs of the busy seaport city of Baltimore—at that time the third city in population in the Union—men were, that spring, flocking to the old Mount Clare estate of the Carroll family, to ride a mile or so upon a car with flanged wheels which ran upon a track leading due west from the city. This was the very beginning of the Baltimore & Ohio Railroad, that ambitious project by which Baltimore hoped to regain some of the commercial prestige that she had lost to New York by reason of the completion of the new Erie Canal. The men of Baltimore had decided that in some way, somehow, they would have to possess their own efficient traffic route through to the Ohio country, and that the National Road, no matter how well it was built, was not the answer to their problem. They had read of these new railways over in England, and they wondered if something of that sort should not be brought to their aid in America.

As a matter of fact there already were railroads in the United States—although not railroads in the sense that we generally recognize them today, as public carriers of man and his goods. The very earliest lines of rail upon this continent were in truth mere utilities, or accessories, to other forms of industry. Thus it was that Gridley Bryant had already (1826) laid down three miles of railroad at Quincy, Massachusetts, from his quarry to the waterside, to bring out the stone that he was cutting for the Bunker Hill monument over at Charles Town; thus it was that up in those same anthracite hills of Pennsylvania there were, in 1830, two short railroads

—one at Mauch Chunk and the other at Honesdale, to bring the coal from those hills down to the loading-wharves of the canals—they thought vastly more practical for carrying than these uncertain new railroads. A railroad company—the Mohawk & Hudson—had been incorporated (in 1826) to build a direct line from Albany to Schenectady (seventeen miles) to cut across the devious bends and climbs of the Erie Canal between those two places. In this way passengers arriving by boat at Albany could be transported quickly by the packet boats upon the canal and the traveler saved several hours of time upon his way.

The men of Baltimore had visions of a far larger enterprise than any of these. Their new railroad to the Ohio would not be less than 300 miles long. And before its rails came to the brink of that stream they would have to cross one of the steep and precipitous mountain ranges of America—the Alleghenies. What a wild idea, all that must then have seemed! And not made saner by the fact that the promoters of the Baltimore & Ohio planned at first to operate their entire 300 miles of railroad by horse power. They were not so sure about those new steam locomotives in England. Horses were dependable. There simply would have to be relay stations each eight or ten miles of the distance to the Ohio. So it was definitely fixed that horses would haul the cars of the new railroad along the level stretches of the line; and then when the hills were reached, inclined planes (also worked by horses, in treadmills) would raise and lower the cars between levels just as canal boats were then being raised and lowered by endless cables on inclined planes, both in Pennsylvania and in New Jersey.

They set out to build their railroad in just this fashion. Following the incorporation of the company in 1827 (it has continued under the same name and charter from that day to this), they began the construction of the road from Baltimore west at almost a dead level—much like a canal. At the first sizable hill, five miles west of the city, they met their Waterloo. They dug a deep cutting into the hill, so deep that it all but

TRANSPORTATION AND COMMUNICATION 95

bankrupted the company and halted the enterprise, and discovered the futility of their entire scheme of levels. There would *have* to be some grades.

And there would have to be a stronger, better motive power than horses. A commission was dispatched to England to look into the steam-locomotives in use there. But when, in May, 1830, the Baltimore & Ohio actually began the commercial operation of its line, it hauled trains back and forth between Baltimore and Ellicott City (fourteen miles) by horses. But before the commission could return from Great Britain, an outstanding New Yorker—Alderman Peter Cooper—who had invested large sums in Baltimore real estate and who was seriously alarmed at the possible future of the place, went there and placed a small locomotive, the *Tom Thumb*, in experimental service. It was hardly more than a toy, an upright boiler set upon a flat-car and with but a single cylinder in its engine, but it worked. It convinced the directors of the B. & O. that they would have to abandon their horse-power idea for steampower and so they advertised and held a locomotive contest outside of Baltimore, which was won by a larger and quite practical engine, the *York*, named after the city in which it had been built and from which it had been transported, with great difficulty, over a hilly turnpike for more than sixty miles.

Neither the *Tom Thumb* nor the *York* was the first steam-locomotive in the United States. Five years before, Colonel John Stevens had operated a tiny toy contraption of the sort around a small circular track in the grounds of his estate at Hoboken, New Jersey, while in May, 1829, the Delaware & Hudson Canal Company imported the *Stourbridge Lion* from the Foster, Rastrick & Company Works at Stourbridge in England and sent it up the Hudson and through their canal to Honesdale. There in August two attempts were made to use it on a level stretch of their railroad over the mountains to their mines at Carbondale. The railroad, however, was unsafe and the locomotive was abandoned.

This really marked the beginning of modern railroading

in America. Swift development ensued. In the South, the South Carolina Railroad and Canal Company, putting down its line from Charleston, 136 miles, to Hamburg (across the river from Augusta), experimented and went into actual operation with a steam-locomotive in the fall of 1830. This engine was the *Best Friend of Charleston*, built by the West Point Foundry in the city of New York, and the first practical locomotive built in the United States. It went into regular service on Christmas Day, 1830. The third engine built by the West Point Foundry was the *De Witt Clinton* (the second engine was the *West Point*, delivered in February, 1831, to the South Carolina Railroad), and in August, 1831, it had the distinction of hauling the first train over the Mohawk & Hudson from Albany to Schenectady.

Other engines, British and American, now followed in quick succession. The *John Bull* came across the Atlantic to the Camden & Amboy, the *Herald* to the Baltimore & Susquehanna, still others to the Boston & Lowell and to the Boston & Worcester. Philadelphia became a recognized center for the construction of American locomotives—a distinction which she never has relinquished. Matthias Baldwin, who had started as a watchmaker in Philadelphia, began his career as a builder of locomotives—a career that was to give his name enduring fame in a great engine-building company. Norris and Eastwick and Harrison, and down in Baltimore, Ross Winans, a New Jersey horse-trader who had gone there originally to sell horses to the railroad, started the construction of a remarkable type of locomotive of his very own. Winans' great fame was to come, however, in the development of the railroad car. He it was who advanced it from the original type of wagon or of stage-coach adapted to the rails by the simple device of adding flanges to the tires of the wheels, to length, carrying capacity, and riding comfort by the use of two trucks of at least four wheels each.

Not long after these earliest builders of locomotives in the United States came another group—Hinckley and Brooks among them—headed by William Mason of Taunton, Massa-

TRANSPORTATION AND COMMUNICATION

chusetts. To William Mason must go much of the credit for the perfection of the American steam-locomotive. He brought it close to its present appearance. In fact, it may be fairly said that Mason gave machines something of a "streamline appearance," as an automobile designer might put it. Despite the ungainly "balloon stacks" of the wood-burning engines of his earliest designs, they were gay-looking craft, bright in color and much decorated in brasswork and other shiny ornamentation. And in the 'fifties and the 'sixties coal was to supplant wood as a fuel, just as it had done upon the steamboat but a few years before. But aside from its huge and bulging smokestacks, the American locomotive was to suffer little change in outward appearance.

RAPID DEVELOPMENT OF RAILWAYS

Swiftly the American railroad developed in those days of the 'thirties, the 'forties, and the 'fifties. In New England, the most thickly populated and industrial area of the United States of those days, it got its first and (for a long time) its greatest foothold. Lines shot out from Boston in almost every direction; to Lowell, to Worcester, to Providence, and then to Portland and still farther down into Maine. Finally a railroad essayed the crossing of the steep Berkshires, and in January, 1842, one finally could go by train all the way from Boston to Greenbush across the Hudson from Albany. At Albany—then reached by a steam ferryboat—other trains awaited the traveler. A succession of nine railroads (afterwards to become the main stem of the New York Central) carried one all the way on to Buffalo, 300 miles distant.

These little roads across New York State had had a rather difficult time. They had dared to put down lines paralleling the Erie Canal, and so they had brought down upon their heads the enmity of the great political forces back of that highly political enterprise. At first they had gone about this railroad construction rather diffidently, building, as we have seen, a short-cut to the ditch, from Albany to Schenectady, and, a little later, a similar short-cut from Rochester to Buffalo. But the original line between Syracuse and Rochester did not dare to parallel the canal. It went ten miles to the south of the waterway, at the expense of a large increase in mileage to itself. Other lines between Schenectady and Utica and Syracuse gradually filled in the gaps, so that in 1843 one could go by train from Albany to Buffalo, although it took twenty-five hours to accomplish the trip and it was necessary to change cars several times on the way.

But in the 'forties there was no rail route between Albany and New York—although a prosperous railroad operated between Albany and Boston. The steamboat-owners of the Hudson, grown fat and strong in the passing of two or three decades, were as difficult a political force as the proponents of the Erie Canal. On this account the first railroad that attempted to reach Albany from New York—the Harlem—chose a route far to the east of the river. The Harlem always was an unfortunate enterprise. It spent eighteen years in completing its route from the chief city of the Empire State up to the capital, and then was beaten by a few brief months, for a group of men had finally been found with audacity enough to attempt a railroad through the valley of the Hudson, close by the river brink. Starting with a well-organized movement at Poughkeepsie in 1846, these men had the satisfaction of seeing their railroad completed all the way from New York to Albany (Greenbush) by October 1, 1851. Moreover, they could have the satisfaction of knowing that their Hudson River Railroad was the best built and most expensive of any road laid down in the entire United States up to that time. This last may not have been a matter for congratulation, because, for a number of years after it was finished, the Hudson River Railroad notoriously failed to earn dividends. Not until a former steamboat proprietor—by name Cornelius Vanderbilt—bought most of its stock did it develop into a highly profitable enterprise.

Mr. Vanderbilt—generally known as the Commodore—acquired the Hudson River by the rather simple process of first buying the bankrupt Harlem and then, by threats of

TRANSPORTATION AND COMMUNICATION 99

using his fine abilities to perfect the road into a first-class carrier, bought control of the Hudson River. This done, he reached west of Albany, acquired the eleven roads west of that city which already had been consolidated into the first New York Central Railroad, and in 1869 united all his rail properties into the New York Central & Hudson River Railroad. So were laid the solid foundations of the present powerful New York Central system.

Commodore Vanderbilt's methods were almost always substantial, far-visioned, creative. Not so much could be said for those of some of his compeers. As he was entering the railroad field a group of men—notably Daniel Drew and James Fisk—were endeavoring to complete, across the southern counties of New York, a railroad that would reach from the Hudson River to the waters of Lake Erie at Dunkirk. This line—from the outset planned as a single company—was for many years to be notorious as the New York & Erie Railroad. This is not the time or place to go into the fantastic financial history of Erie. It is enough here and now to say that in 1851 the road was finally completed and that Daniel Webster was a passenger on its first train from Piermont-on-Hudson to Dunkirk, riding much of the way in a rocking-chair affixed to a flat-car so that he might the better see the country.

The great mountain ranges that cross Pennsylvania at right angles—in contrast to the smooth natural pathway across the state of New York—offered no easy invitation to the early railroad builders. Yet despite these great natural obstacles, a curious combination of railroad and canal was, by 1840, in operation all the way from Philadelphia to Pittsburgh. The traveler now took the old state-owned railroad from Philadelphia to Columbia or to Harrisburg, upon the Susquehanna, and then a canal boat up the valley of the Juniata and beyond to the very base of the forbidding Alleghenies. Yet even then the traveler did not leave the canal boat. The inclined plane and the endless cable again came into play. The canal boat, load and all, was hauled up the plane, transported by rail across the top of the mountain and through the first railroad tunnel in America, and then was lowered into the waters of the Conemaugh to resume its leisurely course on to Pittsburgh. Charles Dickens was one of those who made comments upon this early transport arrangement; he was so impressed by it that he afterwards wrote upon it at length in his *American Notes*.

Reference already has been made to the early start of the Baltimore & Ohio enterprise. It took exactly twenty-five years to reach its goal of the Ohio shore (at Wheeling) and, like the Harlem, it found itself preceded; by the Pennsylvania, which, uniting the various roads and canals across that state, had been completed as an all-rail route from Philadelphia to Pittsburgh in 1847.

WESTERN RAILWAY ADVANCE

So began our railroad structure within the United States. Unfortunately there is not, in a single chapter of this book, space to relate in detail the story of its entire development—sometimes swift, as prior to the great panics of '37 and '57; sometimes slow, as just following those bitter episodes; but for eighty years steadily advancing in mileage. Two railroads—the Michigan Southern and the Michigan Central—raced from the east into Chicago and arrived, in 1852, almost simultaneously, only to find the Illinois Central—beginning of the present extensive system of the same name—already there. The Galena & Chicago (now part of the Chicago & North Western) was the first railroad to reach out over the prairie west from Chicago. Its wooden passenger station stood upon Halstead Street and it is related that its first president—John B. Ogden—was wont to ascend to the tower of that small structure and with his spyglass endeavor to see the smoke of the eastbound trains approaching across the prairie.

The first road to reach the Mississippi and to cross it was the Rock Island. It crossed the great river (at Rock Island) on a many-spanned wooden bridge that at once became a source of great annoyance to the steamboat men upon the Mississippi, with the result that only a few years later the steamer *Effie*

TRANSPORTATION AND COMMUNICATION 101

Alton, burning to the water's edge, drifted against the railroad bridge and destroyed it. This was unquestionably a premeditated act of revenge. In the bitter litigation that followed, a young attorney by the name of Abraham Lincoln made his first great legal reputation. He prosecuted the case for the railroad company—owners of the bridge—with brilliancy and acumen.

The Hannibal & St. Joseph (nowadays a part of the Burlington) was the first road to reach Missouri. Then the railroads again multiplied so fast that it is difficult to keep track of them. The dream of the Pacific Railroad—an all-rail route across the broad continent—a dream which had been in the minds of men ever since 1832, now was acute. From St. Louis they were building the Missouri Pacific, from Kansas City the Kansas Pacific, and from Council Bluffs and Omaha the Union Pacific. Nor was this all.

At the same time a tremendous railroad enterprise was thrusting itself east, from California. Collis P. Huntington and Leland Stanford and Charles Crocker and Judah P. Benjamin were great dreamers—highly practical men, too—and their dreams were taking effect in the Central Pacific Railroad that Chinese coolie labor was constructing east from Sacramento (connected by steamboat with San Francisco) to join the westward-reaching Union Pacific.

In all the annals of railroad construction, no lines have ever been put down quite as rapidly as were the Pacific railroads. A mile a day was frequently reached by the construction gangs, laboring feverishly all the while. As a result progress on the transcontinental route, which for a long time had languished, now went ahead with great force and energy. The Union Pacific, working toward the west, and the Central Pacific, working toward the east, finally came together—on the 19th day of May, 1869. On that memorable day the railheads were joined, near Promontory, Utah, close to the north shore of the Great Salt Lake, and an important chapter in the history of American transport was closed.

Another began. The immediate success of the transcontinental railroad led to similar ventures, to the north and to the south of it. The lines forming the Southern Pacific and reaching from New Orleans and El Paso to Los Angeles (a little later to San Francisco) were completed and opened in 1882; the Northern Pacific in the following year. Thereafter the building of the railroad in the United States was to be, almost entirely, an intensive thing.

LATER PHASES

With no great new worlds to conquer, the iron horse contented himself with poking his inquisitive nose into odd corners where he had not ventured before. The Santa Fe opened up northern New Mexico and Arizona; other railroads, of no mean size, operated here, there, and everywhere.

And when there were no more odd corners to be explored, the business of the railroad-builder was to perfect the lines already set down. A Hill arose to project still another system across the northern tier of the United States. A Harriman took the original Pacific railroad, a rather wabbly and attenuated line in its first location, shortened it, and improved it at a cost of many millions of dollars. A single feature of this task of rejuvenation was to build a railroad straight across the hitherto impassable Great Salt Lake, with a bridge twelve miles in length as its salient feature. This Lucin cut-off cost $4,500,000, but it shortened the main line of the Union-Pacific-Southern-Pacific Overland Route by more than forty miles and it was worth every cent that it cost.

In the East, railroads were similarly improved, new lines, new tunnels, new bridges, better grades and curvatures, with great economies in operating costs coming as a direct result. In the South a railroad feat, rivaling that of the Lucin cut-off in its daring, put a track for seventy miles over the ocean and the keys at the very tip of Florida and made Key West (ferry point for Havana) an important terminal point of the American railroad system.

TRANSPORTATION AND COMMUNICATION 103

ELECTRIC TRANSPORTATION

In 1895 the electric locomotive arrived.

It had been preceded some six or seven years by the electric trolley-car displacing the old-time horse-car in the streets of almost every important American city or town. Horse-cars in the streets of certain of these were nearly as old and as well established as the steam-cars. They were slow and they were incommodious, but they were the best form of city transit that had offered itself.

A variation and considerable improvement came in 1873, when one Andrew Hallidie first met the problem of the hilly San Francisco streets by installing what quickly became known as cable railways—small street cars being hauled at a slightly advanced speed by endless cables in slots underneath the pavement. For a time these cable railways had a considerable vogue. They were installed, at large cost, in St. Louis, in Chicago, in Washington, in Baltimore, in Philadelphia, and in New York. Their day was a short one. The trolley-car, much less expensive to build and to operate, supplanted them—except in San Francisco and certain other cities of the Pacific coast where steep hills still demand the super-safety of the cables.

The trolley-car widened very much the area within which the American city might hope to develop itself intensively. Because of its increased speed and carrying capacity it provided a new periphery for development. It did more. It ran into the countryside for long miles. It aided suburban development and came for a time as a great boon to rural communities.

Until the rise and development of its great antagonist, the automobile, it well filled a great transport need. . . . The coming of the motor-car (including the motor-bus and the motor-truck) has raised sad havoc of late with the electric railways. A great many of them, especially the so-called interurbans, have been torn up or abandoned. In the larger cities, however, the electric street railroad still seems to have a distinct field and service of its own, of which it cannot easily be robbed.

It was quite logical that a form of rail transport which, beginning in Cleveland, Ohio, and in Richmond, Virginia, had leaped into such quick public favor, should have made an appeal to the steam railroaders. They had peculiar problems, problems of short lines and of very heavy passenger traffic, particularly in the suburban districts of the larger cities. So it was that presently many of these services were being transformed from steam operation to electric. The construction, in the first decade of the present century, of the two great passenger terminals in the city of New York—the Grand Central and the Pennsylvania stations—brought with them correlated problems of the operation of many trains in quick succession through long tunnels, of stations themselves built in a large degree underground. Electricity—clean, swift, adaptable—was the successful answer to these problems and so electricity was at once adopted as a sole passenger motive-power in the New York suburban area of the New York Central, the New Haven, the Pennsylvania, and the Long Island railroads. Yet even before then (1895) the Baltimore & Ohio had successfully employed electric locomotives in its tunnels under the city of Baltimore. The New Haven also had pioneered in the use of electricity to a slight degree.

The immediate success of these suburban services of the larger railroads in and about New York has led to similar installations in Philadelphia, in Chicago, in San Francisco and elsewhere. The cleanliness of the electric iron horse has helped him to supplant his fellow of steam-made energy in many of the longer tunnels outside of the cities—the historic Hoosac Tunnel of the Boston & Maine, the tunnels from Chicago to Ontario of the Grand Trunk and the Michigan Central, the long Cascade Tunnel of the Great Northern Railway up in the Northwest. The Northwest is the theater of another large electric installation—the longest in the land and one of the largest in the world—656 miles of main line of the Chicago, Milwaukee, St. Paul & Pacific through the states of Montana,

TRANSPORTATION AND COMMUNICATION 105

Idaho, and Washington. A further field of electrification is upon two important coal-carrying roads over the mountains of Virginia—the Norfolk & Western and the Virginian. And at the present moment the Pennsylvania Railroad is preparing to electrify its busy main stem between New York, Philadelphia, Baltimore, and Washington. Other roads will follow its example in the near future. Electrification upon the standard steam railroads of North America—even after thirty-five years—is still in its beginnings.

THE AUTOMOBILE

The turn of the century brought two other important forms of transport into notice and increasing use.

It was back in the gay 'nineties that the folk of certain American towns noticed a strange new form of vehicle passing through their streets. For lack of a better name, they were being called "horseless carriages." In Kokoma, Indiana, one Ellwood Haynes was driving one of these contraptions; in Springfield, Massachusetts, Charles E. Duryea another; in Rochester, New York, George B. Selden a third. There were a few other audacious pioneers with their uncanny vehicles going down the crowded streets and frightening horses and pedestrians as they pursued their reckless course—at perhaps seven miles an hour.

With the development of the electric trolley-car there had been some effort to apply that heretofore unappreciated form of motive power to a car or carriage that did not have flanged wheels or run upon a pair of rails. The electric carriage was a pioneer almost alongside of the one that was fitted with a gasoline motor. For a time it quite rivaled it in popular favor. Its great difficulty was that it was required to carry a heavy dead load in storage batteries powerful enough to move it for any appreciable distance. Yet it started out quite bravely. In New York and other very large cities it had an early field in public cab service. Gradually it has been supplanted, quite largely, by the gasoline automobile. One adaptation of it which has remained, however, is the trackless trolley, which is used to quite an extent in certain American cities. In that same Rochester, where Selden carried out his first experiments in the gas-driven car that were to bring him both renown and extremely valuable patents, it became necessary a few years ago to build a new trolley line across the north side of the town. In doing this it was absolutely necessary to cross the deep gorge of the Genesee, and the only possible crossing was upon a high arch-bridge which had not been designed for the heavy rails of the standard trolley track. The trackless trolley solved this problem. With no construction through city streets other than overhead trolley wires it was able, and still is, to perform swift and efficient city transit service, without being compelled to carry heavy dead weight in the form of storage batteries. The power taken from one overhead wire and returned to another by trolley is used to run electric motors, at a maximum of convenience and a minimum of cost. It is a form of city transportation that is apt to show increase in coming years.

But let us return to the story of the gasoline-motor automobile as most of us know it today. As far as the United States is concerned its history begins, in a slightly complicated fashion, with the efforts of those three outstanding pioneers, Selden and Duryea and Haynes. The first man of this trio unquestionably had been influenced by two large events: the discovery and production upon a large scale of practicable fuel—petroleum—for a small, energetic engine, and the efforts that were already being made by such men as Otto and Daimler and Benz in Germany, Napier, Royce, and Austin in England, and Peugeot, Renault, Panhard, and Levassor in France to adapt a small energetic oil-burning engine to the propulsion of a carriage upon the highway. Be that as it may, Selden filed his first patents for an automobile as early as 1879, and for sixteen years thereafter managed to keep them alive, without the construction of a single car of any sort. For it was not until 1895 that he astonished the good citizens of Rochester by going forth upon their popular Lake Avenue with his curious horseless vehicle.

TRANSPORTATION AND COMMUNICATION 107

In the meantime, both Duryea and Haynes actually had been building automobiles. Both men were products of the Middle West, although Duryea moved to Springfield before he had advanced very far in his work.

In 1892 Duryea assembled a car that actually carried him through the streets of the old New England town, but rather secretly, at night, in order not to alarm either the horses or the natives. Finding certain weaknesses in his pioneer car, he promptly dismantled it and started work upon a second, which he operated with complete success. It is this machine that is now shown so proudly in the National Museum in Washington. After he had constructed his second carriage, Duryea decided to go out to Chicago and see the automobiles which Europe had sent there, to be exhibited in the great World's Fair of 1893. A Daimler automobile and an electric car from overseas both interested him immensely. He then and there decided that he would build, in America and for Americans, "quality cars," and returning to Springfield he started forth definitely upon this principle. But results came slowly. It was not until another two years that the Duryea Motor Wagon Company had really begun to function. In the winter of 1895-96 it produced thirteen automobiles and was proud of the feat.

In the meantime Ellwood Haynes was also busy. With the backing of Elmer Apperson, who had a machine shop in Kokomo, he produced, in the midsummer of 1894, a practical motor-car. How that pioneer car was built and first began to run is well told by Carl W. Mitman in his *Beginning of the Mechanical Transport in America* (Smithsonian Publications). Mr. Mitman says:

> ... A flat, rectangular tank was installed under the floor-boards, while the water-tank for cooling found a place under the seat-cushion, with a small rubber hose connecting it to the engine. The machine had no radiator. The engine was started by cranking from the side, the crank being poked between spokes of the right rear wheel.
>
> By the first of July, 1894, the machine stood ready for the finishing touches. It had solid rubber-tired wire-wheels and a tiller-handle steering mechanism. On July Fourth Haynes decided to give it a road test. Word got out about his plans and so many people crowded around

the Apperson shop when the much-talked-of horseless carriage was pushed into the street that Haynes decided to hold his test outside of the city. A horse-drawn carriage pulled the machine 3 or 4 miles out into the country. For safety's sake the faithful horse was first driven some distance to the rear. They then cranked the engine. Haynes and Apperson got aboard. Haynes threw in the friction-clutch and the horseless buggy moved out Pumpkinvine Pike. For a mile and a half two delighted men "flew" at an estimated speed of 6 or 7 miles an hour, then turned the machine around and drove all the way into town and to Haynes' house without a stop. . . .

That machine also stands in the National Museum at Washington today.

The experiments of Duryea and Haynes and Selden drew others to tinkering with these new horseless carriages. Alexander Winton of Cleveland, and Herman Buick of Detroit, were among the other pioneers in this new form of transport that presently was to revolutionize the carrying methods of a nation. Up in Detroit also was a hard-muscled young man, a born mechanic, tinkering with a carriage himself. His name was Henry Ford and few people knew of him outside of his immediate small circle of acquaintances. In twenty-five years his name was to be a household word around the world, the product of his brain, the most widely sold and distributed vehicle in all history.

There apparently was much to the horseless carriage idea. Remember that on the other side of the ocean the French especially were also making long strides in its development. They called the new vehicle the automobile, and by this name it now has become known in almost every corner of the globe. Gradually the automobile came to look less and less like the carriage that had been designed for haulage by horses. The body came closer to the ground, grew a little longer all the while. Designers played with what they were pleased to call "stream lines," making the new agent of transport more beautiful as well as more efficient all the while. Henry Ford came into the picture with his idea—radically opposed to the original one of Duryea—of making the automobile in quantity and so low-priced that almost everyone in America could own one

TRANSPORTATION AND COMMUNICATION 109

for himself, and succeeded in capitalizing that idea to the fullest possible extent.

A great new industry went into the making in the early years of the present century. Huge factories, covering acres of space, began to be devoted to the mass production of the automobile. Other factories specialized in the making of its equipment, tires, lamps, electrical devices, and these, in turn, grew apace. Detroit became a city of a million people—the fourth in all this land—while, because of tires alone, Akron developed from a sleepy Ohio town into a bustling and important industrial center.

Side by side with this dramatic development went another. The highway in the United States, after years of comparative neglect and obscurity, came again into its own. The automobile as finally perfected could make its way with some difficulty over roads of almost incredible roughness. But it made its way far better over paved roads, and so paved roads became the order of the day in America. First the trunk highways and then the side roads. Two-way roads, three-way roads, four-way and even six-way roads. Highway traffic increased all the while, by leaps and by bounds. Old-time bridges, amply sufficient for the horse era, were supplanted by much stronger and finer ones for the horseless days. Garaging became one industry; the regulation of city traffic another. So, in brief, has the automobile come into its own in the United States. In capital invested, in labor engaged, and in wages expended it is now the second or third ranking industry in all the land.

OPENING THE AIRWAYS

Now for the newest of man's many pathways—the pathways of the air. Through the ages it has been one of man's pet dreams to soar, birdlike, above trees and houses and hills, but to go straight in a definite direction from point to point—a dream, alas, that seemed incapable of actual fulfillment. Some centuries ago he discovered the principle of the gas-balloon, was actually making ascensions in those queer, pear-shaped, all-but-unmanageable devices, but this could hardly be called transport in any practical sense of the word. As far back as 1859 four men actually sailed in a huge balloon—the *Atlantic* —from St. Louis to a point near Watertown, New York, more than a thousand miles in direct line, but their destination was only a matter of the whims of the winds. If the prevailing wind at St. Louis that day had been from the east instead of the west, they might have landed in Denver instead of in northern New York.

To these balloonists, these early aviators, if you please to call them such, it seemed as if an engine-driven ship was the only solution of their problem. But efficient engines, effective enough to drive an airship straight against unfavorable winds, if need be, were not being produced. Not then. When the internal-combustion gasoline-fuel engine came along, with its vast economies of space and of weight in relation to the power given out, the airplane of today first became even a possibility. And so, in turn, the dirigible.

For fifty years past, men have been experimenting with the so-called heavier-than-air machine; a device which, without relying upon the buoyancy of any form of a gas-bag, would nevertheless be able to ascend, propel itself in any direction that its pilot might desire, and then descend, gracefully and easily, at any designated point. Yet it was not until the 'nineties that Samuel Pierpont Langley, then secretary of the famous Smithsonian Institution in Washington, really seemed to make much real headway with the absorbing and excessively difficult problem. Langley was an indefatigable worker. He was a prolific writer, producing more than two hundred published works, including one most outstanding volume, entitled *Experiments in Aërodynamics*.

Many of his conclusions in this work were attained by a series of interesting experiments that he carried forward in his laboratory at the Smithsonian. At first he worked with exceedingly small-scale apparatus. He felt that if some sort of "plates" could be devised, light enough and strong to hold a man aloft, and then the entire device could be moved fast

TRANSPORTATION AND COMMUNICATION

enough through the air, the entire problem of flying would be largely solved. But Langley also found himself facing that inherent and initial problem of getting an engine light enough and strong enough to accomplish the result. Yet, knowing little or nothing of the coming triumphs of the gasoline internal-combustion engine, he set about perfecting a practical flying-machine.

... Langley is universally accorded the great honor of having been the founder of the science of aviation. ... It remained for the Wright brothers, not many years later, to demonstrate the accuracy of his predictions. ... (Van Wagenen's *Beacon Lights of Science*.)

For all of which Samuel P. Langley was for years laughed at and derided, called "Darius Green" until he must have come to hate the very sound of the words. However, ridicule did not easily discourage him. He went straight ahead. On May 6, 1896, he launched his "aërodrome," as he called his device, on the edge of Washington, and flew it twice, gloriously. It was the first time in the history of the world that such a device had ever sustained itself aloft for more than a very few seconds. It made a flight of a full three-quarters of a minute each ascent, which was a far different matter, and flew a good half-mile, settling down easily and quietly upon the smooth surface of the waters of the Potomac.

This Langley machine, some sixteen feet in length all told, looked, more than anything else, like two giant birds soaring tandem. Two sets of wings were inclined gently upward. The rudders were fixed in advance. A gas-engine being out of the possibilities, Langley had introduced a very small steam-driven one which, in turn, drove two propellors. The entire device weighed not more than twenty-six pounds. It was started upon its flights by being catapulted out into the air by an ingenious special device.

So, the beginnings. For seven years Langley tinkered faithfully with his machine. In the mind of the general public, he seemed to be having no particular success. Yet the fact remains that in these seven years he achieved his greatest results, actually building a man-carrying machine and perfecting a radial gasoline-engine of great efficiency.

In the meantime, some two hundred miles to the south of the national capital, aviation actually was being accomplished. Two young men, brothers, from Dayton, Ohio, were really flying. Their names were Wilbur and Orville Wright and since the days of their young boyhood they had been playing with air devices—box-kites and the like. For a time, discouraged by their friends and neighbors, they had put air transport aside and had conducted a small newspaper in Dayton. But eventually aviation again demanded them. And then it was, at the turn of the century, that they put away everything else and went straight to their great appointed task in life. . . . They chose the lonely Albemarle Sound country in North Carolina as the best available place for trial flights, and prepared to settle there for long months each year, going back in the winter to their Dayton shop to work on the purely mechanical features of the man-bird, very largely problems of the propelling mechanism that gradually they were bringing into being.

They, too, had their disappointments and they were grave disappointments. But they never became entirely disheartened. For three years they worked on, and in December, 1903, they made four true flights with a motor-driven airplane, at Kitty Hawk, North Carolina. In the first of these their machine was up for three and one-half seconds, but in the final one the machine was in the air fifty-nine seconds and it had actually carried a man—Wilbur Wright—852 feet. From that time forward discouragements were fewer, the flights longer and better sustained. The Wrights had hit upon a vital and successful principle, aside from the compact internal-combustion engine: by the seemingly simple, but really extremely valuable device of warping the wings of their machine at the pleasure of its driver, not unlike the "feathering" by which a good oarsman dips his blades into the water, they actually had perfected the principles of the modern airplane. Hencefor-

TRANSPORTATION AND COMMUNICATION 113

ward the problem was merely to better the machine, in all its details, large and small. It was another five years before the Wright brothers were willing to go on record that their machine was entirely practical and win a contract of twenty-five thousand hard-earned dollars from the United States government for an airplane for military purposes. The rest of their achievement rests within the pages of very recent history.

Others followed on the pathway they had blazed. One of the most capable of these men close upon their heels was Glenn H. Curtiss, of Hammondsport, New York, whose early experiments fill some of the most glowing pages of the record of aviation in America. Bigger, swifter, more powerful airplanes, capable of going longer and still longer distances, were devised and built. A great devastating war swept over a third of the world, and the airplane in that great conflict proved its full worth, over and over again. It went into active and practical commercial life on both sides of the Atlantic. It carried not only men, but the mails, swiftly and dependably. And it still, apparently, is just in the beginning of its triumphs.

Meanwhile men began turning their attention back to the possibilities of a gas-balloon, but a gas-balloon mechanically driven and controlled, in the solution of the vast problems of air transport.

The heavier-than-air machine had and still has very great limitations in dependable commercial use. One of the chief of these is the very fact that it is heavier than the air. To keep it safely aloft any time whatsoever, one or more of its motors must be in swift and continuous operation. Why not, an interested group of men argued, make a machine that even in case of mechanical breakdown, particularly of its engines, will sustain itself aloft for an indefinite period, just as a ship at sea might be expected to do under similar circumstances? Their answer to this came in a reversion to the idea of a gas-bag. In Germany, Count Ferdinand von Zeppelin, working upon this fundamental idea, finally built, launched, and successfully operated the giant craft that bears his name. When the World

War was over, the production of zeppelins, or dirigibles, as they soon became known, became a regular and a commercial business, not merely for the purposes of military offense and defense on the part of the several nations, but for actual trade routes as well. The first dependable advertised crossings of the Atlantic have been by dirigible. The great flights of the *Graf Zeppelin* around the entire world in the autumn of 1929 made transportation history and were in no small way responsible for the plans for the further commercial use of the dirigible.

New forms of transport have brought new problems to the older forms. Just as the railroad came a hundred years ago and greatly diminished the importance of the river steamboat, the canal, the stage-coach, and the freighting wagon upon the highway—practically ending the career of the last two of these—it has been predicted that the automobile and the airplane will, in turn, end the brilliant career of the railroad. Wiser folk do not believe this. They saw the telegraph come, and then the telephone, and then the radio; yet there has been room enough and work enough for all. So these wiser folk have come to a belief that the automobile and the airplane will not supplant the railroad, but rather will supplement it, will correlate with its activities. In the last analysis it must be that there is a form of traffic best suited for each of the main avenues of transport—the railroad, the highroad, the waterway, and the airway. Wise railroaders have grasped this idea and already have made great headway with it. One already may make a trip across the continent, riding by night in the sleeping-cars of the railroad and by day in high-speed airplanes, and so go from the Atlantic to the Pacific in thirty-six hours of swift and comparatively easy travel. One large railroad uses busses to pick up and deliver its passengers in the city of New York. A number of others use not only busses, but motor-trucks for short-haul and local traffic in correlation with their long-distance trains, both passenger and freight. The possibilities of such correlation have hardly even been touched.

TRANSPORTATION AND COMMUNICATION 115

ELECTRIC COMMUNICATION

One form of man's communications, so closely allied to the definite problem of transport as to be almost part and parcel of it, is that of communication through the air or over wires—the telegraph, the telephone and the wireless, or radio, as it has come to be known commercially.

First in the order of these, chronologically, is the telegraph. It came as a distinct outgrowth of that day more than a century ago when the story of the battle of Waterloo was being told by semaphore signals from Dover to London (being sadly interrupted at the most important point of the telling by a swift oncoming fog) and the opening of the Erie Canal at Albany was being signaled all the way across New York State by guns planted each six miles between Albany and Buffalo.

The great necessity of man for having swift communication was first met to a satisfactory degree when the electric telegraph came into existence. By common consent, an American, Professor Samuel F. B. Morse of New York, is given credit for having invented the practical electric telegraph of today, although, of course, he had his forerunners. It, too, came into existence nearly an even century ago. For it was in 1832 that Morse, who had already come to no small meed of fame as a portrait-painter, made sketches of his device; he also worked out the telegraph alphabet which has come down to this day as the Morse code, better than which has not been devised.

Morse found practical assistance in Alfred Vail of Morristown, New Jersey, an early scientist and manufacturer. Working together, they perfected the new machine and took it down to the national Congress at Washington in hope of receiving aid from that body. For a time Congress was lukewarm. Gradually it responded to the inventors' appeals and finally it made an appropriation of some thirty thousand dollars to install an experimental line from Washington to Baltimore, forty miles distant. It was over this crude line, placed in a trench alongside the tracks of the Baltimore & Ohio Railroad, that the first official telegraph message was flashed, May 24,

1844. "What hath God wrought!" were the words it quickly spelled, and spelled history at the same time.

From this dramatic beginning, quick growth came to the telegraph. On the day after the line was opened a Baltimore paper printed a news telegram from the Washington Capitol. Two years later another Baltimore paper, (the *Sun*), printed the President's Message, transmitted in full by the "magnetic telegraph" as it was then called. Not only did the use of the telegraph increase, but the actual lines. By June, 1846, the thin strand of copper was in Jersey City, from which messages were carried by boys across the ferry to New York. (The problem of laying a cable across the muddy bed of the Hudson was not lightly to be met—not just then.) By 1847 the telegraph was at Cincinnati; the following year at Boston. A little later at St. Louis and Chicago. It was not until October, 1861, that it finally spanned the entire continent. In the meantime, a most interesting form of swift communication by courier with the far outpost of the magnetic telegraph (at St. Joseph, Missouri) had been put into successful operation, and then (upon the completion of the less expensive and more efficient electric telegraph across the continent) abandoned.

This was the Pony Express, and in the brief two years of its history it formed one of the most picturesque and romantic chapters in the transport history of America. From St. Joseph to Sacramento, where direct communication to San Francisco was by river steamboat, the distance was two thousand hard miles of plain and mountain wilderness. Yet over this the Pony Express operated—slender young men riding at top pace on slender young horses, changing mounts each ten miles, drivers each sixty—and carried messages an average of twelve days for the entire distance. Upon occasion they covered the distance in less time; as with Buchanan's final Message as President (in nine days and twenty-three hours), the flash of the election of Lincoln (in eight days) and his first Inaugural Message (in seven days and seventeen hours). This last remains a world record.

In October, 1861, a brief eighteen months after its incep-

TRANSPORTATION AND COMMUNICATION 117

tion, the Pony Express forever ceased to function. Telegraph wires, built simultaneously east from California and west from the Missouri, met that month at Salt Lake City, and the necessity of the couriers was gone. The telegraph continued to expand, to perfect the details of its practical operation. It came down with its shimmering wires to the edge of the sea, and then, through the genius and tireless energy of one man—Cyrus W. Field—it passed under the waters and across to far-distant Britain. This was the Atlantic cable, laid under incredible hardships, and first opened in 1858. Yet in a few weeks it was inert and dead. A lack of knowledge of proper operating conditions upon a 3,000-mile, unbroken telegraph line had led to the application of too much current to the new cable, and so to the complete ruin of its insulation. The whole work had to be done over.

For seven years thereafter the Atlantic cable scheme was a topic for derision. Yet Field and his associates never lost faith in it. They engaged a colossal ship (for those days)—the *Great Eastern*—which had been a failure in almost every other way, and made her a successful cable-layer. This time they made no mistakes in insulation or in electrical energy, although they had desperately trying times, with the cable breaking, more than once, in mid-ocean. By 1866, however, it was again in fairly successful operation, and thereafter, while it became necessary to make great additions and improvements to the original plant—including duplicate cables—the service continued, without break or serious hindrance. And other cables were laid under other seas, until the day came when a man could send a message by telegraph around the entire world—a feat of no small significance.

The telegraph was a clattering thing, spelling its messages in dots or dashes or printing them out mechanically upon a well-nigh endless paper tape. If only the human voice could be carried in a similar fashion over a fine-spun tenuous copper wire! Many men gave thought to this quite obvious suggestion. One American took it and made it into a device—presently also to become world-wide in its scope—of greatest value to his fellow men. This man was Alexander Graham Bell of Boston, and a little more than a half-century ago he was a teacher of acoustics—particularly as they applied to actual speech in the education of deaf-mutes. Dr. Bell had an unusual interest in the problem of a "voice-telegraph" and unusual ability for its eventual solution. Men before him who had studied the problem had aproached it, invariably, from the purely electrical side; Dr. Bell regarded it very largely as the creation of a mechanical organ of human speech, connected with a mechanical ear. So definitely did this idea become fixed in his mind that some of his earliest experiments were with a human ear, cut from the head of a dead man. This gruesome experience gave Bell a direct pathway of approach to the greatest single fundamental of the telephone, the delicate metal membrane which, upon receiving the human voice, vibrates and transmits it in electrical energy to a complement machine with a similar metal membrane which, in turn, converts electrical energy into human voice.

Upon the development of this principle Alexander Graham Bell won his struggle and took his proper place among the eminent inventors of all time. It was in March, 1876, that he first succeeded in making a real telephone talk—from the basement to the third floor of an old-fashioned house in Scollay Square, Boston. A few months later his device was the center of much attention at the great Philadelphia Centennial Exposition. Among those who came to see it there were the famed Joseph Henry, of the Smithsonian Institution; Sir William Thomson (afterwards Lord Kelvin), the greatest electrical expert of his day; and Emperor Dom Pedro, of Brazil.

For a short time telephone development was slow indeed. The telegraph companies, possibly fearful of its future, were at first intensely jealous of it. They opposed it at every turn. Eventually they were to join forces with it, to see in it a device of communication that supplemented and correlated with their own methods, rather than a thing which fundamentally opposed them. In the meantime, the telephone slowly gained headway. A crude exchange, with but five customers (all of them banks) was set up in Boston in May, 1877. At about the same time a long-distance telephone message was carried forward successfully over a leased telegraph wire from that city to New York. The idea of modern telephone service was firmly established, with almost infinite possibilities for its future development.

The first Boston exchange was quickly succeeded by others, in Bridgeport, in New Haven, in New York, and in Philadelphia. Bell, with the help of Gardiner Hubbard, his father-in-law, and of that tremendously practical man, Theodore N. Vail, who came to him from the United States Railway Mail Service, soon was leasing a thousand telephones a month. The Bell Telephone Company, forerunner of the American Telephone and Telegraph Company of today, was in formation. Progress then became swift. In 1877 there were but five telephones in service in the city of New York; a year later there were 252. Girls, swift and deft and dependable, supplanted boys at the switchboards; pay-stations were installed. Telephones became popular. When the New York system in 1893 —the year of the first great world's fair in Chicago—had come to have eight exchanges and over nine thousand subscribers, it was felt that here at last was a thoroughly developed thing. And yet the process of expansion had hardly even begun.

As the telegraph in its day and generation had begun to push itself across the country, so did the long-distance telephone now begin to reach out. For a time its progress was comparatively slow. There were large practical difficulties connected with its expansion. Gradually these were solved, and by 1895 it was possible to talk from New York to Chicago. That day in 1911 when, for the first time, one could pick up his telephone in New York and converse with Denver was made an occasion for a large celebration in the capital of Colorado. Yet how it paled compared with the one, four years later, when long-distance telephony was at last established between New York and San Francisco.

And after continental telephony came transoceanic. But this is recent history. On January 7, 1927, the president of the American Telephone and Telegraph Company, sitting at his desk in New York, talked rapidly and easily with the head of the British postal service in London. That very day commercial service began between the two chief cities of the world. By night, when the circuits were closed, a total of thirty-one business calls had been handled. The service was firmly established. A little later it was much extended; not only between cities, near and remote, both of North America and of Europe, but between other continents as well. It was only the other day that the King of Siam, sitting in his palace in Bangkok, lifted the receiver off his telephone and talked quickly and easily with New York in regard to a forthcoming visit there.

One other feature of telephony—highly dramatic and spectacular—remains for consideration here.

This is television. As these paragraphs are being written, this device still is in experimental stages, yet rather advanced ones. Already men are going to especially equipped stations, both in the city of New York and elsewhere, and are talking to one another, facing each other as they speak. The President, sitting in the White House in Washington, has both spoken and shown himself to an audience in New York, two hundred miles away. Plans are being made for the commercial adaptation of this fascinating new super-telephony. How far it may go no man may easily predict.

How far any form of transport or communication may go in the future it is difficult, even for men possessed with the widest possible range of imagination, accurately to predict. We have seen the stage-coach and the sailing-craft, followed by the railroad train and the steamboat; these followed by the motor-vehicle and the airplane; the telegraph opening its field to the telephone (in turn greatly extending that same field) and then to the radio—even though this last is to be regarded even more as a method of national amusement and education than a commercial means of communication. A great nation, a whole world of great nations, crossed and crisscrossed by all these avenues—of transport and communication!

This is but a single chapter that endeavors in a brief fashion to summarize the progress in these avenues in a single nation through a single hundred years. Yet a hundred years that can with good reason take that glorifying name—a Century of Progress.

CHAPTER FIVE

AGRICULTURE

Every city in the world could be leveled to the dust and countless millions could still live on the earth in a fair degree of comfort and with a good deal more security than the masses of the cities enjoy today. On the other hand, if agriculture were destroyed, every city would sink down in hopeless decay from its own weakness. Its buildings and its counting-houses, its tenements and its factories, would be the tombs of its inhabitants. And all mankind would have to return to primitive savagery. This is a commonplace too often forgotten by those who have occasion to write and speak in these times.

Agriculture

By Frank O. Lowden

BEFORE the rise of agriculture, human life was of necessity nomadic, following the wandering herds wherever sufficient herbage could be found. Then came the discovery of agriculture, pursuit of the civilized, crowned by towns and cities where the arts and sciences flourish. Without urban interests, civilization as we conceive it today could not exist. But it rests on the extraction of food from the soil, one of the most ancient of industries.

Notwithstanding its antiquity, however, more progress has been made in agriculture during the last hundred years than in all the centuries before since the dawn of time. Recent excavations in Egypt and other places have disclosed proof that agriculture was practiced in much the same way two thousand years before the birth of Christ as it was when our forefathers landed on the Atlantic coast. I have myself seen upon the pictured walls of recently discovered Egyptian tombs, built more than four thousand years ago, scenes which represent with fair accuracy the most advanced agriculture of all the centuries until the present one. The *Georgics* of Virgil, whose two-thousandth birthday we celebrated recently, was perhaps as good a textbook upon agriculture as any we had a hundred years ago.

THE HISTORIC SYSTEM OF AGRICULTURE IN THE UNITED STATES

During the Colonial days and even in the early years of the Republic the practices in agriculture were not different in the main from the methods that had prevailed theretofore throughout the civilized world. The farm was a self-sufficing home and little more. The farmer produced wheat primarily to feed his family. He hauled his wheat to a near-by mill and had it ground into flour, paying for this service by a portion of the produce. He raised cattle and hogs chiefly to furnish produce for his own table. He had a few sheep upon his farm to supply wool for his family, which was manufactured into clothing in the household. His farm implements were few and inexpensive and largely fabricated upon the farm. His taxes were low. His cash outlay, therefore, was very small; competent authorities have asserted that it did not exceed on an average a hundred dollars a year.

From the beginning the land was occupied largely by farm owners. The feudal system never gained a foothold in America. There was from time to time a battle between land speculators who sought to purchase land in large tracts and the actual tillers of the soil. The battle in one way or another usually resulted favorably to the actual occupant. This policy found its final expression in the Homestead Act, which opened up vast areas of rich and virgin land throughout the west in small holdings to actual settlers without charge. Thus America on the whole became a great agricultural nation, with land ownership vested largely in the actual occupants.

An exception occurred in some portions of the South, where cotton was the principal crop. The invention of the cotton-gin, just before the end of the eighteenth century, made cotton culture upon a large scale possible and slave labor profitable. In fact, the invention of the cotton-gin was the most significant factor in the evolution of American agriculture within the last hundred years. Large plantations, consisting of several thousand acres, became numerous in the cotton-growing states, and constituted the beginning of commercial agriculture in our country. The old idea of the farm as a self-sufficing unit was still cherished in the main, however, even by the large cotton-planters of that time. They largely produced their own food for man and beast and manufactured the clothing which was required upon the plantation. I myself have talked with former slaves upon a cotton plantation I own in the South, who described picturesquely the vast warehouses for food products, the great smokehouses for curing meats, the large amount of spinning and weaving that went on to supply clothing upon this plantation in antebellum days. They have told me that their old master used to say that there were only two things that they should buy upon the plantation. One was coffee, and the other sugar, as I recall it.

In recent years the cotton farmers generally have depended upon outside sources of supply for food for themselves and their livestock. It is interesting to note that the highest authorities are now urging the cotton farmers of the South to return in this respect to the older practice.

With this exception, however, the great body of American agricultural land was owned by those who lived upon it and cultivated it with their own hands. I know of no other instance in which the farmers of a nation have had so fair a start. In any other of the modern nations these lands would have been apportioned among the chief nobles, who would have imposed a heavy charge upon those who cultivated them. Thus from the beginning we have had an independent, self-respecting, and, upon the whole, relatively prosperous farming population.

The dangers the pioneer farmers faced, in their westward march across the continent, from hostile Indian tribes and no less dangerous privations gave them courage and strengthened their self-reliance, so that they have been the most potent factor in the swift upbuilding of our great country. It is from the farming population in the past that we have largely drawn our leadership in public affairs and in all the city vocations of men. Hardships, of course, they had, and yet there is a brighter side to the picture, which we are likely to overlook in these mechanistic days. There was an independence which is hardly possible to any class now. There was an abundance of game and fish. There were rural sports, of which we become aware when we read the life of Abraham Lincoln. There was often a finer community spirit than is possible in these days of the swift automobile. The people lived closer together and shared their common joys and sorrows more sympathetically, and, as many now think, builded character more securely. The awful fear of losing one's job or one's farm was not so immediate and pressing then as now. Of course, as measured by the wealth of today, these farmers of an early time were poor, but so was everyone else; no one went hungry and seldom did any face actual want of the necessities of life.

THE MECHANICAL INVASION

I have already mentioned the cotton-gin invented by Eli Whitney near the close of the eighteenth century. This great event marked the first step in the transition from the age-old, self-sufficing farm to modern commercialized agriculture. For the first time, I think, in history an important crop was produced on a large scale primarily for sale rather than for home consumption. Before the invention of the cotton-gin one man could separate only about three pounds of lint cotton from the seed in a day. The cotton-gin soon developed a capacity of a thousand pounds a day. As a result the growing of cotton expanded rapidly. Within a quarter of a century the cotton produced by the South had increased by 2,000 per cent. It continued to expand until very recent times, and for more than a century has furnished considerably more than half the cotton consumed by all the world.

Outside of the cotton areas, however, it is only a hundred years since the earliest invention of those farm implements which have revolutionized the agriculture of the United States. It is just a century since Cyrus Hall McCormick, of Virginia, devised the first entirely successful reaper, though many efforts had been made before. The elder Pliny, in the first century of the Christian Era, recorded the fact that "in the vast domains of the provinces of Gaul a large hollow frame, armed with teeth and supported on two wheels, is driven through the standing corn, the beasts being yoked behind it; the result being that the ears are torn off and fall within the frame." None of these inventions had proved practical, however, and throughout the centuries the harvest-

AGRICULTURE

ing of cereal crops had been largely by the sickle, supplemented in some instances by the scythe and cradle. The amount of labor required was enormous and was the limiting factor upon the farm in the acreage sown to cereal crops. McCormick's reaper, with the subsequent developments and improvements made by a long line of inventors, has transformed the agriculture of the nation. One hundred years ago every bushel of wheat represented the expenditure of three hours of human effort. Now we produce a bushel of wheat with about ten minutes of labor.

It is significant that this first of modern farm machines was invented at a time when the vast fertile reaches of the Mississippi Valley were being opened up to settlement. The wide and comparatively level prairies were particularly suitable to machine cultivation. It is important, too, that at about the same time the great transportation system of America, due to Watt's invention, was being initiated and vigorously pushed forward. Simultaneously, England repealed the corn laws, and the way was opened at once for a tremendous foreign trade in American farm produce. These factors interacted upon one another and ushered in a new age of agriculture such as the world had never seen before.

Another invention made a little later than the McCormick reaper, though not so spectacular, is almost as important. The plow of a hundred years ago was very like the plow of ancient times. It was built largely of wood with a metal share. It is true that the cast-iron plow had partly taken the place of this ancient tool, yet while more durable than the plow of antiquity, it was hardly more efficient.

While McCormick was working upon his reaper in Virginia, a blacksmith moved from Vermont to Grand Detour, a little town upon the Rock River about ten miles from the place in which I write. The pioneer farmers were finding great difficulty in plowing their rich black fields. The old plow had done its work after a fashion in both heavier and sandier soils, but it was found almost impossible to operate one successfully in the very fertile black prairie soils after the sod of this new country had once been broken and subdued. This young blacksmith, John Deere, set about the task of solving the problem. He took an old steel saw blade and fashioned from it the first steel plow ever made. It was found, as he had hoped, that the new plow took on a high polish and easily scoured in the most difficult soil. The invention of the steel plow was as important, I think, in the successful cultivation of the soil as the invention of the wheel in transportation.

The reaper and the plow have been followed by a long line of new mechanical devices, culminating in the combine and tractor of today. The industrial revolution which has created the modern world has not passed by the farm. The modern farmer, therefore, has become of necessity a mechanic, as well as a cultivator of the soil. The machines of which I have spoken substituted horse power for man power. They enabled the individual farmer to cultivate a larger area than he had cultivated before. Indeed, they made this almost necessary because these machines required a larger investment than the old agriculture. They resulted in a greater degree of specialization than had theretofore been known.

EFFECT OF SPECIALIZATION ON AGRICULTURE

The farmer discarded the spinning-wheel and loom and purchased his textiles from the factory which could produce them more cheaply than he. In many sections of the country he gave up his sheep in favor of other sections perhaps better suited to the growing of sheep, for he no longer needed the wool for his own household. The farmer who specialized in grain found it cheaper to buy pork and beef produced in the regions specially adapted to meat than to produce it himself. The smokehouse has largely disappeared from the modern farm, for the farmer's meat comes chiefly from the great packing companies that have grown up as a result of the machine age. The picturesque little flour-mill which a hundred years ago was found upon every small stream, to which the farmer carried the wheat to be ground for his own use, has given way to the mammoth mills in Minneapolis and elsewhere. The

AGRICULTURE

farmer of olden times who found the village blacksmith equal to the task of making the few simple implements he required has yielded to the great farm-implement manufacturing plants employing thousands of men.

Thus the farmer, hardly realizing it himself, has found himself an integral part of the great commercial world. The old self-sufficing farm has passed into history. Commercialized agriculture has taken its place. While we recognize the revolution that has occurred in industry since the beginning of the machine age, we are likely to overlook the fact that almost as complete a transformation has been effected in agriculture. Indeed, the industrial triumphs of which we boast would have been impossible if the machine age had not given the same momentum to agriculture that it gave to industry. A hundred years ago 90 per cent of our population lived upon the farm. They were producing only enough food for themselves and the remaining 10 per cent of our population, with exports negligible as compared with today. At the present time less than a quarter of our population is living upon the land, supplying its own needs, feeding our great and growing cities, with surplus so heavy as to reduce the price of farm commodities generally below the cost of production.

The application of steam power was the initial step in the industrial revolution. Steam power could be most economically produced by large units and it had to be applied at the place of its origin. As a result, we have witnessed a constant increase in the growth of population in the industrial centers. Our cities have grown ever larger and larger. Now electric power is rapidly supplanting steam power in the industrial world. Though electric power, too, can be generated more economically in large units, it differs from steam in that it can be widely distributed from a central generating plant. The trend in the past has been toward great centralization. It is altogether likely that the opposite process of decentralization will set in as a result of the great progress being made in the electric field. This from every standpoint is much to be desired. Perhaps the greatest beneficiary of this movement will be the farmer.

All through this agricultural depression in which we find ourselves the farmer would not have been so badly off if the prices of the things he has had to buy had not been out of all proportion to the prices of the things he has produced for sale. I am wondering if we have not carried the process of specialization too far. It is possible that many things we now purchase might better be produced upon the farm. It may be that this new and marvelous power we call electricity is going to enable the farmer to retrace some of the steps toward specialization which he has already taken. Isn't it likely that the farm of the future, equipped with electric current, will supply many of the needs of the family for which in recent years the farmer has depended upon others? To the extent he produces and fashions his own products to his own use he escapes the heavy loss which the large spread between the price at the farm and in the retail markets entails.

AGRICULTURAL RESEARCH

During the last century we have evolved a highly technical science of the soils. Justus von Liebig, who began his experiments in 1840, was the first soil scientist, and a few years later Sir John B. Lawes began his world-famous work at Rothamsted, England. The chemical composition of the soil was unknown before. It was then discovered that potassium and phosphorus and nitrogen were the chief elements of fertility in the soil. Lawes made the momentous discovery by actual experiment in the fields that soil fertility cannot be maintained by mineral fertilizers alone.

Virgil recommended to the farmer that he plant a crop of beans, vetch, or lupine before wheat. The beneficial effect upon the soil of these crops, now called legumes, was therefore known at least two thousand years ago. And yet the reason was discovered only near the end of the last century, when it was learned that nodules on the roots of the leguminous plants contained bacteria which are able to fix nitrogen from the atmosphere, convert it by unknown chemical action into protein, and pass it on to the plants. A little later it was found that there are certain bacteria in the soil itself which obtain nitrogen directly from the air. In view of the fact that in most soils nitrogen is the limiting factor in production, the importance of these discoveries may be realized.

It is only a few years since soil was conceived of as hard particles worn down from rocks and mixed with decaying animal and vegetable matter. It is now recognized to be a complex structure of organic and inorganic matter in which micro-organisms play an essential part. As Dampier-Whetham says, in his recent *History of Science*: "The ground beneath our feet is living, not dead; the function of the soil and its multitude of inhabitants is to break up the raw materials it contains, or which it gains from without, and to supply them in forms available as food for the plants that grow upon it."

With the rapid progress of invention and science as applied to agriculture, it was inevitable that agricultural education should develop to facilitate and disseminate this progress. Washington, the first farmer of his time, was not only distinguished by his feats of arms and his wise statesmanship, but he was also famous for his contributions to an advanced agriculture. It was fitting, therefore, that in his last Message to Congress he should have advocated the "establishment of a board of agriculture for the collection and distribution of information on farming methods and the development of a spirit of discovery and improvement." This suggestion, however, was ignored completely for more than forty years. In 1839, Congress created a small division of agriculture, which was tucked away in a corner of the Patent Office, of so little importance was it deemed. During the Presidency of Lincoln, a Bureau of Agriculture was established, from which time the contributions of the government to the theory and practice of agriculture have become increasingly important. In 1889 this bureau was expanded into a department with a Cabinet officer at its head. From that time until the present the work of the department has been rapidly extended, until it now has 22,000 employes, with a range of research, study, and distribution of information covering practically every phase of farm operation and management.

THE WAR ON PESTS AND DISEASE

Especially in the field of research has the department accomplished results of untold value to agriculture. It has sought out in every part of the globe varieties of grasses and grains best suited to the varied and diverse conditions of soil and climate which exist in the United States. It has been particularly successful in discovering and transplanting the natural enemies of imported insects which have threatened entire industries, for one of the dangers which confronts modern agriculture is that of some new insect appearing, deadly to some product of the forest or the farm. These are usually imported from other lands, where in the long course of the centuries enemies have developed, making them less deadly in their native habitat than they are here. I recall that an eminent botanist once said that if it were not for the war which insects of different species waged upon one another, they would soon possess the earth and man would go the way of the mastodon. Perhaps the best example of the kind is the white fluted or cottony scale which once threatened the complete and speedy extinction of the orange and lemon growing industry of California. It was in the 'seventies that this insect came by ship from Australia to California. The insect attacked fruit trees of all kinds, but soon developed a preference for orange and lemon trees. The trees attacked were ruined. Since this insect had left all its enemies behind in Australia it had a free field for action. The Department of Agriculture discovered in Australia, whence the scale had come, a particular ladybug whose diet was the scale. The entomologist of the department secured a number of these ladybugs in Australia and brought them to California. They multiplied rapidly and put an end to the cottony scale.

Equally important are the achievements of the department in stamping out livestock diseases. Three times in three decades the whole animal industry of the United States has been all but overwhelmed by a major disaster. First it was the Texas-fever tick which afflicted the range cattle of the Southwest. A few years later the hog industry of the corn belt suffered enormous losses from cholera. About the same time the dreaded foot-and-mouth disease arrived on our shores from Europe and struck down hundreds of thousands of farm animals within a few months. Upon each occasion the federal authorities succeeded in bringing these devastating invaders under control and finally in eradicating the fever tick and the foot-and-mouth disease. Hog cholera has lost its old terror because of the perfection of control measures. The triumphs of sanitary science have nowhere been greater than in the conquest of the scourges of the livestock world.

AGRICULTURAL COLLEGES AND STATIONS

The largest single contribution, I think, which America has made to scientific agriculture is the college of agriculture. America was the first nation of the world to recognize the most ancient of all pursuits of civilized men by establishing higher institutions of learning devoted to the study of that pursuit. The first agricultural college was founded in Michigan in 1857, and a few years later the forward-looking Morrill Act was passed by Congress, by which a grant of public lands was made to insure the training of youth in the science of agriculture. Every state now has an agricultural college. And from the agricultural colleges of America the cause of agricultural education has spread beyond the boundaries of our own country, until similar schools and colleges are now to be found throughout the civilized world. Our own colleges of agriculture, however, have maintained their preëminence, and students from every quarter of the world are found within their walls.

Experiment stations have played an important part in the progress of our agriculture. The first of them was founded in Connecticut in 1875, and twelve years later the federal government, through the Hatch Act, assured the continuation of these scientific agencies. The experiment stations are devoted to research in agricultural problems and exist alongside the agricultural colleges, with which they are closely correlated. Scattered, as they are, throughout the country, they are able to devote themselves largely to special problems affecting the several states. There is hardly a phase of agricultural production which these experiment stations have not considered, to its great benefit.

A notable instance of their achievements is the invention of the butter-fat test, by Stephen M. Babcock, of the Wisconsin Experiment Station. By Dr. Babcock's test the exact amount of butter fat, the most important part of milk, could be simply and accurately determined. In all the centuries that preceded Dr. Babcock there had been no radical change in the methods employed. Everything was by "rule of thumb." After him the dairy industry began to take on its present highly scientific form. The processor was at last dealing with raw material which could be expressed in precise terms. The dairy farmer found himself in a position in which he could accurately determine the relative value of his cows. He could then for the first time begin to breed intelligently along those blood lines which the Babcock test had shown to be superior. The coöperative creamery and the cow test association, which give a greater promise than any other I know for an improved economic status of the dairy farmer, were made possible by the Babcock test. I have already spoken of the cotton-gin as ushering in a new era in the great cotton industry of the world. I have also pointed out that the McCormick reaper likewise transformed that large area devoted to the growth of cereals. In no less a degree did Dr. Babcock's invention revolutionize the dairy industry of all the civilized world.

The colleges of agriculture and the experiment stations have from the beginning been uncovering new and better methods for agriculture. It proved difficult, however, to translate their newly discovered learning into actual practice upon the farm. Therefore, nearly forty years ago, they founded what is called extension work, by which direct contact was

AGRICULTURE

established with the farmers throughout their several states. In recent years their effectiveness has been greatly increased by the creation of county farm bureaus, which, with a trained farm adviser at the head, are now found throughout all sections of the country. These agents are the direct connecting link between the farmers of the county and the agricultural college and experiment station of the state.

Another great agency of education in agriculture is the farm press, which also is the product of the century. The oldest existing farm paper was founded in 1841. The farm press has a number of unique qualities in the field of journalism, of which perhaps the outstanding is the tremendous hold which leading editors have upon their readers. In the old days the identity and personality of the editor of the general newspaper were well known. Today, however, anonymity is the distinguishing character of the great metropolitan press. Many regret this change. It is interesting to note, however, that it is one which has affected the farm press in only a small degree, for the leading farm papers are readily identified in the farmer's mind with the names of the editors who control their policies.

We have seen that the past century has been characterized by marked improvements in crops, in practices in orchard and field, in soil management. Livestock improvement has been equally notable. The old fashioned long-horn steer has gone the way of the buffalo, and quick-maturing baby beeves now supply the markets. Similar changes have eliminated the razorback hog and developed his slow-maturing successor in the Miami Valley.

The increased efficiency of American livestock is best measured in the field of dairy production. A survey by the Department of Agriculture established the fact that the average milk production per cow in 1850 was 1,436 pounds annually. This rose to 3,646 pounds at the end of fifty years, and to 4,600 pounds in 1927. Thus the efficiency of the cow has increased more than 200 per cent since the middle of the last century.

If today we were obliged to go back to the farm animals of even two hundred years ago, famine would threaten the world.

AGRICULTURAL ORGANIZATIONS

Agriculture was the last of all great industries to employ the principle of organization. There seemed no need of organization with the self-sufficing farm. The farmer's commerce with the outside world was hardly more than the barter of ancient times. In effect he exchanged his surplus products in the near-by market place for the few things he was compelled to buy. He traded butter and eggs for his sugar and coffee and tea. He paid for grinding his flour, not with money, but with a portion of the product. He supplied the blacksmith and the carpenter and the mason with food of various kinds for the services which they rendered to him. The moment agriculture passed from this simple stage to the present commercialized form, all this was changed. Transporting his surplus products to a far distance, even to foreign lands, the farmer became interested in transportation rates. Purchasing his machinery, not from the local blacksmith, but from the great farm-implement factory, and buying those other commodities no longer fabricated at home, he suddenly found himself a part of the great industrial and commercial world. He then began to see that he too must organize if he were to secure justice for himself in a highly organized world.

Organization is a most powerful factor in human progress. It means the difference between the mob and a highly progressive society. The psychologists tell us that in an unorganized mass of men the primal instincts have full sway. Therefore he who appeals most strongly to the passions of the unorganized mass influences them most. Reason, justice, mercy, all the finer qualities which civilization has evolved, are swept aside by the rising tide of the primitive passions of men. The voice of the scholar, of the humanitarian, of the moralist, and of the preacher is drowned. Now, if we take this same unorganized mass and succeed in giving it a thorough organization, the wisest and best men within it rise to positions of leadership. The result is that its conduct, instead of inclining to the plane of the lowest, will tend to rise to the level of the highest. Through organization only does progress come.

In the modern world, the farmer has been the last to realize the value of organization for its own sake. Therefore it happens that when the farmers in any community organize for any one purpose, they soon find that there are other benefits to be derived in addition to the one which was their special aim. A finer community life, a widening of sympathies with their neighbors and associates, a broadening of their outlook upon the world, a new sense of the dignity and worth of their calling, an elevation of the ablest and worthiest among them to places of leadership, are among the by-products of farmers' organizations.

It was not until the last half of the last century that farm organizations of importance came into being. The first of any large influence was the Patrons of Husbandry, commonly known as the Grange, which was organized in 1867. It adopted an ambitious program by which it was hoped the farmers collectively would have a larger voice economically and politically. Though it failed in many of its objectives, it did exert a powerful influence. To its efforts were largely due the policy of political control of public utilities and other measures which have now become a part of the settled law of the land. Though it has abandoned some of its earlier projects, it still has great vitality. There followed the Farmers Alliance, the Wheel, the Equity, and the Farmers Union. The youngest of the great general farm organizations is the American Farm Bureau Federation, an organization resting primarily upon the county farm bureau, of which I have already spoken. Each of these organizations has from time to time exerted a genuine influence toward agricultural improvement. They have all fostered coöperative dealing in some form or another. Indeed, antedating any general farm organization there were local organizations through which groups of farmers disposed of their produce. The reason for this must be obvious to everyone.

MARKETS AND PRICES

In the industrial and commercial world we have been undergoing a silent revolution as to methods of control of production, of selling and of determining prices, of which we are hardly conscious. This has come about in various ways. In some of the important industries there is a single corporation so large and powerful as practically to dominate the field. It has competitors in a way, and yet, by common consent, this large corporation is permitted initially to make the price. Its rivals in practice follow substantially the price thus made. There may be no agreement in violation of anti-trust laws, but the lesser rivals have learned from experience that it is perilous to engage in price-cutting with the larger corporation. In other industries the same result is obtained through the trade association, which has become so popular a device of modern industry.

Through these and similar practices the manufacturer has acquired some measure of control over the price of his products. He may sometimes err as to the price at which his commodity will move, and thereby have to revise his price downward, but he himself in the first instance sets the price. It has followed from this trend that brokers and middlemen in all other fields have been gradually disappearing. The producer has assumed the burden of marketing his product. And so the farmer finds himself in a business world in which the prices of the things he has to buy come to him ready made. He can, of course, refuse to buy for the time, but in the end he must either yield or go out of business.

Upon the other hand, when it comes to selling under present market conditions, he is not in a position to have a voice. The price which is offered to him he, acting as an individual, must take. He wonders how long he can survive in a world which decides for him what he shall pay for what he buys and also what he shall receive for what he sells. He is groping for

AGRICULTURE

some way by which he shall have the same voice as to the price of his products as other people have in determining the price for theirs. The most promising remedy for this unbalanced condition seems to be coöperative societies owned and controlled by the farmers themselves.

AGRICULTURAL COÖPERATIVES

The first American coöperatives were dairy associations in Wisconsin and New York and fruit associations in New Jersey. For many years the coöperatives made little progress, rising in periods of distress and declining in prosperity. Gradually, however, the fundamental principles of producer coöperation were established and the success of the California Fruit Growers Exchange, founded in 1893, caused the application of these principles to practically all commodities produced on the farm. There were some spectacular developments, followed by several disastrous experiences, but the volume of product handled through coöperatives is now close to the highest figure in history. The farm coöperatives of one kind and another now market a substantial and growing percentage of all the products of the American farm.

Organization of the farmers for the purpose of marketing their crops collectively is progressing. Denmark has shown how, under the most adverse circumstances, it can transform the agriculture of a people. Wherever coöperative marketing is farthest advanced, either in the United States or abroad, there you find agriculture in its best estate: violent fluctuations in the markets lessened; better prices to the producers without an increase in cost, and sometimes with an actual decrease to the consumer; an approach to standardization of product; a more intelligent effort to adjust production to probable demand; a finer and more satisfying community life.

It is doubtful, however, if the coöperatives of the staple farm products will ever be sufficiently organized to take care of the ever-present problem of seasonal surplus unless some way be found by which the cost of handling it is borne equally by all producers of the particular commodity. If the producers of any staple farm product are only partly organized and attempt to take care of the surplus, the producers of that commodity who are not members of the coöperative receive the full benefit of the improved price which the increased bargaining power of the coöperative may secure without bearing any burden incident to the surplus. To illustrate:

The tobacco coöperatives were very successful for a number of years. When farm prices broke in 1920, the tobacco growers were among the severest sufferers. Tobacco was selling far below cost of production. And then coöperative marketing associations were formed. Through their largely increased bargaining power these associations were able to sell the bulk of their crop at remunerative prices. To accomplish this it was necessary to withhold a surplus temporarily from the market. That entailed a necessary expense. The non-member, therefore, who was able to avail himself of the better prices which the association had established without bearing any part of the burden of handling the surplus, profited even more than the members of the coöperatives.

It is impossible to maintain the *morale* of an organization when outsiders receive its benefits in a larger measure than do the members themselves. For this reason the tobacco coöperatives and some others were driven out of business.

THE LARGE CORPORATION IN AGRICULTURE; ITS FUTURE

Owing to the continuous improvements in labor-saving machinery, the outstanding farmers of the country have gradually enlarged the acreage they have cultivated. This has led to a quite prevalent belief among city people that the family-sized farm is soon to give way to large-scale corporate enterprises. I do not believe there is sufficient evidence to justify this assumption. It is commonly recognized that for ten years agriculture has not participated in the prosperity which most other industries have enjoyed. It has indeed been a difficult time for the farmers to make both ends meet. It is interesting to note that during this period the larger farm enterprises have suffered more than the comparatively small, one-family farms.

In fact, the two most widely advertised corporation farms in the United States are now in severe financial distress, and a considerable portion of the others have suffered greatly. This recalls the fact that a generation ago there was a great development of large-scale farming in such districts as the Red River Valley, the Sacramento, and the Palouse, but in the course of a few years nearly all were overtaken by vicissitudes which most family-sized farms survived. No other form of farm enterprise has as yet demonstrated its superiority under American conditions to the family farm.

In England experience seems to have led to the same result. In a recent book on *The Future of Farming in England*, by C. S. Orwin, the author states that only the "family farm" has escaped the disaster which has come to agriculture generally in that highly industrialized country.

It is probable that the ablest farmers will continue to expand their acreage somewhat as farm machinery improves, but I see nothing that warrants the belief that corporate farming is ever to take the place of the family-sized farm. Indeed, while in the corn belt, the cotton belt, and the wheat belt of the Southwest and the Northwest, there may be many larger-sized family farms, it is not at all unlikely that smaller farms than we have heretofore cultivated will become the rule. The prices of farm products at the present writing are lower than they have been for many years. And yet on the smaller farms, where cash outlay is reduced to the minimum, there is a sense of security and well-being today that we do not find in the great industrial centers of the land. These farmers are obliged to practice a rigid economy, but they have no fear of going hungry or of suffering for the actual necessities of life. Indeed, there are millions of workingmen in our great cities who look longingly to the little farm. It is entirely possible that the serious problem of unemployment which now afflicts us could be best met if some of our surplus lands which are now devoted to commercial agriculture were to be utilized as nearly self-sufficing homes for the millions of unemployed. Particularly will this be possible, I think, when industry has become more decentralized, which seems to be the tendency of the present time.

Those who think the corporate farm on a large scale is to supplant the family farm overlook an important factor in the situation. There are millions of men—and always will be—who love the farm, who love to work in the open and close to the soil, and who are willing to take a part of their compensation in the satisfactions which they can get out of such a life, its independence, its freedom, its guaranty of a home life. A home, in the old sense of the word, is more easily achieved upon the farm than anywhere else. The husband and the wife and the children are all engaged in a common task. There is a mutuality of interest among the members of the family in the country which is impossible elsewhere in the complex civilization which the machine has brought. The family is more closely knit because its members are all engaged in the same enterprise—that of making a home. If large-scale corporate farming were to take the place of the present system, wages would have to be paid commensurate with wages paid in industry.

THE DARKER SIDE OF THE PICTURE

I have spoken of the achievements of American agriculture. It is an inspiring story. We come now to a less pleasant task. The triumphs of American agriculture have been due primarily to our great abundance of land of unsurpassed fertility. Because of this we have been more wasteful of our fertile soils than any other nation of which I know. In the early days of the Republic the pioneers who had taken the first fruits from their fields moved on toward the west where other virgin fields awaited the plow. This process continued until there were no longer good new lands to be had. It is only recently, therefore, that we have come face to face with the problems of the conservation of our soils. For the future we shall have to depend, as the older nations do, not upon new lands to be

AGRICULTURE

opened up, but upon the maintenance of fertility upon the lands already under cultivation.

The land of any country is the basis upon which its civilization must rest. Industrial structures may multiply without end, but the land has limits which nature has fixed. The common brown dirt from out of which all vegetation springs, the minerals found therein, and the trees that grow thereon compose the natural resources of a nation. And these resources are the limiting factors in a nation's growth and progress. Most important of all is that portion devoted to agriculture. For upon this the food of the people and an important part of the raw materials of industry depend. Less than any other civilized nation have we given heed to these primary considerations. We have destroyed our forests and given over the land to farming, where it is suited only to a new growth of forest products. We have established farms in semi-arid regions where the land was suited only to grazing purposes. In an age when system and science have powerfully impressed every other of our major activities, we have employed our agricultural resources without either. Isn't it time we made an inventory of these resources? Isn't it possible to make a survey of these millions of precious acres, and to classify them according to the best usage to which they can be put? If this had been done a half-century ago we would not now be humiliated to find huge devastated areas in the Great Lakes region upon which the forests are entirely destroyed, lands which have reverted to the states, or which have been sold to deluded farmers, only to be abandoned when their owners had been starved out. We would not have seen those immense migrations to semi-arid lands fitted for grazing and nothing else, where only bankruptcy in both purse and spirit awaited the farmers who settled upon these lands.

All students of our agriculture agree that the depletion of our soils is going on at a dangerous rate. At a time when farm surpluses are often driving the price of farm products below the cost of production, we are adding to those surpluses by robbing the soil. If we had the care for future generations which our great heritage demands of us, we should find means to stop this double loss. There is probably not an agricultural college in the land which does not teach that in order to maintain the soil some legume crop should be grown at least one year in three or four. And yet it is safe to say, I think, that upon the farms of America as a whole our lands are in legumes much less than one year in ten. Upon most of our lands which have been farmed for any length of time it is necessary to apply lime before a satisfactory legume crop can be raised. The average farmer knows this, but this costs money. And when he is straining every nerve to pay fixed charges, he puts off these needed things until a more prosperous year.

Not only are the soils being gradually depleted by constant cropping, but erosion of the soils proceeds at a progressive rate. Indeed, Dr. H. H. Bennett, of the United States Department of Agriculture, makes the startling statement "that the productivity of our agricultural and grazing lands is being depleted and even destroyed by erosion at a rate many times greater (certainly not less than twenty times faster) than that by the crops taken off."

We commonly think of erosion as occurring only on hilly lands, where gullies greet the eye. This is but a fraction of the damage which erosion causes. For there is another form which it takes. This is called sheet erosion and is slowly at work upon even comparatively level fields which are constantly under cultivation. For if the land is overworked through constant cropping, the humus or organic matter in the soil grows less and erosion moves forward at an increasing rate. An experiment made by the Missouri Experiment Station showed that a field with less than 4 per cent slope, which is probably near the average for corn belt lands, in corn continuously, would sustain a complete loss of the top soil in fifty-seven years, but that with a three-year rotation of wheat, corn, and clover, it would require three hundred and twenty-four years to bring about this result. If one acre out of three could be in clover or another legume, it would not only revolutionize the movement for soil conservation, but at once the

AGRICULTURE

farmers of the United States would be relieved largely of their burdensome surpluses, and future generations would be insured an adequate supply of food at a reasonable cost.

THE NEED FOR FORESIGHT AND PLANNING

After the needs of agriculture and forestry are satisfactorily met, there will be great areas scattered all over the United States not best suited for the growth of either crops or trees. Some of the land contained within these areas is now being farmed without profit. I believe that this land can be utilized to a better purpose.

Owing to the advance of science and invention, the tendency, which would seem to be a continuous one, is toward shorter working-hours. The problem now confronting us is whether we shall be able to use the increasing leisure promised to the advantage of mankind. In the past, the character of a civilization has depended largely upon how men and women disposed of their time at work. In the future, isn't it possible that the quality of civilization will depend largely upon what we do in our leisure time?

I can conceive of no higher use to which we can devote our superfluous lands after the needs of agriculture have been met than as great playgrounds upon which the people from our cities and towns may find that communion with nature which seems to be an elemental need of the human heart. Our mountains and lakes, and even deserts, speak a lofty language to the tired souls of men which is never heard in the most splendid cities of the world.

In all the past good land has been one of the objects most keenly sought. Not only have individuals so thought of land, but nations as well have regarded themselves happy when they could add to their domain some area somewhere of fertile soil. A year ago last winter I visited Egypt. The historic valley of the Nile contains only about ten thousand square miles of delta land, and yet every empire of ancient times hazarded war to gain possession of these ten thousand square miles. Rome's possession of this area probably prolonged the life of the Roman Empire for a century or more by furnishing food to the Roman populace when it was no longer able to feed itself.

In the Mississippi Valley alone we have about four times as much delta land as Egypt contains. This single resource ought to distinguish us among nations. And these lands are but a small part of our agricultural heritage. All at once, however, we seem to have abandoned our idea as to their importance and value. Perhaps in earlier times we attached too much importance to land. If so, we have now gone to the other extreme. If we were planning intelligently for the long future, I think we should find that we have not a single acre beyond our needs. Land again would recover its ancient prestige in the estimation of mankind.

Progress has heretofore been effected largely along the lines determined by the exigencies of the present, but in late years there has been a notable trend toward conscious and comprehensive planning for the future. One of the most notable instances is city planning. It was formerly thought that a city could be left to develop naturally in response to the needs of the individual owners. Factories were welcomed to the city. They were permitted to select their own sites. Apartment houses were needed and their builders were allowed to erect them wherever they wished. Everyone was given the liberty to use that little portion of ground which he called his own for any use he desired. As cities became larger and larger, it was found necessary to restrict the liberty of the citizen in the use of his land. There followed building restrictions of many kinds. City zoning was introduced. Today city planning is one of the recognized professions. Something like this must come to the rural regions of the United States.

The industrial age of which we have boasted so much is only a little over a century old, and yet it seems to be sadly out of gear. We are in the midst of a world-wide depression. We are producing too much of wheat, too much of cotton, too much of raw materials of all kinds, and too much of manufactured products for the world to consume. So in all fields men are discussing a broader and more comprehensive planning in industry than ever before. Nowhere is this needed more than in the proper utilization of our lands.

Science and invention have transformed the modern world. They have achieved mighty triumphs, but they have created many problems. In no field of human activity is this more true than in agriculture. That these problems are pressing for solution is now commonly recognized. The future welfare of our country depends upon their wise solution. For though industry looms larger and larger all the time, it becomes clearer as the years go by that a country can remain in a healthy condition only by maintaining a proper balance between industry and agriculture.

CHAPTER SIX

LABOR

"One of my most vivid early recollections is the great trouble that came to silk-weavers when machinery was invented to replace their skill and take their jobs. . . . Misery and suspense filled the neighborhood with a depressing air of dread. The narrow street echoed with the tramp of men walking the street in groups with no work to do. Burned into my mind was the indescribable effect of the cry of these men, 'God, I've no work to do. Lord, strike me dead—my wife, my kids want bread and I've no work to do.' Child that I was, that cry taught me the world-wide feeling that has ever bound the oppressed together in a struggle against those who hold control over the lives and opportunities of those who work for wages. That feeling became a subconscious guiding impulse that in later years developed into the dominating influence in shaping my life. Up and down Fort Street men were walking, wringing their hands, and striking their heads."—SAMUEL GOMPERS, *Seventy Years of Life and Labor*, I, p. 5.

Labor

By William Green

THE past hundred years has recorded the change from conditions when an individual relied upon his own activities to supply his wants to an era of social and economic interdependence where the material necessities of life are supplied by various groups and even opportunity to work is controlled by group agencies. When free land was no longer open to all, control over work opportunities passed into the hands of employers and then to paid management as ownership became collective. With the development of factory production and mechanical power, ownership of tools even passed to employers. With standardization of parts and progress in perfecting machines, manual skills were transferred to machines. These fundamental changes have swept through many industries and partly invaded others. The trend is everywhere apparent, although the degree of development has been unequal in various geographic areas. Industry had forgotten that labor was a partner in production.

Formerly when an individual selected his occupation he supplied capital, tools, and labor. But as the function of supplying these elements became differentiated and owners of capital had to act first, the capitalist made himself a dictator.

The fundamental economic problem has been for labor to recapture its functional partnership in production. As workers had to act collectively to function normally under the changed conditions, labor was the first economic group to develop the practices and technique of associated activity. The pattern which labor evolved is now being followed by industries through trade associations, technical groups, and financial groups. Associated activity has made necessary coöperation by individuals with similar interests. Interrelation, the outstanding fact of the past quarter of a century, makes imperative coöperation between groups within an industry, between industries, between producers and consumers. Progress with and not against other groups is a conception that has grown out of the century's experiences.

Labor is an integral part of national life in its daily practices and its efforts to supply its varying needs. A labor movement is, therefore, a national institution whose policies and character take their set from the spirit and the habits of the nation. The immediate ideals of the labor movement are determined by the industrial situation, the degree of social progress which the community has reached, the development of democratic governmental institutions sensitive to the needs and problems of workers as well as other groups.

THE ECONOMIC BACKGROUND OF THE LABOR MOVEMENT

In 1840 only 44 towns had a population of over 8,000. Of a population of 17,069,453, a little more than one-third were gainfully employed; 3,719,951 in agriculture; 117,607 in commerce; 791,749 in manufactures; 15,210 in mining; 56,021 in ocean navigation; 33,076 in inland navigation; 65,255 in learned professions and engineering.

The 1930 census reports a population of 122,775,046, of which 48,800,000 were gainfully employed; 10,500,000 in agriculture; 7,500,000 in commerce; 14,300,000 in manufacturing and mechanical trades; 1,200,000 in mining; 171,000 in navigation; and nearly 3,000,000 in the learned professions and engineering. There were 93 cities with a population of 100,000 or over and 982 with a population of 10,000 or over. In this hundred years the United States has changed from a rural to an urban nation.

With inventions and technical improvements, costs of production per unit were decreased and output increased. The less expensive processes of puddling and rolling multiplied the output of iron. The sewing-machine reduced the costs of

LABOR

factory-made clothing to one-fourth that of hand-sewed; the cost of shoes to one-eleventh that of hand-sewed. Within this hundred years we have substituted electric dynamos for hand power, central heating for fireplaces, the telegraph and telephone for the post-rider, electric lights for candles, tractors for hand-plows, sewing-machines for hand-needles. Electric cranes, steam shovels and dredges, power presses, assembly lines, and combustion engines revolutionized industries as well as modes of life. Our use of mechanical power has widened until we now employ as much as the rest of the nations combined. During the past ten years use of power has grown nearly four times as rapidly as the population. Such changes as these made possible larger-scale production housed in factories.

With large-scale industries came the development of recording, sales, production, and financing as distinct functions. Planning and operation of routines have become indispensable to the orderly flow of work. At first it was thought that management was a special function to be exercised by a central department. Later we found that management was a method of work that needed to be participated in by all on the job as well as the central office. Labor, which handles materials and tools, has a special knowledge of production that is necessary to efficient management. With the massing of workers in the employ of a single company individual relations were no longer possible. The need for organizing to develop a representative basis for participation in planning and decisions directly affecting wage-earners was voiced by trade unions. Unions have adapted methods and principles to varying problems and conditions.

One hundred years ago homes were still the center of hand industries. Professional craftsmen were employed principally in construction work, the metal trades, printing, the leather industry, textiles, carriage- and wagon-building, flour-mills, and a few other industries. Employers managed their own businesses, knew their employes personally, and sold their products within the community.

Labor organizations were local in scope, mainly benevolent and fraternal in character. In the evolution of fundamental economic functions essential to stable development of industry and society, the spirit and the practices of the labor movement are strikingly interdependent on general levels of thinking and attainment, spiritual and material.

The changes that have occurred in American industrial life in the past one hundred years are more revolutionary than have come in the span of any thousand years of previous history. With the winning of manhood suffrage, non-property-holders had the right to participate in civic affairs. The whole trend toward a wider basis for the organization of business activities, the increasing importance of the federal government, and the development of the agencies to reduce the barriers of space welded geographic divisions into one nation and wiped out divergences in customs coming down from Colonial times.

HISTORICAL DEVELOPMENT OF UNIONS

The earliest forms of labor organizations were societies for mutual relief. What inter-craft organizations existed were concerned with reform issues or specific projects. Early federated efforts pertained to land reforms, suffrage, public schools, a ten-hour law, and greenback currency. These unions were designed for temporary purposes.

In the larger centers, however, unions often formed local federations. Until a national system of transportation by rail and a network of communication binding communities into a national market brought the need for national standards for an industry, local craft unions felt no impulse to unite in fraternal trade organizations, extending throughout the country.

The steadily increasing use of machinery which established the factory system, together with those unifying forces which were welding geographic districts into a nationally integrated economic organization, made necessary union developments paralleling industrial and social trends. In the 'fifties came the formation of national trade unions in a number of crafts, though but few survived the Civil War. The war itself brought sharply rising costs of living and widespread movements for wage increases, greenbackism, or proposals for democratic control of credit and capital. Then came the development of standard trade-union practices and policies. Many unions joined in these movements. Politics was the method by which workers approached their larger problems. The National Labor Union, the first attempt at union federation on a national scale, organized in 1866, gave promise of valuable work, but was diverted into politics and died.

After experiences with coöperative production, secret organizations, and successful resistance to radical propaganda, American trade unions began to find their distinctive problems and to discover how to deal with them. Unions rejected both socialism and anarchism and set themselves to make economic institutions function more equitably for wage-earners. There were no traditions of fixed social classes to lead them to a program of labor domination.

While national trade unions were increasing in number among the several crafts, various efforts to secure national labor legislation, together with the interrelation of the problems confronting all wage-earners, emphasized the need for national coöperation among all unions. The need for this coöperation was first evident in connection with such problems as competition in national markets, control over child labor, and the restriction of immigrant labor. To meet these demands, two organizations were founded in due course— the Knights of Labor (1869) and the Federation of Trades and Labor Unions of the United States and Canada (1881). The first was a secret organization which brought together workers and those sympathizing with their cause, and the second sought to unite wage-earners by a federation of their union organizations. The Knights of Labor, following established customs, admitted employers and labor sympathizers to its membership. Although it chartered some trade-union groups as separate organizations and attempted to perform union functions, it did not allow these groups trade autonomy and could not furnish the economic administrative control and leadership which became increasingly necessary.

THE AMERICAN FEDERATION OF LABOR

To secure the mobilization of labor strength necessary to secure more favorable work agreements and higher standards of living for workers, the national trade unions in 1886 reorganized the Federation along its present lines, changing its name to the American Federation of Labor. From that date, the Federation concentrated on economic problems. It held that the best interests of wage-earners as well as the best solutions for industrial relations problems could be reached through joint conferences of representatives of workers and employers. Believing that voluntary agreements afforded the best basis for group activity and progress, the Federation has applied principles of voluntarism in all its policies and relationships. Mutual interests are the proper force to maintain an organization and keep it flexible and industrious in accomplishing its purpose.

The labor movement gradually transformed itself from a sporadic, crusadic effort into a stable business organization with established agencies and practices for dealing with wage-earners' problems of life and work. To the end of the nineteenth century the Federation's development was characterized by the slow extension of union organization and a merging of unions so as to define union responsibility and jurisdiction more precisely. The union became a stable institution with well-defined business methods and financial resources. Then followed two decades of more rapid growth, checked temporarily by business depressions and the outbreak of the World War. Post-war readjustments brought a slump, succeeded by steady gains. At its formation in 1881 the Federation had an affiliated membership of 50,000; its peak was 4,078,740; its average membership for 1930 was 2,961,096. There are 105 national and international unions affiliated to the Federation. For the last fiscal year, these international unions paid in benefits to their members $32,242,440.40. We estimate that the benefits paid directly by local unions duplicated this amount.

While earlier unions were composed of manual workers and persons in wage-earning capacities, other workers, such as stenographers, office-workers, and bank clerks, saw that the principles of unionism applied equally to their problems. In the field of federal employment, unions of workers in the various branches of the postal service came first, and those working in the executive departments of the federal government organized to affiliate with the Federation. The membership of these unions covers a wide range of occupations—professional persons to charwomen. These unions have a constitutional provision against strikes. In the professional field, the first important union to identify itself with the labor movement was the musicians. Years later came the actors and the teachers.

The most notable organized groups not identified with the Federation are the four railway brotherhoods. Their industry developed separately and evolved distinctive problems. But in legislative undertakings there has been most cordial coöperation between the railway brotherhoods and the Federation.

While the craft has been the foundation around which unions have been formed, the jurisdictions of unions have changed notably as changes have come in industry. The Federation has found that it is often necessary to make a whole industrial group the basis. Therefore it has not advocated either principle to the exclusion of the other and maintains that neither shall be dogmatically advocated. In each case the basis of organization best suited to the specific conditions should be chosen.

Reflecting large-scale business organization and a resulting concentration of economic power, some craft unions have found it advantageous to coöperate through departments so as to operate on an industrial basis. Three such departments have been organized—the building trades, metal trades, and the railway shopmen's trades.

The craft basis was particularly essential when the work experience of a craft needed to be handed on to the next generation. As technical changes and machine tools materially alter the nature of work skills, bases for union structure must be developed which will function efficiently under the new conditions.

ORGANIZED LABOR AND IMMIGRATION

The changing make-up of the wage-earner group has been an important element in the growth of the labor movement. The Colonial population consisted chiefly of English, Dutch, Germans, and Swedes. These "Nordic" immigrants had fundamental characteristics in common so that all were readily welded together in the making of a nation to conquer the continent. In the decade between 1831-1840 the total number of immigrants was 599,125. Then immigration increased rapidly until 1860. After the Civil War the total mounted again until it reached 8,705,386 for 1901-1910. After the 'eighties the source of the heaviest immigration began to shift to southeastern Europe. The Latin and Slav nationals had little in common with the Nordics. They came from agricultural communities and their standards of living were low. They were able and willing to underbid American workers. Limited resources and language handicaps tended to concentrate them in the poorer urban communities. They flooded the market for unskilled or semi-skilled labor.

Newer waves of immigrants tended to push American-born workers or earlier immigrants to other jobs or industries as the newer immigrants poured into the ranks of common labor and certain immigrant industries.

In the decade 1881-1890, 5,246,613 immigrants were admitted to the United States. The number from Russia practically equaled that from Germany. But the workers from Russia and other Eastern countries had no knowledge of a labor movement and no understanding of liberty with its rights and obligations. The majority of the earlier immigrants brought with them understanding of industrial work and experience in the labor movement. These craftsmen helped to build the American labor movement. The later immigrants

LABOR

had less capacity to adapt themselves to their new conditions. They had to be industrialized, Americanized, and unionized.

It was the labor movement that first realized the necessity for a national immigration policy. On the Pacific coast wage-earners had their first experience with a labor group which threatened their welfare because it persistently retained the low living standards of another country. The first step to control immigration was the Chinese Exclusion Act of 1882. The principle was the exclusion of a nationality that was essentially different in race and standards. Later the principle was extended to the Japanese.

When large employers of labor, such as the steel companies and the coal operators, adopted a deliberate policy of importing large blocks of different races in order to create barriers to unionization, the labor movement began to give serious attention to the need for a controlled labor market. The way in which immigration complicated union problems may be illustrated by a single example: in one organizing campaign it was necessary to translate labor literature into eleven languages.

In 1896 the American Federation of Labor suggested a literacy test as the first immigration regulation. This measure was not made law until 1917, when the need for regulation could be denied by few. After the war, quota provisions made it possible to control the source of immigrants in order to maintain a better balance between national groups so as to provide time for the necessary assimilation. The war brought out the unwisdom of creating barriers to the assimilation of wage-earners into the national group conforming to American ideals and standards.

FUNCTIONS OF UNIONS

Originally the workday in industry followed the standards in agriculture. Wage-earners were spending practically all their waking hours drudging away in the shop until they undertook to get better conditions for themselves. They had to plan for their own progress and organize to carry out their plans in order not to be pinned down by the machinery of industrial progress. We have succeeded to the extent of winning the highest wages and the highest standards of living in the world. Progress was made possible by these facts: (1) The unusual natural resources of the United States with openings for self-employment that made wage-earners scarce; (2) the determination on the part of labor to have a hand in controlling certain industrial decisions so as to promote their own interests; (3) technical progress which increased productivity per individual; (4) educated workers with individual initiative.

In the early days there was no way for workers to get a hearing for their opinions except by public declaration and strike. Individuals working side by side did not know what was in their fellow workers' minds and what action they would be willing to take collectively, until they had an organized policy-making institution.

The beginning of participation by wage-earners in those industrial decisions that affect them so vitally came through organization and conference. Employers resisted attempts of wage-earners to secure a voice in labor decisions—first on the ground that they would tolerate no interference in their business and then that they would tolerate no interference by outsiders. But the labor movement represented a determined effort to secure for wage-earners a status in industry—a status predicated upon identification with our industry and functional service. It implies rights as well as obligations.

Wage-earners carry on the actual processes of production. They invest their labor and personal capacities in industry and make just as necessary an investment as do those who invest capital. Because labor investments were intangible and workers no longer owned the tools of production, employers at first found it possible to resist workers' claims to a status as a producing partner in industry.

Yet, collective bargaining became necessary when the individual worker could no longer make an equitable work agreement with his employer man to man. Representation of the whole work group alone has an authority that puts them on an equal footing with capital and management. It makes work relations matters to be discussed in the light of facts and decided intelligently. Such conferences result in balanced decisions, taking into account the welfare of all interested.

A few examples will illustrate. In 1876 the steel-workers agreed upon a sliding scale of wages for their industry following a practice established by the Sons of Vulcan in Pittsburgh in 1866. In the early 'eighties workers in the various glass industries persuaded their employers to agree to the practice of annual conferences to negotiate work agreements—the flint-glass workers, the glass-bottle blowers, the window-glass workers.

One early collective agreement dealt with specific issues, such as the ten-hour agreement between the operatives and employers in manufacturing industries of Delaware County, Pennsylvania. In the late 'eighties collective agreements became more general. Where the industrial conditions permitted and unions were strong, there was a trend toward covering whole industries.

The International Typographical Union claims to be the first union to adopt a wage scale and submit it to the employers of the industry. In 1891 the International Iron Molders and the Stove Founders National Defense Association established the practice of conference and joint labor agreements covering the industry. This is still maintained.

The collective agreement protects the individual worker and gives him a real work contract. When an employer hires a large number of workers, rates of pay and work rules must be standardized. Shall standardization be delegated to the employer or shall it be jointly determined? There are but a very few persons who believe employers are sufficiently disinterested or omniscient to set rates without the coöperation of those who handle the tools and materials of production. The individual worker cannot bargain with the employer, but a spokesman for the whole group can meet the employer on an equal footing. The agreement negotiated is the work contract. It usually specifies rates of pay, hours of work, work rules, sometimes conditions of work, provisions for interpretation of the contract, and arbitration of differences. Under the safeguards of collective agreements some unions have extended the field to include provisions for regular conferences on the problems of daily work. Such provisions make possible continuous coöperation between management and workers.

The progress and scope of collective agreements have depended upon developments within industries and competitive areas. In some industries, such as construction, local agreements prevail; in coal-mining and longshoremen's work, agreements cover geographic districts; in railroading, the railway system is the basis.

Collective bargaining makes it necessary for workers to study the problems of the industry and for management to consider the progress of wage-earners. The result is an appreciation of interdependence of interests and the realization of the need of joint planning and action to put the industry in the forefront of progress.

The first form of joint-relations machinery was usually the grievance committee, designed to adjust personal difficulties. The second step was usually to provide a method for deciding between different interpretations of an agreement. Another was for arbitration of differences that arose. More recently provisions for coöperation in the problems of production have been added.

Wherever workers have a status through collective bargaining, a new feeling grows up in the work. With guaranties of fair dealing, many barriers to putting their full coöperation into the daily job disappear. Coöperation for better and more efficient production benefits management and workers alike. There is hardly a collective agreement that has not resulted in some coöperative development. More formal provisions and machinery for coöperation have recently grown out of collective agreements. These are usually classified under the head of union-management coöperation. This type of agreement has developed since the World War.

Out of the necessities of war needs came a new vision of hu-

LABOR   161

man relationships and joint endeavor. A developing spirit and technique of coöperation, carried over into the post-war period, gave a new perspective to many labor problems, revealing the interdependence between the interests of all groups contributing to production, and the interrelation between progress for wage-earners and stockholders.

At the same time came a new emphasis on industrial planning. Management became a separate function and its responsibilities were vastly increased by large-scale production, mass production, and the rapidity of technical change. "Efficiency" systems were advocated as panaceas. Workers pointed out that efficient management must be done on the job and challenged the efficiency of management that does not plan to coördinate the thinking of those engaged in all the processes of production. One of the conditions which union-management coöperation disclosed as essential to unrestricted efforts to improve production efficiency is security of employment. Modern management has definitely begun to plan to increase stability of employment. This function of unions will be increasingly important as managements assume more and more responsibility for providing assured employment for all attached to the industry.

This new approach to fundamental problems puts labor in the position of challenging the efficiency of management, pointing out that no management can decide wisely upon production policies without considering the experience of those who carry on the processes of production or fabrication. Labor maintains that each functional group can help to make industry more efficient by participation in policy-making, so that correct understanding of the labor field shall be integrated in the information upon which decisions are based.

The various national and international trade unions provide services for their memberships such as benefits, employment bureaus, apprentice training, and education for the journeymen workers. Benefits provide against such emergencies as sickness, accidents, death, unemployment, old-age pensions. Not all unions provide all types of benefits. The death benefit is the most general. Some organizations have put their benefits on an insurance basis. The Brotherhood of Electrical Workers has its own insurance service, while a number of organizations have combined to organize and conduct an insurance undertaking to serve their unions.

Many unions assume responsibility for training those who work in the craft. Where unions do not provide their own vocational education, they make coöperative arrangements with public schools. Among the latter are building-trades unions and printers. Printing-pressmen and photoengravers have given especial thought to educational opportunities through their unions. Practically all unions supply their membership with a trade journal.

The crucial problem before modern industry is stabilization which involves the application of the best practices in all fields and relations. Good management no longer is conceived as stopping with the individual plant. It involves linking up all activities in that plant with broader bases of information so it can buy, produce, plan, and sell its output to the best advantage. This means, of course, organized sources of information and policy-making.

One labor phase of industrial stability is security of job. Producing workers have a right to expect work opportunity from the industry in which they have invested their labor. Regularized production is necessary to job security. The responsibility resting upon industries has increased with their control over forces conditioning social welfare. It is a poorly managed industry that expects the community to feed its workers when they are laid off because production runs now large and now small. Management must plan production and sales instead of letting orders dictate production plans.

Not only must industry put security into employment for its workers, but there must be larger social planning to provide against technological unemployment and unemployment due to business mishaps.

Improved machinery and technical progress should perform their real function—labor-saving. They make possible shorter workdays and workweeks. Increased productivity means that the individual creates more wealth and he should thereby be able to make use of things he helps to create.

Labor holds that wage-earners must share in industrial progress or become a drag on it. Higher wages and shorter hours of work are essential to balanced growth, which alone can sustain prosperity. At the same time labor, where it has status, is concerning itself increasingly with problems of production and enlarged output, realizing that higher standards of living can be established and maintained only through greater productivity.

Workers through their unions are offering to industry an essential agency for good management and sustained progress. By providing the machinery for adjusting industrial relations, by participating in the problems of production, and for keeping the wage-earners' progress abreast the general level, the American labor movement is a constructive, stabilizing force in national life.

LEGAL STATUS OF UNIONS

The development of an effective labor movement has involved the changing of legal principles and precedents as well as the setting up of new economic practices and the modification of management principles. The forces of tradition in law have frequently united with established economic powers against recognition of new principles of justice under changed conditions. When old legal patterns are applied without regard to realities of life, injustice is done. Law cannot remain constant while the relations and agencies of living change—it must constantly conform to justice.

On the whole the doctrines of conspiracy and restraint of trade have dominated court decisions in labor cases. The tradition of the English common law made strikes criminal conspiracies and strikers liable to indictment and imprisonment. Early cases against unions were usually tried in the lower courts and not appealed. Consequently there was little discussion of the principles involved. The gist of the conspiracy lay in the unlawful federation. When the labor conspiracy cases at length reached higher courts, the first step toward justice came when it was definitely settled that a combination of workers for a lawful purpose was not illegal (Commonwealth *vs.* Hunt 1840). Later decisions, while recognizing the necessity for unions as a legitimate factor in the nation's industrial system, did not always find, however, that the means employed by unions were legal. The enactment of the Sherman Anti-Trust Law (1890) resulted in litigation charging unions with a new form of conspiracy—restraint of trade—and in the federal sphere was a setback for labor. A local federal court issued an injunction against the longshoremen of New Orleans on the ground that the strike was an illegal restraint of interstate commerce. This was the first injunction under the Sherman Act. The Sherman Act was held applicable to unions in the case of Loewe *vs.* Lawlor and later in Buck's Stove and Range Co. *vs.* Gompers, Mitchell, and Morrison. In the case of the Hitchman Coal and Coke Co. *vs.* Mitchell, the United Mine Workers of America was held by a district court to be an unlawful organization. This ruling was reversed by higher courts, but an injunction forbidding organizing activity among the miners was upheld.

Only the government could apply for injunctions under the Sherman Act. When the Clayton Act (1914) was enacted, this restriction was removed. The labor provisions of the Clayton Act were supposed to relieve labor from injunction abuses and to remove the stigma of illegality from unions and their normal functions. But the first Supreme Court decision after the enactment of the Clayton Act held that it was unlawful for machinists to boycott presses made by a company that had refused to permit its employes to organize and overruled the decision of the two lower courts refusing the company an injunction. This and following decisions have practically nullified the labor provisions of the Clayton Anti-Trust Act and have made union activities criminal under the Sherman Act.

The most reactionary decision of recent years is that of the

Supreme Court in the case of the Bedford Cut Stone Company *vs.* Stone Cutters Association. The stone-cutters unions were unable to negotiate an agreement with employers in the Bedford Limestone District of Indiana to replace an expiring contract. The employers set up company unions and refused to hire union men. The constitution of the union contained a clause requiring members not to handle stone cut by men working in opposition to the union. Since employers refused to hire union men and the union refused to let its men handle non-union stone from the Bedford Cut Stone Co., building operations came to a standstill. The builders and the Bedford Company asked for an injunction declaring the union was conspiring to stop interstate commerce. The two lower courts thought the request not well founded, but the Supreme Court granted the injunction, enjoining the stone-cutters from refusing to set the stone. In this case there was no violence and no rioting. The stone-cutters had simply quit work.

In a dissenting opinion Justice Brandeis said:

> If, on the undisputed facts of this case, refusal to work can be enjoined, Congress created by the Sherman Law and the Clayton Act an instrument for imposing restraints upon labor which reminds one of involuntary servitude.
>
> It would, indeed, be strange if Congress had by the same act willed to deny to members of a small craft of workingmen the right to cooperate in simply refraining from work when that course was the only means of self-protection against a combination of militant and powerful employers. I cannot believe that Congress did so.

This record makes it plain that unions have not yet succeeded in freeing themselves from the application of the common-law theory of conspiracy to its normal activities.

Labor's difficulties have also been increased through the judicial practice of giving legal protection to "yellow-dog contracts," that is, pledges not to join the union as a condition for getting a job. Injunctions have been issued forbidding unions from interfering with these "contracts." The result is to deny workers the right to organize and to deny unions the right to extend the effectiveness of their work.

The legal difficulties of labor are also multiplied by our federal form of government—necessitating relief by state as well as federal legislation and the education of public and legal opinion.

SOCIAL SERVICE OF UNIONS

During the past century life has been lengthened by approximately twenty years and the possibilities of health, comfort, and efficiency for the great majority of our citizens vastly increased. A considerable factor in this achievement has been a better-educated nation. This period has seen the establishment and development of our public-school system. The labor movement took the initiative in putting educational opportunity on a democratic basis. In 1928, the latest figures available, 25,179,696 students were enrolled in elementary and secondary schools—68 per cent of the school population five to twenty years of age. In 1840, 9 per cent of the school population, or 468,268 pupils, were in public schools. Labor has steadily urged the extension of public educational opportunities through state universities and a wider use of educational plants. In 1840, there were 16,233 students enrolled in colleges and universities; this number had grown to 919,381 by 1928.

To keep our children in schools and out of factories, we have urged compulsory school-attendance laws and child-labor restriction. Eight states have raised the school age limit to eighteen years; seven to seventeen years; twenty-eight to sixteen.

Unions have helped bring about a fairer distribution of wealth resulting from joint work by raising wage rates. The average money rates per hour in 1929 were seven times those of 1840; the average rate for 1840 was 10 cents per hour; for 1929, 72 cents. Over a long period the wage trend has been upward. A general upward trend in real wages has helped to raise standards of living among wage-earners.

Higher wages bring comforts into the wage-earners' homes. In 1830, wage-earners' homes were bare of conveniences;

LABOR

water and toilet facilities were outside. There were no stationary bathtubs or kitchen sinks. Candles or oil lamps supplied the light. By 1926, 68 per cent of urban homes had bathtubs, 84 per cent kitchen sinks, 82 per cent flush toilets. By 1930 there were 20,000,000 domestic customers using electric light—many of these must be wage-earners; 12,000,000 families have telephones—many must be wage-earners; there are 10,000,000 radio sets—some of these are certainly in wage-earners' homes; in 1930 there were 23,000,000 registered passenger-cars—probably half of these are owned by wage-earners.

Expenditures for free social services, including education, libraries, recreation, and health in cities of 30,000 population, were $2,790,000,000 in 1926.

Many homes of industrial workers now contain rugs, curtains, living-room furniture, table linen, dishes, victrolas, radios, etc. Money income determines opportunities for wage-earners and their families. Wage increases, therefore, enable wage-earners to share in the increasing national wealth and social progress.

By reducing the hours of work, unions enable wage-earners to share in the leisure made possible by technical progress. In 1830 the workday was from 12 to 14 hours long, while in special industries from 16 to 18 hours were spent at work. A movement for a ten-hour day began in that decade. Laws were enacted in several states. In the 'sixties the eight-hour standard was raised.

By the 'eighties practically all skilled craftsmen had a ten-hour day. The American Federation of Labor launched an eight-hour movement in 1884. It is significant to note that a new standard appears before the old standard has found universal application and that union hours are about six hours per week lower than the average workweek.

Increasing productivity has made possible these gains in leisure for workers. Between 1899 and 1919, the time necessary to produce equal portions of the country's manufacturing products was decreased by 12 hours, and further decreased between 1919 and 1929 by 18½ hours. But the workweek decrease was 6.8 hours in the 1899-1919 period and 2.2 hours in the last decade. Obviously these two trends were not balanced in industry. Unless hours of work decrease in proportion to technical improvements in processes, workers are displaced and unemployment results. Improved work methods and technical progress should reduce manual labor and give more leisure for the other interests necessary to a balanced living.

Through higher educational opportunity, higher incomes, and more leisure, Labor has helped to provide the means for higher social standards. It has steadily set its face against rigid class barriers, insisting that Labor has a right to equality with all other groups. The labor movement has, therefore, advocated the extension of the ballot to all citizens regardless of sex or ownership of real estate. It has done much to widen the concept of equal protection before the law adapting methods to varying conditions. We have tried to write into statutes the recognition of workers' rights and the protection of their interests. We have persistently urged the establishment of bureaus of labor statistics so that there might be available information showing trends and measuring-rods of progress. We have sought to assure wage-earners the legal protection of their rights. Every state, for example, now has a mechanic's lien law. Forty states and the District of Columbia have established employers' liability for accidents to workers; forty-five provide compensation for industrial accidents. To assure greater equality of opportunity, all states have compulsory school-attendance laws and all states have some form of law regulating child labor. Forty-four states, two territories, and the District of Columbia have mothers' pension laws.

All states have factory-inspection provisions and all have regulations for special industries. These provisions are not of uniform excellence, but they represent definite advances. All states have laws to prevent the exploitation of convict labor. All states have established the principle of protective legislation for women, though the laws vary. All states have laws regulating the payment of wages. Thirty-four states and the

LABOR

federal government have provided some type of employment service. Seventeen states have old-age pension provisions.

Every group must have political influence if it needs to have its interests considered in the formulation of public policy. After discouraging experiences with efforts to solve its difficulties by legislation, labor found it wise to separate its problems into two groups: (a) the work problems of those employed in private industry; and (b) the problems of those employed by governmental agencies and the issues of social policies. For the latter, we use political methods. As it rejects the principle of class domination and makes use of practical expedients, the American Federation of Labor consistently has followed the policy of non-partisan political action. We formulate labor principles and seek to get them written into laws by one or both political parties. Our unions mobilize the labor vote as a flexible force to create a balance in favor of candidates pledged to our principles.

Workers were the first to call attention to insanitary work conditions and living conditions. In some industries insanitary conditions in workers' homes endanger the health of all the customers of the industry. Protests of workers led to the development of industrial medicine and sanitation. Protests of workers resulted in efforts to abolish tenement work and to secure the housing of industry under controlled conditions with factory inspection. Higher wages have spread opportunity for better homes and better living conditions among the rank and file of American citizens.

It is characteristic of the spirit and the methods of the trade-union movement that it has consistently sought to extend the basis of law and institutions to include protection and aid for labor. In this period of one hundred years, the restriction of the suffrage to property-owners has been eliminated from the definition of civil liberty. This change is itself a symbol of a new concept of rights and interests, developing through urbanization and industrialization. Workingpeople, whose investment of personal capacities went into the processes of industry, found it necessary to secure a legal recognition for new practices and relationships. Wage-earners who functioned in collective undertakings had to get their collective agencies and practices accepted by public opinion and judicial thinking. This period has also brought a revolution in the forms and concept of property. Labor has sought to incorporate in law the changes that have taken place in industry and society. If, instead of continuing to raise wheat and sell it, workers in the new order make bread of flour made from wheat, their relationship to what they create is less obvious but none the less real. Organized labor is simply helping to define these new rights and new investments. In other words, the trade-union movement holds that workers are partners in production and that this relationship is most accurately expressed in union-management coöperation.

Within the past hundred years the forces that control labor welfare have changed from a community to a world-wide basis. Mechanical power has revolutionized craftsmanship and mass production has introduced standardization and dependence upon mass consumption. With interdependence tying together all groups for common fortunes in depression or prosperity, it becomes necessary to extend the principles of planning and organization to cover all areas. To enable labor to fashion and develop its functions in the new economic world, unions are indispensable.

CHAPTER SEVEN

BANKING AND FINANCE

In 1846 there were 707 local banks in the United States with $105,600,000 in circulating notes supported by $42,000,000 in specie. The merchant or traveler, in attempting to do business outside his own community or with visitors from other sections, had to deal with scores, perhaps hundreds, of varieties of notes of uncertain value in many cases and liable to undergo amazing fluctuations. If, for example, a traveler from Indiana dined with friends in a New York hotel and tendered in payment for the meal a crisp, new note on a bank in an Indiana town, the waiter would accept it gingerly and refer it to the proprietor. The latter, uncertain whether it was worth forty or seventy or ninety cents on the dollar in New York bank currency, would send a messenger on foot, perhaps half a mile to the nearest bank, to get the latest exchange news and then, duly informed in high finance, would give the guest the balance due him. In those happy days of fiscal liberty, banks issued notes with impunity in many parts of the Union, cunningly deceived bank inspectors searching for specie, and failed in cold blood, leaving note-holders to heal their own wounds. Moreover, great states issued bonds with indiscretion and a few repudiated them without remorse, making American finance a chaos of folly.

Banking and Finance

By H. Parker Willis

A SURVEY of the past hundred years of American history brings to the front no series of developments more striking in their significance or more pronounced in their change of character than those which have to do with finance and banking. Owing to the great expansion in size and wealth through which the nation has passed, the figures of its financial experience, which were at first written in units, are now written in their thousands. Change in size always brings with it some change in method and in type; but the transformation through which American banking and finance have passed in the course of a century of development has been immensely more significant from the standpoint of theory and technique than that of mere amount or volume.

EXPERIMENTATION IN FINANCE, 1830-1860

There has been from the beginning a close and intimate relationship between our banking system on the one side and our plan of tariff levies and of public finance generally on the other. This intimate connection has sometimes been injurious in nature to the sound growth of one or the other element in the composite whole. Be this as it may, the two branches of financial practice—banking and public finance—are closely interwoven, and it would be impossible to separate them, even for purely technical purposes of classification. Hamilton himself, in his early foreshadowings of American policy, regarded the two as closely interdependent, and thought of the national debt in its political rather than in its economic aspects; and in his plan for the reorganization of a national bank, later to

BANKING AND FINANCE

become the First Bank of the United States, he expressly provided for the uniting of banking with the public debt by authorizing subscriptions to the bank stock in the form of government securities.

What Hamilton thus foreshadowed, and the lines of reasoning by which he arrived at his conclusions in this matter, turned out later to be characteristic of the entire trend of American development in the field of finance. This no doubt was much more than chance, for Hamilton's projections owed much of their success to the accuracy with which he analyzed American business and commercial prejudice. His firm policy, which was embodied in proposals that led to the introduction of central banking on a plan somewhat patterned after that of the Bank of England, resulted in a forty years' experiment with centralized finance (omitting only an interregnum of a few years during and after the War of 1812). In this period the First Bank of the United States was brought to life and ran its course, and its successor, the Second Bank, fully developed, rendered considerable service to the nation and, curiously enough, was facing in 1830, when our story begins, some of the same problems of central banking with which we, a century later, still find ourselves concerned.

The refusal of Congress to grant a second charter to the Bank of the United States in 1836 represented in effect the determination of the young country to experiment with banking upon its own responsibility and to see whether it could or could not work out for itself an independent system of banking which would not follow too closely either the European precedents or the desires of the Eastern financiers who were supposed to have obtained control of the financial machinery during certain periods of time when the First and Second Banks were running their course. It was true that, during the brief intermission between the closing of the First Bank and the opening of the Second, there had been some little experimentation with uncontrolled state banking; and this experiment had been anything but encouraging to those who believed in strictness and conservatism as factors of financial management. But what had then been done had, after all, represented a very limited experiment, while keenness of appreciation for the lesson of it had been largely dulled with the passage of time. State banking in that brief interval increased to a total of 788 banks (in 1837), with a circulation of $149,200,000. This great growth had, of course, occurred for the most part by and with the consent and approval of the several states which had chartered the new institutions. They acted, doubtless, with the best of intentions, but with an ignorance of the subject which rendered the first banking enactments wholly inadequate to the requirements of sound finance. Unwittingly, they laid a foundation for insecurity and disaster which was soon to be disclosed by bitter experience and to serve as a warning for the future. The transfer of public deposits from the Second Bank to the "pet" banks chartered by the state governments, which had been selected for the service (numbering 29 in 1835 and increasing to 89 at the end of 1836, with $50,000,000 of public funds) was the preliminary for a great expansion that led to a general business crash in 1837.

By this time, however, there had been a great deal of careful analysis of banking theory on the part of specialists, and much harsh experience with faulty banking practice. During the first thirty years of the nineteenth century, New England banks had had their own share of difficulties and had finally worked out among themselves certain principles of operation which were being studied (and in some cases adopted) by banks in other parts of the country. Among these principles, we may mention (1) the rule that the capital of banks should be paid up in cash at the time of their organization; (2) the establishment of a definite limitation upon the total amount of notes to be issued by any one bank; (3) the requirement of special security in the form of bond deposits or other provisions for segregation of assets; and (4) limitation of bank loans to transactions not likely to involve a complete "freezing" of assets.

Moved by the unsatisfactory conditions existing in many

BANKING AND FINANCE

parts of the country and by the depreciation of state bank notes, the Massachusetts banks had in 1824 established the so-called Suffolk system, which provided for the par redemption of country-bank notes at an agency in Boston conducted by the Suffolk Bank of that city. The system had proved very satisfactory in its constant maintenance of parity and in its steady testing of the ability of each issuing bank to redeem its notes. Few other regions had actually adopted it, even in part; yet all recognized the merits of the idea and were disposed to accept principles taken from it—at least in those cases where such acceptance was not too costly.

The first period of really unrestricted state banking that began with the collapse of the Second Bank of the United States was now, however, to develop certain additional fundamentals which by the outbreak of the Civil War had attained a fairly general acceptance. Even with the unrestricted liberty which had been accorded to the state banks, there were many persons who still felt that centralized banking under close state supervision, perhaps with state ownership, was advantageous; and the period accordingly witnessed the establishment of several such institutions which were now perforce organized by the state as such, since the Federal government itself had resolved no longer to charter or to countenance any bank. Among such institutions may be mentioned the State Bank of Indiana, formed in 1836 under a twenty-year charter, with its branches scattered throughout the state; the State Bank of Kentucky; and the Bank of Missouri. In each of these institutions, an effort was made to reproduce on a local scale the services which had been rendered by the Bank of the United States, although, of course, without the restoration of any evils, real or imaginary, which had accompanied the older form of banking.

As the country passed into the state banking period, and as the weaknesses of such unrestricted state banking became more and more evident, there was a search for types of banking management that would cause as little deviation from the independent system as might be, yet would in some way protect the funds of the customer. Efforts in this direction were furnished by the "safety-fund" system of New York State, established in 1829, and by the system of free banking with secured currency, established under the auspices of the same commonwealth in 1839. In the one case, the idea—later to be accepted by the Canadian banks—of creating a joint fund out of which to pay the notes of banks that might fail, constituted the central thought; while in the other the underlying principle was the plan of investing the assets of the banks in good bonds and mortgages up to an amount equal to the note issue, such securities then to be placed in trust with a state official. In some measure it might be said that both experiments had a moderate success, but their achievement was not sufficient to command widespread imitation, although something of the same kind was instituted in sundry of the Middle Western states, best represented perhaps by Illinois. In these instances, however, the currency experiments were less successful on account of the poor character of the securities which the law permitted banks to purchase and hold for the protection of the notes. Great losses accordingly resulted.

While different regions of the country were thus making trial of almost every system of protecting and issuing bank notes, the nation was slowly effecting the introduction of the check and deposit system upon a much larger scale than ever in the past. With the growth of towns and cities, a better scope for the use of checks and deposits was revealed, and there was a general effort on the part of local business men of any considerable importance to substitute for note payments the process of settling by check.

Discontent with the unsatisfactory conditions growing out of the new local and widely divergent plans of banking was quick to make itself felt, while the Federal government, accustomed to safety and security for its funds, formerly furnished by the Second Bank of the United States, now found itself subject to all the hazards of bank failures and the resulting losses which necessarily came through the distribution of public deposits primarily on a political basis among the dif-

BANKING AND FINANCE

ferent banks and communities of the country. The federal government itself was thus moved to search for and introduce new banking devices. Having already tried, with little satisfaction to itself, the plan of employing the state banks as depositories and being debarred by reasons of political consistency from experimenting further with the central banking idea, federal administrators naturally began to think of developing their own method of handling and disposing of public money. The outgrowth was a new experiment involving the use of expedients previously not thought of, certainly not attempted upon any such scale by any other modern country. They crystallized into what was known as the sub-Treasury or Independent Treasury system, first recommended in 1837 by President Van Buren, then actually provided for in 1840, and finally reaching a completed form in the Act of 1846.

The central thought in these efforts was the notion of placing the government upon a cash basis: both its receipts and its payments were to be in hard money—a policy first established by the famous "specie circular" of 1836 and then enforced with some intermissions as a permanent policy, though with modifications. Underlying the whole experiment was the thought that the operations of the federal government could be regarded and treated as the business of an independent entity and could be handled as if the well-being of the administration at Washington were quite distinct from, or independent of, that of the people at large. It was a conception that could be realized only by keeping the receipts and disbursements of the government within a small compass and by limiting the total of the borrowings to nominal figures. Could these conditions be fulfilled?

One of the major reasons for creating the First Bank of the United States and for renewing the experiment by the chartering of the Second Bank had been furnished by the existence of a very large public debt. Remarkable success had accompanied the disposition of this debt. Although at the opening of the First Bank in 1791 the government was in debt to an amount then regarded as extraordinary, namely $75,400,000 at most, this obligation had been reduced to $45,200,000 by careful management and steady redemption by the middle of the year 1812, notwithstanding increases from time to time during the period, due to new undertakings. The War of 1812 had augmented the burden and had left the government at the time of the rechartering of the Second Bank with an outstanding debt of $127,300,000. But this had been entirely disposed of by 1835, and the extinguishment of the debt was followed by a popular distribution of surplus revenue before the panic of 1837. Although greater necessities grew out of the panic or were presented from time to time, the government henceforward had at no period before the Mexican War in 1846 an outstanding total of obligations amounting to more than $15,000,000. Some $63,000,000 were added by the Mexican War, but with good tax yields the total outstanding in 1861, at the opening of the Civil War, was only $75,000,000.

This state of things made it much more possible to carry into effect the Independent Treasury plan, as first developed in 1841 and brought to maturity in 1846. The plan provided for the actual payment to the federal government of all taxes and other obligations in cash, neither checks, drafts, nor bank notes being acceptable, while in the same way the government was to pay its debts in hard money. Practice, of course, soon modified this original intention and permitted the use of "vouchers" and "warrants" which represented money in the Treasury and obviated the necessity actually of paying in cash. But the essential idea of having as few relations as possible with the banks was generally adhered to and lay at the root of the new system.

It would not have been possible to make a success of the Independent Treasury system had the government at the time been a large borrower, or a large collector, of money. As later experience showed, operations in the debt or withdrawals of money in large amounts from circulation invariably mean serious disturbances of the circulating medium and result in changes of business conditions as well as alterations in borrowing power at the banks, which prove a severe economic handi-

BANKING AND FINANCE

cap. This, however, was not perceived at the time or at least was recognized only in a very limited degree. It grew plainer as time went by and as the influence of the system upon the economic organization of the country became more obvious.

The Independent Treasury system, including the parent Treasury at Washington and the nine sub-Treasuries at various points throughout the country, nevertheless, proved sufficiently flexible to permit its continuance through all the years until a return to the central banking method under the Federal Reserve system of 1913 permitted the abandonment of what had been a useful and powerful and yet obsolete and harmful type of government financial control.

DEVELOPING A NATIONAL SYSTEM, 1860-1900

By the end of the decade 1850 to 1860, the United States had thus settled down upon a basis of local banking under various forms and guises with the government's management of public funds developed under its own organization and without the help or oversight of a central banking system. The total number of banks in the country at the outbreak of the Civil War was probably about 1,600; the circulation of currency notes, varying in soundness and constantly fluctuating, was perhaps $200,000,000; and the amount in the Treasury was, on the average, in the neighborhood of $1,716,000. A new and striking financial phenomenon was now to make its appearance as a commanding factor in American policy.

Since the opening of the Civil War had thus found the federal government with but little in the way of resources on hand in the Treasury, and since, as usual in such cases, the prospect of raising money through taxation could be realized only after long experimentation, a new expedient was sought. The first efforts to borrow were barely successful. Only a few months, therefore, after the outbreak of the war, Congress confronted a situation in which the Treasury would probably be empty within thirty days, and seeing no way to fill it without resort to some new device. The result was the adoption of an inconvertible paper policy: the government printed legal tender treasury notes to the nominal sum of $150,000,000, in small denominations, which were at once used to pay off the government's clerical and official staff and then came into general acceptance as a circulating medium. Inevitably they drove out the specie and the local bank notes which had previously existed, and substituted for the latter a cheaper medium without specie foundation—with only the government's designation of worth as an indicator of value. The new notes, usually referred to as "greenbacks," were issued under three successive acts, and finally reached the total of $431,000,000, a sum which apparently provided a temporarily adequate basis for business and banking. Before the close of the Civil War, the notes had thus become the standard of value of the United States and their worth, like that of all such currency, fluctuated in accordance with the prospects of victory on the part of the Northern armies. When the war was finally over, the value of the greenbacks continued to vary in proportion to the changing prospects of an eventual redemption in gold.

The final passage of the Resumption Act in 1875, with provision for redemption of the greenbacks in gold after 1879, seemed to many the equivalent of a restoration of the gold standard in full force. But as Treasury policy unfolded itself, it was soon discovered that the law was nothing more than a mere act of stabilization since the greenbacks were ordered paid out as fast as they were redeemed. They had thus come to constitute a permanent obligation of the government, which, in times of financial depression—as during the years from 1893 to 1896—involved serious danger because the Treasury had no positive or assured way of obtaining gold, although it was obliged to pay specie to holders of the greenbacks whenever they might choose to present their notes for actual conversion.

Notwithstanding this dangerous character of the situation with respect to greenbacks, the Gold Standard Act of 1900, adopted at the instance of the Republican party for the purpose of improving the entire currency situation, did not retire the legal-tender notes although it might beneficially have done so. On the contrary, it simply established a separate gold fund

BANKING AND FINANCE

in the United States Treasury. This fund was to amount to $150,000,000; and, in case it became depleted through the constant presentation of greenbacks for redemption, it was to be restored by issues of short-term government obligations, sold for that particular purpose. Such sales were ordered to take place whenever transfers of gold from other funds in the possession of the Treasury proved insufficient. The greenbacks, however, gave comparatively little real trouble after the year 1900; and, during the past quarter-century, have gradually been absorbed into the lower denominations of circulation, where they are held as part of the supply of currency needed for fulfillment of the expanding needs of the population. We thus still retain, as a souvenir of our Civil War necessities, the fiat currency issued in 1862, which has now made a definite place for itself as a medium of circulation.

Civil War requirements, however, were only temporarily satisfied by the issuance of the greenbacks. Although the government speedily exerted itself to raise money through taxation and through the sale of government bonds, it at first had great difficulty in marketing the latter. As the machinery of bond distribution became better organized, bonds passed slowly into the hands of the real buyers, but the movement was at times very tedious and unsatisfactory. Accordingly the question occurred to Secretary Chase and his advisers, whether it might not be possible at the same time to improve the banking situation and to furnish a means for disposing of a large quantity of bonds. One result was the National Bank Act of 1863, shortly remodeled and improved upon by the Act of 1864.

Perhaps the most interesting feature of the financial side of most wars has been the extraordinary unpreparedness for serious conflicts shown by governments. This was true at the opening of the Civil War, certainly by the end of 1861. The long-continued controversies which had preceded the outbreak of the struggle had not resulted in laying the foundation for any definite plan of financing, North or South; so that, both at Washington and at Richmond, early war finance was an improvisation. Reluctance to levy new taxes in time of war has almost always been characteristic of governments, and was never more strikingly illustrated than during the Civil War. In 1862, the brittleness of the social and economic structure of the United States, coupled with the inevitable hesitation in devising new applications of taxation, explains the practice of resorting to all sorts of expedients as alternatives. Among such, the national banking system stands out as a striking example. It would probably not have been thought of had not many persons already recognized the shortcomings of the state banking system; and it is likely that the government would not have had the courage to attempt it, or perhaps any other far-reaching banking experiment, had it not been for the urgent needs of the Treasury.

Experience in New York and other states had laid the basis for a plan designed to protect currency by specially deposited bonds, while the inability of the Treasury to sell bonds in large quantities inclined the statesmen of the day toward a particular plan that promised to provide a demand for such bonds. Hence the national banking system, based upon the Act of 1863 as amended in 1864, which provided for the creation of national institutions supervised by an official called the Comptroller of the Currency and authorized to issue notes up to an amount equal to their capital stock, provided that they first bought and deposited with the Treasury of the United States an identical amount in the bonds issued by the government in meeting the costs of the war. The system was at first unpopular, and banks neither organized, nor exchanged their state charters for national, with any rapidity. By the close of the war a membership of only about 584 banks had been developed.

Then Congress decided on March 3, 1865, to make the issue of bank notes a monopoly in the hands of national banks by passing an act which imposed a tax of 10 per cent upon all transfers made with the issues of state banks. This had the desired effect, and state institutions in much larger numbers entered the national system, although there were still many

BANKING AND FINANCE

which preferred to continue as banks of deposit, tacitly surrendering the power of issue, inasmuch as the burden to which they were subjected as note-issuing banks would prove impossible to bear. Thus a genuine system of nationally supervised banks was brought into existence, and has continued to the present time, constituting in its later days the foundation of the Federal Reserve System.

Of the national banking system it may be said that it was the first effort in American history to provide a single national currency, everywhere redeemable at par, and not subject to exchange, while guaranteed as to soundness and freedom from exchange charges, even though the method employed for that object (bond deposits) could not be very highly recommended in theory. No note-owner of a failed national bank has ever lost money; for the notes have been redeemed by the government under the terms of the National Bank Act, which provided for the sale of the bonds deposited with the Treasury; while immediate funds for taking them up were furnished through the deposit of a 5 per cent redemption fund with the Department as the notes themselves were gradually issued.

The attainment of a safe, even if not otherwise very scientific or satisfactory, system of note issue, after long experience enforced by war necessity ought, in the abstract, to have saved the nation the pains of a monetary controversy. But the national banking system had provided no real elasticity in the currency, and there was no satisfactory buffer to take up the shock to public convenience and equity caused by the steady decline of post Civil War prices, with their usual reflex upon the relations of debtor and creditor. The panic of 1873 and a subsequent series of failures and breakdowns served to accentuate the injustice always flowing from great changes in prices.

One result was a popular movement for the free coinage of silver—a metal said to have been "surreptitiously" demonetized in 1873. The post-panic agitation led to the so-called Bland-Allison act of 1878, in which a compromise was reached at the cost of the purchase by the Treasury of not less than $2,000,000 nor more than $4,000,000 worth of bullion per month and its manufacture into coin, which could be represented by silver "certificates." Despite various efforts to suspend this artificial output of dollars, the coinage continued down to 1890, when, after the minting of 371,000,000 silver dollars, the Sherman Silver Purchase Act was adopted July 14, 1890. This measure permitted the purchase of 4,500,000 ounces of silver bullion monthly and the issue of an equal amount of legal-tender treasury notes in payment therefor. The consequence was an effort to increase the outstanding greenbacks; and confidence in the government's ability and will to maintain parity between silver, gold, and paper was hardly restored by the repeal of the Sherman Act on October 30, 1893. Decline of the government's gold holdings followed; and two large issues of bonds were required to restore them.

The hard-fought election of 1896 resulted in the defeat of the silver advocates and the adoption (in the gold standard law of 1900) of a plain declaration respecting the monetary standard of the United States put the subject to sleep for a time. With the coming on of the World War, another outbreak of the controversy proved inevitable; and again, in the Pittman Act of 1918, the government undertook to buy silver in large quantity—this time to replace the silver bullion formerly held by the Treasury to support certificates but temporarily lent to Great Britain for the strengthening of her silver reserves in India during the war. Final purchase of the entire amount was not completed until 1926, and the conclusion of the operation left the market unsupported, prices finally dropping, as the result of a complex of causes, to 27 cents an ounce in 1931.

The national banking system, correcting as it did a great many evils and uncertainties in connection with note issues, and gradually providing a widely diffused and convenient system of small banks available to the public was soon to show its weaknesses shortly after the Civil War had ended. They clustered around the variations of the public debt, inasmuch as the note issue itself was so closely intertwined with the existence of the bonds, which the constituent law had ordered to be used as a basis for protecting the new currency. As the bonds of the Civil War, issued at high rates of interest, appreciated in value, rising to substantial premiums, the banks found it less and less profitable to issue notes, for the amount of notes which they could obtain was limited to the par of the bonds. Thus as the debt increased in value, forces were set to work tending to contract the notes, and such contraction usually took place at the precise time when business was expanding and was making an enlargement of currency more necessary.

As the wealth of the country increased, and as Treasury surpluses developed and were applied to debt reduction, it became doubtful whether the amount of bonds in existence would continue to be sufficient to furnish an adequate notecurrency. This doubt presented itself more and more urgently as years passed, and by 1900, through the gradual absorption of the outstanding bonds by purchase for trust estates as well as by withdrawal in other ways from the field of active financing, the supply of such securities available for the protection of the national currency became very small. This inadequacy, and the consequent difficulty or impossibility of ever enlarging the supply when a sudden stringency required it, led to a reexamination of the old question of "currency reform." After the panic of 1907-08, the demand for some remedial legislation broadened, and an important result was seen in the Federal Reserve Act of 1913, which provided the present basis of our currency and financial system and has permitted the reorganization of the national banks as elements in the general structure of national finance.

Before attempting any account of the Federal Reserve System and its relation to present-day problems, it is necessary, however, to examine, in brief historical survey, the national revenue conditions which preceded the World War. Upon these was built the war system of taxation and indebtedness which, together with the Federal Reserve Act, have reshaped the entire financial outlook of the country. As indicated at an earlier point, the United States had entered the Civil War with the scantiest preparation. Whatever might be done through such expedients as the sale of bonds to banks, temporary borrowing, or Treasury notes, the issuance of greenbacks, and other expedients of a similar sort, it was apparent from the outset that the only successful method of dealing with the war situation must be found in the introduction of a system of taxation and a supplementary system of borrowing, upon which the government might fall back for the provision of funds to carry it through a long war, should such a conflict prove inevitable. This system of war financing, however, was, as is true in most such cases, reluctantly and rather poorly constructed, one feature after another being added as fiscal necessities impelled a further resort to measures which, it had been hoped, might be avoided.

For many years previously the government had relied to a preponderant degree in its plan of taxation upon customs duties. From these, during the year 1861, it obtained not less than 95 per cent of its entire receipts, amounting to $50,000,000, while most of the remainder came from the very moderate revenues produced by the sale of public lands. A mild tariff enacted in 1846 had been still further reduced by the more liberal measure of 1857. Not until the opening of the Civil War did the tariff, originally regarded, in the main, as a producer of revenues, afford to industrialists who desired a better economic status a rare opportunity to get the kind and amount of protection they demanded. Before the war was fairly underway, the Morrill Tariff Act was passed, to assure protection to certain industries and bring in the maximum revenue. It was speedily followed by other provisions of the same tenor, which had, at all events, the merit of accomplishing the object in view, for the year 1866 found the United States with an income of more than $85,000,000 from tariff duties. During this crisis the income tax also made its appearance, although on an uncertain footing, and the internal revenue system was materially expanded. The total yield of income and internal revenue rates in 1865 was $209,000,000.

Notwithstanding the wide range of the measures to which

BANKING AND FINANCE

the government resorted in its dilemma, almost every expedient for raising revenue through the issue of securities was also adopted.

The close of the war found the United States with what was, for those days, an unheard-of outstanding debt of $2,700,000,000. To clear away this immense indebtedness and to bring it into what was considered manageable limits now became the care of every administration. Such an accomplishment necessarily implied the continued maintenance of the structure of revenue that had been erected, along the lines already described, and it was accordingly natural to retain high-tariff duties and to eliminate only the more unpleasant of the internal-revenue rates. Income taxation, always the most obnoxious method of collecting the sums needed by a government, was the first to be done away with—the War Act being repealed in 1872. Further modifications in the taxation system were soon necessitated because of changing international trade conditions not peculiar then to the United States but prevalent throughout the world. Alteration of the war tariff taxation accordingly took place on a substantial scale in 1883 and again in 1890, followed in 1894 by a distinct reduction which sought to do away with some of the inadequacies and unfairnesses of the older measures. But soon the tide turned. The so-called Dingley Act of 1897 brought tariff taxation to its peak for the nineteenth century, while the Payne-Aldrich Act of 1909, undertaken nominally with the idea of reform, gave tariff rates still another push upward. President Wilson and his administration, coming into office in 1913 with the pledge of tariff revision, made moderate reductions in the Tariff Act of that year, yet left the essential structure of duties very much as before. In the meantime the internal-revenue system of the Civil War had been greatly curtailed, especially in the law of 1883, article after article having been cut off as the economic condition of the country improved and the Treasury grew stronger.

The year 1913 found the country with an income for that fiscal period which amounted approximately to $725,000,000, including about $320,000,000 from tariff duties, $345,000,000 from internal-revenue rates, and the remainder from miscellaneous sources, including the sale of Indian lands and a variety of minor levies or fees. A small income tax of 1 per cent on individual earnings had been imposed in 1913 in order to make the receipts from other sources rather more adequate; yet it could still be said that, for practical purposes, the government of the United States did not resort at all to direct taxation. Among the several states, various kinds of tax systems had been built up. After the Civil War, the improvement and cost of local government rapidly increased and state, city, and county governments sought sources of income from which they could draw what they needed, while the federal government by tacit consent sought to adhere as far as possible to its "constitutional" revenues, chief among which were the tariff and internal revenue duties. This left for the states the land tax, usually divided between the state, county, and municipal governments, taxes upon a variety of acts and processes, corporation taxes, taxes on personal property, and, after the year 1890, in an increasing number of cases, taxes upon incomes and inheritances. The opening of the World War found the country with a general tax system in which the state and municipal governments probably collected not less than $370,000,000.

WORLD WAR TRANSFORMATIONS, 1900-1930

The World War brought with it the necessity for complete reconstruction of the financial system of the United States. Although prior to the entrance of the United States itself into the war, there had been but slight change, it was evident that participation would immediately involve great alterations in method. Early in 1917, therefore, after war against Germany had been declared, the problem of financial reorganization was taken in hand. It was recognized that the situation called for three distinct lines of change, the first being an alteration of taxation, the second a revision of methods of borrowing, and the third an adaptation of our banking system to new con-

BANKING AND FINANCE

ditions. Of the three, the latter was naturally the first to be undertaken. It was fortunate that the Federal Reserve System was at hand, although it had not been adopted with a view to its use as a method of war finance, but solely with the idea of the reconstruction of banking in the United States. Prior to the World War, popular attention had been almost entirely concentrated upon the banking problem to the exclusion of other national financial issues.

The panic of 1907-08 convinced the business public that an extensive revision of banking legislation must be provided for. Accordingly, the National Monetary Commission was appointed, and developed the "Aldrich bill," first made known to the public early in 1912. The Federal Reserve Act, which took the place of the Aldrich bill, was an outgrowth of the same general movement which had become strong in both of the major political parties. During the years 1912 to 1913 this Act was in process of formation and preparation, finally becoming law on December 23rd of the latter year.

Its general object was that of organizing the existing banks into a compact system, built up around twelve institutions called Federal Reserve Banks, the whole group being supervised and directed by a Federal Reserve Board appointed by the President and Senate. This board was given large powers with respect to approving and vetoing the acts of the banks, appointing three out of the nine directors in each of the twelve institutions (the remainder being elected by the banks of each district, one vote to each member bank), approving discount rates proposed by the banks themselves, and issuing regulations designed to define commercial paper, indicate its scope and terms, and otherwise controlling the discount activities of the reserve banks themselves. The latter were to have the primary function of discounting for their members (the national banks and any others which might join the system), issuing notes based upon commercial paper, holding the reserves of these member banks, and granting bank credit as the result of rediscounts designed to provide the members with funds when they might need such accommodation.

With these large powers, the banks and the Reserve Board came into existence in the autumn of 1914, almost simultaneously with the opening of the World War. The first duty of the reserve system was, therefore, that of furnishing a rallying-point and assisting the financial institutions of the country to concentrate their resources, thus protecting themselves against the losses and dangers incident to the struggle in Europe. In this they were successful, but the first two years of their lives were largely spent in keeping their organization together through protective measures designed to safeguard the other banks and, finally, in establishing reserves of notes and currency available for use in the event that the nation might be drawn into the war.

The opening of the year 1917 found the Treasury in a depleted condition with little in the way of surplus. Tariff revenues had been naturally greatly reduced as the result of the original outbreak of the war, and the interruption to trade. The Treasury might thus have been obliged to borrow in any event, even had we ourselves not become participants in the conflict. Almost immediately upon the decision of the United States to enter the war, it called on the reserve banks for a short-term loan of $50,000,000. It soon became apparent that we should be obliged not only to take a part in the conflict, but to finance the allies, and that the Treasury would not be able to wait for the more or less gradual effect of new taxation.

Consequently the very first policy of the department was that of mapping out a loan system for continuous use, and of enlisting the aid of the Federal Reserve Banks in putting it into effect. Thus began the creation of the immense war debt, eventually amounting to $25,000,000,000, which constituted the larger part of our financial contribution to the World War. It could not have been placed without the organization provided by the reserve banks and their branches. Fundamental to the new plan was the notion of a series of great loans, running for long terms and widely distributed among

BANKING AND FINANCE

the people. Five such loans were floated, furnishing the country with about $20,000,000,000 in the aggregate.

It was not possible, however, to wait for these loans to produce their results; and consequently the Treasury department began to borrow upon short-term certificates of indebtedness from month to month, placing these certificates with the banks. In turn, the banks themselves were permitted to borrow at reserve banks on certificates, or on long-term bonds, as collateral, at a rate equivalent to the rate borne by the obligations themselves, so that borrowing cost them nothing. From time to time, maturing certificates were funded into long-term bonds, the latter thus taking the place of these certificates. In this way, the reserve banks were able to carry the banking system of the country practically through the war, and after the conflict was over, they initiated a policy of gradually eliminating the government obligations from the banks and shifting them into the hands of the community.

Meantime, the revenue act of October 3, 1917, had given an immense extension to the income tax, greatly broadened the system of internal revenue taxation, and established a substantial inheritance tax. There had been urgent requests from many persons that the war should be financed on the basis of taxation but, as has just been noted, this proved out of the question, due to the fact, among other things, that the necessities of war were so vast, while a successful working of the tax system could only be effected by a slow process. The maximum yield of the new tax system was, however, very great; and at its peak in 1920 the government was able to derive from all forms of taxation, despite the decline in customs duties, a revenue close to $6,700,000,000. Incomes thus obtained supplemented the enormous proceeds of the loans and produced the funds with which the European conflict was carried on both by ourselves and by the allied governments.

With the war over, the problem of paying off the debt and retiring the burden of taxation naturally presented itself as the first duty of the government. Expenditures were gradually cut, and in order to assist in their more effective management a budget system was installed in Washington under an Act of 1921. The idea of a budget system, as known in European countries, had always involved giving Cabinet officers seats in the parliament called upon to make appropriations. This, of course, was out of the question in the United States with its system of divided powers and checks and balances; yet the establishment of an administrative budget, designed to oversee and control outlays, was entirely feasible. The creation of such a budget system had often been suggested and introductory plans had been made under the direction of President Taft in 1910-12. At last the administration of President Harding took up the idea and, in the course of the year 1921-22, installed the Bureau of the Budget. In this bureau are collected and consolidated all departmental plans for appropriations, with the idea of adjusting them to one another, curtailing them where possible, and especially eliminating duplication and waste. This budget plan unquestionably has operated to bring about a more rational control of federal expenditures and has permitted a careful planning for the adjustment of income and outgo to an extent never before possible.

The introduction of a budget system for the national government did not stand alone. State finance had passed through many years of local experimentation and error, and had profited from the costly lessons of local extravagance. The process of education had been slow. Pre-Civil War finance had witnessed many curious episodes. Among these perhaps the most spectacular, if the least creditable, had been the financial collapse of several commonwealths, during the years 1837-44, followed or accompanied by debt repudiation. Hasty borrowing and unwise application of funds, with the privilege of taking refuge behind the "principle of sovereignty," which protected a state from being sued save with its own consent, account for this episode. Down to the present day it has been an unerased blot on the record of several states.

Greater burdens and responsibilities during the Civil War brought, in the North, the steady expansion and systematization of state methods of taxation, followed or joined by similarly advanced methods in municipalities and local governmental units generally. In the South, the entire change of economic organizations caused by the alterations incident to the war swept away the older primitive systems of local taxation and finance, and these were eventually followed by the gradual reconstruction of systems analogous to those of the Northern commonwealths. An underlying structure of land and general property taxes, accompanied by the provision of licenses and special taxes of various sorts, was gradually built up during the post-Civil War years. And in the first few years of the nineteenth century these were supplemented by a wide use of inheritance and income taxes. World-War needs expanded the system as Civil War requirements had transformed the earlier property-tax systems of the several commonwealths. By the close of the World War, at least a dozen of the states had installed an income tax of some sort, and the following decade saw this number nearly doubled. The close of the War, too, found all but three of the states resorting to the inheritance tax in some form.

This systematization of income, and the constant study of tax problems by state commissions transformed the credit of the several governments. Today the strongest of them can borrow more cheaply than the federal government. In the meantime municipal and state governments began to adopt more or less complete budget systems. The former started on this line of reform soon after the opening of the twentieth century, and the latter quickly followed in the wake of the cities.

Parallel with this development of a better mechanism for financial planning ran the growth of a quick and vigorous, if unsoundly developed, prosperity. The great war taxes speedily produced a large surplus. Although steadily cut down by successive acts of legislation in 1921, 1924, 1926, and 1928, which first attacked the huge surtaxes, then emancipated the lower contributors and relieved the highest taxpaying groups in the community, the surplus remained a continuing factor. This was due to the growing wealth of the country that supplied an increasing yield from income and inheritance taxes, notwithstanding the fact that the total number of reporting incomes fell from 7,000,000 in 1922 to 5,000,000 in 1930. In the latter year, the income and profits taxes yielded $2,410,000,000 as against $4,000,000,000 at the opening of the decade, while the total number of individual incomes over $5,000 reported as of 1930 was 763,357 as against 525,000 in 1921. About 99 per cent of the total taxes were paid by less than 1,000,000 persons.

The productiveness of taxation and the continuance of the surplus at all events permitted a reduction of the national debt to little more than $16,000,000,000 at the close of 1930— itself probably a record in debt reduction, representing as it did a steady cancellation of debt after 1921 at a rate not far from $800,000,000 a year. The great advantage of this rapid curtailment of debt was, of course, seen in the fact that prices were themselves conclusively, if irregularly, falling, with the result of reducing their average level, by the close of the decade (end of 1930), to a figure not far above the level at the close of the pre-war period. The United States had once more passed through a world and national crisis, shown its ability to cope with the financial and industrial problems growing out of it, and emerged with a structure of finance, banking, and credit more efficient and stronger than that with which it had entered the conflict—no matter what its defects and no matter what disasters might later be incurred or invoked.

THE FUTURE

No growing nation ever reaches a static condition of its development, in any branch of its life. The year 1932 finds the United States still deep in the process of growth, its future, more than at any time heretofore, a "function" of its past, its ultimate goals hidden. Yet the main lines of early growth and development are perhaps clearer now than they have been at any point for two decades past.

First of all, the present position of American banking and

BANKING AND FINANCE

finance compels a further expansion in international trade, and an extension of the process of exporting capital, if we are to continue along the lines of our recent progress or even to protect, in any adequate way, the position already gained. During the years between 1920 and 1930 it was possible to expand the plant capacity of American factories to a point probably 20 per cent ahead of the capacity that can be occupied by domestic requirements. To "run full" without unemployment beyond the normal we must export one-fifth of our total output. And this result must be accomplished in a world which has been alienated by a tariff policy that has sought to prevent the payment of our foreign claims in goods—the only medium in which they can be finally liquidated.

The working out of a commercial policy which will open to American products all possible doors is thus a matter of immediate necessity, and one which underlies, moreover, the retention of that commanding position in the world which has been attained at so heavy a cost both in wealth and in human life. The opening of such doors to our trade may be accomplished by the application and constant pursuit of a consistent foreign trade policy, or it may be brought about by a general adoption of trade liberalism, with confidence in the ultimate triumph of a nation whose business is today unquestionably conducted at lowest cost, and with the best facilities for economical financing that exist in the world. Whatever may be the course that is chosen, it will involve the adoption of a world conception of trade, and of the principles governing its movement. The education of our leading spirits in business and manufacture has already carried them to the point at which they fully recognize this necessity. It must now be extended to public men and to all classes of voters.

A trade policy of the sort thus indicated will not be successful without a better organization and the systematic pursuit of the plan of exporting capital when needed abroad, in judicious amounts. The experience of the past has shown not only that to sell largely a nation must buy widely, but also that to do either it must also lend freely, and without fear that the expansion of well-being and prosperity in other countries will harm the prospects of its own citizens. Sheer necessity on the part both of ourselves and of the world at large has made us the greatest creditor and the largest lender in history—, if perhaps, the least wise.

The present temporary check to the process of "rationalizing" and recapitalizing foreign countries is partly a result of a realization that injudicious advances have been made. It would be strange if they had not, for men and nations learn by experience, and the paths of our post-war development were, to say the least, unfamiliar to our feet. Partly, too, the present uncertainty has resulted from defects in our foreign banking mechanism. This, in contradistinction to that of foreign countries, fails to make provision for the holding of securities in the portfolios of issuing banks, and leaves the sponsors to seek the immediate disposition of new issues among the public at large. It counts upon the public to "carry" the costs of foreign trade. While the costs thus indicated, in so far as they are represented by actual losses, were small at the opening of 1931 (only 1 per cent of foreign securities held in the United States having defaulted by that time), the eventual costs are already showing themselves so much greater. In any case they should be more directly borne by those who profit most from the trade. Our bankers, in discontinuing the extension of advances to foreign borrowers, act, no doubt, as they think necessary under present methods of financial organization, but the time must soon come for providing a more flexible and more effective method of caring for the needs of our foreign customers and of carrying them through times of crisis and depression—just as a skillful and foresighted banker carries his domestic borrowers over "rough spots" in the business terrain, in the expectation that they will speedily return to their old condition of liquidity and solvency. The improvement of foreign-trade finance is thus a condition of the broadening and strengthening of trade to which reference has already been made.

Such an improvement is, without doubt, intimately dependent upon the development of a decided preference for genuine business financing as against speculation, and the prevention of the latter from making undue inroads upon the preserves of legitimate business finance. It was this task, among its chief duties, that was imposed upon the Federal Reserve system at the time of its establishment, and it is precisely this responsibility which up to date has proved hardest to fulfill. Even in the short space of the fourteen years since the close of the World War, the United States has passed through two major crises; one centering about the overfinancing of, and speculation in, commodities and lands; the other based upon the speculative over-expansion of dealings in securities. This speculative over-development has retarded the growth of industry and trade, caused unprecedented unemployment, and inflicted incalculable losses to capital upon the nation.

A central banking system better calculated to exert the necessary restraint upon speculation is called for, and must probably be undertaken first among the chief tasks which the nation faces in its reorganization for the future. Some useful steps in that direction, no doubt, will prove to be possible in the early future, but the complete working out of an effective plan for the division of available banking funds between investment and commercial purposes—between the financing of long-term entrepreneurs and the credit support of short-term business requirements—will call for effort carried over a long period.

The gradual development of such a central banking system may be possible without any great change in the structure of ordinary commercial banking, but, as experience shows, it is likely to react upon, and be affected by, the latter. It is sufficiently evident that the drift in American banking is toward a greater degree of centralization. This may take form as an extension of branch banking along the lines which seemed to be making their way a few years ago, or it may move more decisively along the lines of group and chain banking, perhaps with a considerable development of the holding company principle applied to the ownership and management of banks.

Careful study does not as yet reveal with certainty the form of banking organization which can be counted upon as characteristic of the financial future in the United States. One of the most important and immediate tasks before us will be, without doubt, the planning and upbuilding of a new banking structure to be gradually completed as the years pass. The immediate danger of the situation today is found in the chance that conditions may be allowed to shape themselves, and will do so, in a haphazard way which will bring uncertainty and conflict between the various systems of bank organization— branch, chain, and group types, or some combination of all. Whatever new line of organization may be pursued, the chief object in the scheme of development must clearly be that of providing improved access to credit for the ordinary borrower who now is in danger of being less favorably treated than the larger business concern which today is more and more getting its credit through the medium of stock-market financing. Far better protection of savings, and greater safety in banking, which will insure termination of the epidemic of bank failures that has now lasted for near ten years, is possibly the most urgent change in banking methods demanded by recent experience. Action in this matter will be required by public opinion with the least possible delay—a fact now clearly apparent to members of Congress.

The close connection already established between the Treasury and the Federal Reserve System, involving, as it does, the constant issue of short-term certificates which are placed upon the market only with the aid of the Reserve banks, has already attracted the serious attention of students of national problems. Further reduction of the public debt, following the great curtailment that has already taken place since the end of the war, is desirable when circumstances allow it to occur without undue sacrifice. The depression started in 1929 has already seemed to set a period to the continuance of active debt reduction, on the great scale which has been characteristic of the past decade; and, while such reduction may be resumed, it is indicated that the pressure of taxation must not be unduly increased.

There is without question a widespread restiveness under present burdens of taxation, national, state, and local, which affords a warning not only against random additions, but even against the mere maintenance of the present load. Hence the nation confronts the serious question of readjusting its tax burdens in such a way as to relieve both the individual and industry in general. It is a question that involves, first of all, a more positive and definite decision as to the rate of debt reduction, and then a funding of that portion of the debt whose redemption is deferred to the future, into long-term bonds, which can be taken up and held by investors rather than by the banks. Such a policy will help most extensively in freeing the banks for actual business financing on the scale that is demanded by the other conditions already reviewed, while it will relieve the Treasury itself from dependence upon the chances of the money market and its rate changes.

The plan so suggested naturally presupposes a reorganization of the income-tax and other government revenue sources and the establishment of them upon a footing that will cause the least inconvenience and burden to the community as a whole, while promoting fairness so far as practicable among individuals. There is, no doubt, a call for a more equitable adjustment of tax rates in such a way as to spread the burden of taxation more generally, not only with a view to easing its weight upon certain classes, but also for the purpose of linking, in the public mind, the cost of social experiments such as unemployment emergency relief, bonus distribution, and the like, with the rate of tax levies. Above all, the early future seems to call for the division of tax sources between the national, state, and municipal governments, and the ending of double taxation both at home, and, as soon as possible, internationally, through suitable agreement with other countries. As the demands for social legislation to be enacted either by Congress or the state governments grow more insistent, the need of a careful delimitation of objects of taxation and of wise apportionment of objects to means, inevitably grows more imperative. Tariff revision, if undertaken along the lines indicated in connection with the discussion of our foreign trade requirements, will probably make our customs duties more productive, and to that extent will allow the easing of tax burdens in other directions. A more scientific management of the borrowing activities of all types of government, with a smaller and more reasonable reliance upon tax exemption as a means of selling national and local obligations, will obviously be an integral part of our new revenue policy.

CHAPTER EIGHT

GOVERNMENT AND LAW

"If slavery is to be the destined sword in the hands of the destroying angel which is to sever the ties of this Union, the same sword will cut in sunder the bonds of slavery itself. A dissolution of the Union for the cause of slavery would be followed by a servile war in the slave-holding states, combined with a war between the two severed portions of the Union. It seems to me that its result must be the extirpation of slavery from this whole continent; and, calamitous and desolating as this course of events in its progress must be, so glorious would be its final issue, that, as God shall judge me, I dare not say that it is not to be desired."—JOHN QUINCY ADAMS, *Diary*, November 29, 1820.

"In 1830 five-sixths of those incarcerated in the jails of New England and the middle states were there upon complaints of creditors, the majority for debts of less than twenty dollars. In this oppression New York held a sad preëminence."—DIXON RYAN FOX, *Decline of Aristocracy in the Politics of New York*, p. 353.

Government and Law

By Charles A. Beard

STUDENTS of government and law are wont to see in the wild welter of politics and legislation an eternal rhythm of progress and reaction, and not without reason. In a mighty effort, the English overthrew their monarchy in the seventeenth century, but within little more than a decade it was restored. With a kindred zeal the French, more than a hundred years later, uprooted the old *régime*, sent their king to the scaffold, and proclaimed a republic; but when the gamut of democracy and dictatorship had been run, the Bourbons returned to the throne of their ancestors—for a brief spell. So remarkable has been the swing of the pendulum in politics that Macaulay, brooding upon its inner meaning, came to the conclusion that this everlasting swaying of forces has its origins in human nature: mankind is divided by some mysterious alembic into conservatives and progressives. The former instinctively cling to the old; the latter are instinctively venturous and eager to experiment with the new. Yet the figure of the rhythm is an illusion. Monarchies have been restored, but never reëndowed with pristine powers and privileges. Progressives proclaim revolutions and always fall short of the perfection announced as the goal. Still underlying the whirling eddies and tides of politics, there are to be discerned streams of tendency, sometimes obvious on the surface of things, sometimes hidden far in the deeps out of sight. Keen is the eye that discovers them; bold is the thinker who can determine whether particular incidents represent gains or losses for mankind. At all events, not until the tendencies are revealed and the incidents set forth can a judgment be made.

GOVERNMENT AND LAW

In this spirit we may look over a hundred years of development in American government and law.

THE PRESERVATION OF THE UNION

At the head of all the achievements to be written down on the credit side of the ledger of our history is the preservation of the American Union. A century ago, its fate was uncertain. The father of the Constitution, James Madison, was still living, but his closing days were profoundly disturbed by the thought that nullification might undermine the edifice which he and his colleagues had so laboriously erected. It had barely escaped dissolution during the war of 1812, when New England defied the laws of the federal government, and twenty years later it was again put to the test by the ordinance of South Carolina nullifying certain acts of the United States. On all sides were threats of withdrawal and disruption. It was no theatrical cry, but an agonizing fear that Webster voiced when he exclaimed in his reply to Hayne: "When my eyes shall be turned to behold, for the last time, the sun in heaven, may I not see him shining on the broken and dishonored fragments of a once glorious Union; on states dissevered, discordant, belligerent; on a land rent with civil feuds, or drenched, it may be, in fraternal blood! Let their last feeble and lingering glance rather behold the gorgeous ensign of the Republic, now known and honored throughout the earth, still full high advanced, its arms and trophies streaming in their original luster, not a stripe erased or polluted, nor a single star obscured."

In this exclamation there was only hope, not assurance. In Europe advocates of monarchy, church, and aristocracy believed and confidently expected that "the Yankee experiment" in democracy would fail, that the Union would be dissolved, and that the resultant chaos would prove the folly of all such popular undertakings. Even an observer on the whole friendly, de Tocqueville, who surveyed the United States in 1831, entertained grave doubts about the future of the American system. "I am strangely mistaken," he wrote, "if the Federal Government of the United States be not constantly losing strength, retiring gradually from public affairs, and narrowing its circle of action. It is naturally feeble, but now it abandons even the appearance of strength. . . . I do not see anything for the present which can check this general tendency of opinion: the causes in which it originated do not cease to operate in the same direction. The change will therefore go on, and it may be predicted that, unless some extraordinary event occurs, the government of the Union will grow weaker and weaker every day." The future, he confessed, concealed the final results of the tendency and he did not feel able to remove the veil that hid them, but there was no doubt in his mind that the Union was an accident, in process of decline. Certain that dissolution was ahead, he proceeded to discuss the future of republican institutions on the assumption that they might survive the downfall of the Constitution. To Americans of this generation the very thought of the disruption of the Union seems so remote that it awakens only the palest reflections, but a hundred years ago it was a contingency so pressing and burning that none could escape it.

Tried by the fire of the Civil War, the Union emerged triumphant, to the dismay of the English Tories who had confidently expected the Republic, supported by "horny-handed" farmers and "greasy mechanics," to go down to the ruin appropriate to all such violations of the law of nature. It survived, stronger than ever, with its future fixed in the firmament. The dream which the Fathers had cherished, the plan which they had put forth with doubts and misgivings, stood the test of civil conflict; and in the process the fatal disease of particularism had appeared in malignant forms in the weaknesses of the Southern confederacy as disclosed by the resistance of its component states to the government they had erected. The disastrous policy which divided the countries of Latin-America into petty and warring factions was thus avoided in the American Union and discredited beyond all possibility of revival. Even those who fought against the Union, contending for the right as they were given to see it and powerfully supported by law, fact, and logic, came to realize that in some mysterious way the outcome of the lost cause was not all bitter. Viewed in the light of the ages, recalling the long quest of mankind for self-government, the preservation of the Union alone was enough to make the nineteenth century in the United States a century of enduring distinction.

THE ABOLITION OF SLAVERY

Interwoven with the preservation of the Union was another achievement which constituted a landmark no less noteworthy in the history of the nation—the abolition of slavery. A century ago this institution seemed as secure as the Union seemed transitory. It was as old as civilization; States and federations had arisen again and again, only to sink into ruin, but slavery had survived them all. When Webster made the peroration quoted above, slavery stood foursquare in the law of the land. It had, to be sure, practically disappeared in the North, but the Constitution guaranteed its protection in the states which sanctioned it. By press, pulpit, and statecraft slavery was defended with all the engines of human intellect. Once questioned by doubters of the Jeffersonian school, once timidly treated as an evil to be endured, it had now come to be supported by Southern planters and their affiliates as beyond criticism. "I take higher ground," declared Calhoun in 1837; "I hold that in the present state of civilization, where two races of different origin and distinguished by color and other physical differences, as well as intellectual, are brought together, the relation now existing in the slaveholding states between the two, is, instead of an evil, a good—a positive good." And in the light of this willful philosophy pallid hopes for "the gradual disappearance" of slavery shriveled up and blew away.

At that moment this powerful institution—an economic interest running into the billions—was vigorously assailed by a mere handful of noisy agitators, with William Lloyd Garrison in the lead. It was in 1831 that he flung to the winds his *Liberator*, a mere pigmy shooting mere words at a giant. For thirty years no hope seemed more forlorn than emancipation, no cause more futile. Free soil parties arose but never polled more than a handful of votes. The Republican party in the hour of its triumph advocated nothing beyond the exclusion of slavery from the territories, pledging itself at the moment to the protection of slavery in the states. Indeed, after Lincoln's election, with his full approval, it pushed through Congress an amendment to the Constitution making slavery immune from federal interference forever. It was a destiny beyond the design of the grand contestants that reversed the verdict of history and struck down the institution of chattel bondage. Lincoln's emancipation was announced and defended as a war measure, and the Thirteenth Amendment abolishing involuntary servitude, adopted in 1865, was a great stroke of state as well as an act of liberation. Likewise the Fourteenth and Fifteenth Amendments, designed among other things to confer civil rights on the newly emancipated slaves and political rights on Negro men, were measures of policy, no less than justice. Yet there they stand, with all their limitations, landmarks in the history of mankind's long martyrdom. Few there are, if any, who would fain bury them in oblivion and return again in a tide of reaction to the state of affairs prevailing a century ago.

THE EXTENSION OF DEMOCRACY

The flooding years that witnessed the gathering of the forces which preserved the Union and abolished slavery saw the spread of democracy, the extension of the suffrage, and the removal of property and religious qualifications from the right to vote and hold office—a movement continuing through the Civil War and culminating in the enfranchisement of women. Contrary to popular impressions, the founders of the American Republic did not believe in the democracy of one man one vote, and the early state constitutions limited voting and office-holding to the propertied classes or at least to taxpayers. In several states Jews, Unitarians, Catholics were denied all share in government; South Carolina would not intrust political power to citizens who denied the existence of hell. Soon,

GOVERNMENT AND LAW 207

however, a war was started on all such limitations, in the name of democracy. Did not the Declaration of Independence rest government on the consent of the governed? Then why not let all the governed vote and hold office? With growing insistence the question was asked. There were practical results. State after state swept away property and religious tests, and when Andrew Jackson was inaugurated President in 1829, the victory for white men was more than half won. Spurred to greater effort by the triumph of Jackson, advocates of manhood suffrage continued the struggle until, on the eve of the Civil War, religious and property tests for white men had been abolished in nearly every state; there were exceptions, and some still remain, but by 1860 the principle of white manhood democracy was accepted beyond challenge, and in a number of states Negro men could vote and hold office, sometimes under special restrictions.

From manhood suffrage it seemed to ardent democrats a logical step to confer equal political rights on Negro males throughout the Union as a corollary to universal emancipation. Indeed, the Fourteenth Amendment, adopted in 1868, went beyond this intention and provided that whenever a state deprived any adult male citizen of the right to vote in certain elections, its representation in Congress should be reduced. Only the Fifteenth Amendment, proclaimed two years later, took Negroes as such into account by forbidding states to deprive citizens of the right to vote on account of race, color, or previous condition of servitude. It would be idle to contend, of course, that the principles thus announced have been and are now enforced throughout the Union. In many states they are defeated by ingenious devices, but in wide reaches of the country they are applied without discrimination. It was the federal amendments, not popular will, which struck out of the constitution of Ohio the provision of 1851 excluding Negroes from the polls, to cite a single illustration. Whatever may be said with respect to the practical manifestations of the legislation granting the vote in principle to Negro men, it stands written in the fundamental law of the land, a declaration of catholicity in politics.

While the Fourteenth Amendment was under discussion women appeared in the lobbies of Congress, asking a pertinent question: If it is necessary to give Negro men the right to vote in order that they may defend their civil liberties, do women not need the ballot for the same reason? The query was not new. It had been put squarely in the eighteenth century by Abigail Adams, wife of the indefatigable John, and by Charles Brockden Brown, the novelist. During the flowering of Jacksonian democracy it became an issue of agitation, and in 1848 a Woman's Rights Convention at Senaca Falls had announced a new declaration of independence. Swiftly the idea of equal suffrage spread through the North. Garrison indorsed it; Whittier approved it; and it was on a fair way to serious consideration when it was overshadowed by the slavery debate and the Civil War. Shortly after the close of that grand conflict, an amendment to the federal Constitution conferring suffrage on women was introduced in Congress and a national suffrage association was formed under the leadership of Elizabeth Cady Stanton and Susan B. Anthony to carry on a general campaign in support of universal democracy.

Gains were slow. Wyoming conferred the ballot on women while still in the territorial stage, but that was more of an accident than a triumph. Members of Congress were unmoved by the logic of the cause, and its advocates then directed their energies to winning their victory state by state. That, too, proved to be uphill work. Not until Populism engulfed the West in the early 'nineties did Colorado, Utah, and Idaho adopt equal suffrage, in spite of strenuous agitation. And those gains were followed by a long lull. Then came the progressive surge at the opening of the twentieth century, and by 1915 California, Oregon, Kansas, Arizona, Nevada, and Montana had come under the new roof. The next year woman suffrage became one of the major issues of the presidential campaign, for candidates could not appeal for feminine votes without expressing an opinion on the challenge. Within a few months other states enfranchised women, among them New York, and by militant methods at Washington the question was dramatized. In September, 1918, with a congressional election at hand, President Wilson, who had long opposed the national suffrage amendment, went before Congress to urge its passage, as a measure "vital to the winning of the war." In the summer of the following year the requisite two-thirds vote was mustered and the resolution sent out for approval. Within little more than twelve months it was ratified by a sufficient number of states and proclaimed a part of the law of the land. The fruit of a hundred years of discussion had been garnered; equal suffrage rounded out the long struggle for universal democracy in the United States.

Through all these years the leveling forces of democracy worked out in other directions. Fearing popular agitations, the fathers had provided that the President and Vice-President of the United States should be selected by electors, chosen as the legislatures of the states might decide, and in the beginning the legislatures exercised the prerogative themselves; but before long the growing democracy demanded a transfer of that right to the voters. When Jackson was inaugurated the second time in 1833 presidential electors were chosen at the polls in every state in the Union except South Carolina, which clung to the historic method until the crash of the Civil War. In the meantime a new scheme for nominating candidates was devised. After the country had divided into two parties, candidates for the Presidency were selected by congressional caucuses, each composed of party members in the national legislature. But this autocratic institution, called "old King Caucus," was assailed by Jackson's followers, and when the election of 1832 came around they substituted for it the national party convention made up of delegates chosen by local assemblies of partisans. Immediately it took root and flourished, and by 1840 it seemed as firmly established as the Constitution itself. Presidents were to be nominated by "assemblies fresh from the people," and to be chosen, indirectly, to be sure, by popular vote. No wonder Carlyle could write in lament that caucusing, ballot-boxing, and universal democracy were at hand.

Having brought the election of President out into the popular forum, democracy then turned to a consideration of the provision of the Constitution which vested the power of choosing United States Senators in the hands of state legislatures. The framers of that document had intended to establish a conservative upper house which would represent the "substantial" economic interests of the country and act as a check on the "popular distempers" of the lower chamber. As things turned out their plans succeeded beyond all expectations, but only to arouse jealousy on the part of democratic champions. Early in the Jacksonian era, popular election of Senators was proposed for debate; and in 1868 a spiritual heir of Andrew Jackson, Andrew Johnson, then President of the United States, suggested to Congress a constitutional amendment designed to effect this "reform." From time to time thereafter the idea was broached until at last in 1893 the House of Representatives passed the amendment, only to find the Senate obdurate. Again and again the House insisted, gaining heavier support in the upper chamber as more and more Senators were returned from states having the direct primary in force. Finally in 1912 the necessary two-thirds vote was mustered in both houses in favor of the Seventeenth Amendment establishing popular election of Senators; with amazing alacrity it was ratified by state legislatures; and the next year it was proclaimed in force by its veteran advocate, William Jennings Bryan, as Secretary of State.

CIVIL SERVICE REFORM

As Jacksonian democracy spread, widening the suffrage and insisting upon popular election of the President and Senators, it was accompanied by the rise of the spoils system in politics— the use of public offices to reward partisan workers, with little or no regard for fitness or competence. By the middle of the nineteenth century the practice had become universal in national, state, and local affairs. Whenever a new party came into power it proceeded to sweep out of office all public employes under its jurisdiction, save perhaps a few necessary to keep the machine running. Even a change in administration with no change in political complexion usually brought with it a "house-cleaning," the expulsion of old henchmen to make room for a hungry multitude at the gates. In the midst of the harassing perplexities of the Civil War, President Lincoln was more distraught at times by the beggary of office-seekers than by the calamities of battle. It seemed, as he put it, that he was living in a hotel all on fire, surrounded by a clamoring horde demanding an assignment of rooms in the crumbling structure. With the increase in the number of federal positions after the war, the spoils system became worse than ever, and politics in the Republic appeared to be little more than an angry scramble for office and pelf.

Then came a reaction. As Carlyle remarked, when it gets dark enough we can see the stars. In the early 'seventies a vigorous movement was launched in favor of civil-service reform—a program of administrative reconstruction which demanded that the major portion of the public offices should be open only to persons of competence as tested by examinations, that tenure should be during good behavior, and that partisan influences should be eliminated in the field of technical performance. Supported by leaders of undoubted power, such as E. L. Godkin, George William Curtis, and Carl Schurz, the new ideal of public service made headway slowly against the scorn of practical politicians who called its advocates "wolves in sheep's clothing" and laughed at what they called "snivel"-service reform. The agitation came to a climax in 1881 when President Garfield was done to death by a disappointed and demented office-seeker. A shot fired by an assassin rang throughout the land, driving into the head of the most hardened henchman the notion that there was something disgraceful in reducing the Chief Executive of the nation to the level of a petty broker in petty jobs. Within a year a committee of the Senate brought in a report which denounced the spoils system in stinging terms, picturing the President giving audiences to beggars and flinging public employments to "a hungry, clamorous, crowding, jostling multitude."

Driven into action by public opinion, Congress reluctantly passed in 1883 a civil-service law which still remains the basis of federal administration. The Act authorized the President to appoint a commission of three, not more than two from the same political party, and empowered him to introduce the merit system in a wide range of federal employments. Henceforward, within the specified area, political influences were to be prohibited, and the enumerated offices were to be open only to candidates who had demonstrated their fitness by passing appropriate tests. Although the law at first covered only a few thousand positions, the number was steadily increased, and at present three-fifths of the federal employes, amounting in all to approximately half a million, come within the scope of the merit system. Meanwhile sickness and accident insurance and a scheme of contributory pensions have been provided to make the federal service secure and attractive to those who devote their lives to it. So firmly intrenched is the idea that "a public office is a public trust" that even the heads of numerous bureaus and establishments nominally under the spoils system are appointed and retained with reference to competence rather than political affiliations. Although the ideal arrangement has not been attained, the national government has resolutely turned its back on the Jacksonian doctrine that the spoils of office belong to the victors, and is working toward the establishment of its functions on a high level of efficiency.

The example of the federal government was followed in 1883 by the state of New York in the enactment of a civil-service law modeled on similar lines. Other states yielded slowly and reluctantly, and at the moment only nine all told have civil-service commissions or agencies—California, Colorado, Illinois, Maryland, Massachusetts, New Jersey, New York, Ohio, and Wisconsin. In the cities, however, greater gains have been made and in all the leading metropolitan centers the public service now rests, to some extent at least,

GOVERNMENT AND LAW

upon a merit basis. The hands of state and municipal civil-service officials have been strengthened by the formation of a national Civil Service Assembly composed of representatives engaged in personnel administration. This Assembly has a staff agency, known as the Bureau of Public Personnel, which carries on expert investigations and publishes a monthly journal reporting events, advances, and the results of special researches in the field. And outside the official world an association of private citizens, the National Civil Service Reform League, promotes studies and seeks to develop a favorable public sentiment. Moreover, forty or fifty associations of federal, state, and local officials, for example the Society of American Foresters, the Conference of State Sanitary Engineers, and the Association of Highway Engineers, hold conventions, carry on investigations, issue publications, and attempt in other ways to establish and maintain ever higher standards of efficiency in the public service—a guaranty of the perpetuity of American institutions more fundamental in character than the labors of most professional patriots, although less known in the popular forum.

THE QUEST FOR ADMINISTRATIVE EFFICIENCY

A similar quest for administrative efficiency is revealed in the recent movement for the reorganization of state governments. Near the close of the nineteenth century, as the functions of government multiplied and expenditures mounted, it became evident that the chaos of independent offices, boards, and commissions which had come down from the agrarian age was wholly unadapted to modern requirements. Waste, confusion, friction, and lack of coöperation characterized state administration throughout the Union, and everywhere responsibility was dissipated so widely that the governor sank into a mere figurehead. Millions were annually squandered and no official could be held accountable to the public.

As early as 1891 the Governor of Massachusetts analyzed the situation in a message to the legislature, pointed out glaring defects in state administration, and suggested a consolidation in the interest of economy and efficiency. Soon the idea appeared in the writings of students and critics, such as Frank J. Goodnow and Herbert Croly. In 1915 the New York Bureau of Municipal Research prepared for the constitutional convention of that year a complete survey of the state government and proposed a scheme of reorganization and consolidation. Two years later Illinois, under the leadership of Governor Lowden, broke the way by abolishing more than one hundred statutory offices, boards, departments, and agencies and consolidating their functions under the direction of a few great departments responsible to the governor. Other states followed in rapid succession and within fifteen years nearly one-third of them had revamped, more or less thoroughly, their inherited systems of administration, united functions of a kindred character, drawn lines of accountability from the base to the apex, and prepared themselves to handle with more competence the responsibilities devolved upon them by the public requirements of the modern age.

Meanwhile American cities, stirred by the shame of rings, bosses, boodle aldermen, and corruption, confronting prodigious problems made inevitable by technology—the supply of water, gas, electricity, and transportation, the paving of streets, the safeguarding of public health, and the administration of immense educational systems—began to search for more efficient methods of government. Civil-service reform had made a contribution, but that was not enough. Some way had to be found to introduce order and competence into the historic chaos. If a particular point of beginning must be made, it may well be the reconstruction of Galveston after the great storm of 1900 which laid waste whole sections of the city. Since the trying questions of restoration were too much for the old political machine, Galveston swept it all away and vested the entire government in the hands of five commissioners elected by popular vote. This was the beginning of commission government for cities which spread with amazing swiftness to all parts of the country; within fifteen years approximately four hundred municipalities had adopted it, in the hope of secur-

GOVERNMENT AND LAW

ing a better quality of public officers and a more business-like administration of their affairs. And on the whole their hopes for large improvements were realized.

Discovering weaknesses in the commission plan and yet undaunted in their search for competence, American cities took another step in the direction of concentrating power and responsibility, by turning to the city-manager type of government. The essential elements of the plan are simple. All executive functions of the city are vested in the manager who is chosen and removed by the municipal council or commission, as the case may be. He in turn appoints and dismisses department heads, who are few in number, subject to the limitation of the civil-service law. It is his duty to direct the various branches of municipal administration, to initiate the budget, and lay working plans before the city legislature for consideration and action. The manager may attend meetings of the council to defend and explain his proposals, and he may be called before it to render an account of his stewardship. The business of the council is to scrutinize and to legislate; the business of the manager is to plan and to execute. Such are the outlines of the system.

Its history really dates from 1913, when Sumter, South Carolina, adopted it, although Staunton, Virginia, claims the honor of having tried the project five years before. Since that beginning the system has spread rapidly until it is now in effect in nearly four hundred villages, towns, and cities. While most of the municipalities are places of minor rank, among them are also to be found Dayton, Springfield, Cincinnati, Kansas City, and Rochester. In electing their managers these cities are not limited to local politicians, but may choose talent wherever it may be found. Thus the plan overcomes many of the difficulties involved in obtaining technical ability by the democratic process and broadens the field of choice to include the whole country, placing all talents at public disposal. Indeed, city managership has become a profession, with an organization, a magazine, and a research staff, thus helping to lift municipal government from a morass of corruption and to place it on a high plane of technical performance.

Correlated with civil-service reform, state administrative reorganization, and the reconstruction of city governments has been a movement of equal importance, the introduction of the budget system of financing. Under the traditional régime which prevailed until the opening of the nineteenth century in the Congress at Washington, the state legislatures, and municipal councils, the expenditure of public money proceeded without plan; appropriation bills were introduced pell mell, referred to committees without order, reported in chaos, passed in confusion, and not until the end of a legislative session was it possible to discover how much money had been voted away. Deficits accumulated, debts piled up, and occasionally, very rarely, surpluses were collected. That the public should have tolerated such methods is one of the seven wonders of democracy; but it did until, near the opening of the twentieth century, mounting taxes suggested a halt, a survey, and reform. Leadership was taken at this stage by the New York Bureau of Municipal Research, founded in 1907, which advocated in season and out the adoption of a budget system for all legislatures, national, state, and local--appropriation procedure on the basis of a balanced financial plan covering outgo and income systematically presented, debated, and enacted. Within a short time cities and states began to accept the device, and in 1921 Congress yielded to the new demand by passing a national Budget Act. While the history of this development belongs elsewhere (Chapter VII), it must be mentioned in connection with the quest of the American democracy for efficient instrumentalities and methods of government.

THE TRANSFORMATION OF AMERICAN POLITICAL ECONOMY

Striking as may be the changes in civil rights, suffrage, and the machinery of government, they are no more significant than the transformation of its functions and spirit during the past hundred years. When Andrew Jackson entered upon his

GOVERNMENT AND LAW

second term in 1833 the doctrines of classical economy were everywhere dominant in the English-speaking world and they fitted with strange precision into the aspirations of the agrarian democracy of the United States as well as the requirements of manufacturers in Great Britain. According to these doctrines the prime business of government was to defend the country and keep order within its borders. It should not engage in any economic enterprises itself, interfere with the conduct of private undertakings, favor business by subsidies, bounties, and tariffs, intervene in the "natural distribution of wealth," promote social welfare by positive legislation, or seek to protect individuals against the buffets of disease, accidents, and misfortune. With respect to internal affairs it was the prime duty of the government to uphold private property, protect it in gainful enterprise, and sustain freedom of contract between private parties. Although the classical theory of *laissez-faire* had not been carried to its logical conclusion in England or the United States, it was on the whole the guiding principle of American government a century ago and was supported by the strict construction of the Constitution to which Jacksonian Democrats generally adhered. Broadly speaking, they insisted on the reduction of the tariff to a revenue basis, the transfer of banking functions to the states, and the discontinuance of the "protective" system inaugurated in the age of Alexander Hamilton. Jackson himself even vetoed a federal appropriation for highway construction on the ground that it was unauthorized by the Constitution.

According to this philosophy every individual was the best judge of the situation in which he moved, could protect himself against evil, and select the good with almost unfailing accuracy. The purchaser could pass upon the quality and price of the commodities he bought (let the purchaser beware), avoid adulterations and frauds, and rely upon competition to keep prices down to a "just" level. The capitalist in his legitimate quest for profits would automatically serve the needs of society and under a competitive régime his prices would be kept reasonably near to the cost of production. Each workman was free to employ his talent wherever he could find work, must rely upon himself in avoiding accidents, hazardous occupations, and the distresses of business depression. Government interference with the conditions of industry, prices, wages, the quality of goods, the operations of business enterprise, the lot of the poor, sick, and unfortunate, the social order, in short, was, therefore, a violation of "natural law," evil in itself and bound to do harm in the long run to the supposed beneficiaries. In other words, the State was to be limited to the rôle of a "police constable."

Even the most superficial examination of modern government reveals nothing short of a revolution in practice, however far theory may lag behind political conduct. There is not a department of economy in which the government of the United States does not intervene. It protects private industry against foreign competition by means of a high tariff—the history of tariff legislation since 1861 is in the main a record of progressive increases in duties, occasional reductions forming striking exceptions to the rule. It gives bounties to the American merchant marine and to aviation in the guise of generous mail subsidies. It favors the formation of corporations for promoting foreign trade by the broad terms of the Webb Act of 1918. It utilizes the engines of diplomacy and war—the State Department, the Navy, and the marines—in promoting investment and commercial opportunities abroad and safeguarding them against inroads by governments, bandits, and revolutions.

While thus encouraging private enterprise on the one hand, it intervenes in its operations on the other under the Sherman Anti-trust Act of 1890 and the Clayton Act of 1914. Although the formation of large-scale corporations proceeds apace, the federal government proclaims every combination in restraint of trade illegal and subject to prosecution, pains, and penalties, apparently on the assumption that destructive competition makes for the general welfare. Where it does not dissolve by legal process, it seeks to regulate through the Federal Trade Commission and to guarantee the maintenance of "fair practices" in trade and manufacture. In addition it attempts to fix and enforce national standards in food and drugs, thus reversing the ancient rule that the individual purchaser must protect himself against frauds, poisons, and adulteration. And when in time of depression business fails to provide employment for its workers, the federal government, especially under the planning law of 1931, proposes to help take up the slack by the construction of public works.

Agriculture, the most highly individualistic form of enterprise in the United States, likewise turns to the federal government for aid and protection. It enjoys the benefit of tariffs on certain competing commodities. On the one side the government assists in increasing production. It subsidizes agricultural experiment stations, studies the diseases of plants and animals, discovers new plants, improves the breed of animals, analyzes soils, makes recommendations with respect to fertilizers, excludes unsound and adulterated fertilizers, livestock feed, and seeds from interstate commerce, combats pests, large and small, establishes and maintains standards for the grading of agricultural produce, and supplies farmers with technical information on every conceivable branch of their industry. On the other side, the federal government assists them in marketing their produce under innumerable Acts of Congress, among which the Agricultural Marketing Act of 1929 stands out as the boldest departure. In this field it helps in the formation of coöperative societies, encourages coöperative selling, makes experiments in the stabilization of prices, regulates warehousing and packing, penalizes fraudulent practices on the part of commission merchants, collects and distributes market reports, and in short seeks to rationalize agriculture, the most primitive and emotional of the technical arts. In addition the federal government, through the agency of land banks and intermediate credit banks, facilitates the extension of long-term and short-term credits to agriculture, supplying it with capital at a lower rate of interest than would prevail "in the natural course of things."

In one field of business enterprise, public utilities, the federal and state governments intervene and regulate within their respective jurisdictions on principles exactly contrary to the rules of *laissez-faire*. These economic institutions are declared to be "affected with public interest" and separated from the main body of industries. Here it is assumed that something akin to monopoly is involved and that competition cannot operate as a control over rates, services, and charges. Here, it is believed, capital is not entitled to all the profits that the traffic will bear and must be content with a fair return. On such suppositions billions of dollars' worth of property are placed on a peculiar basis and subjected to governmental control.

Although the principles applied vary widely from jurisdiction to jurisdiction, a clear tendency is discernible in the process. The capital invested in plants must bear a close relation to the physical value of the property, and exorbitant profits due to peculiar circumstances cannot be capitalized as a burden upon the future. The rates charged for services must be reasonable— that is, merely high enough to afford a reasonable return to the vested interests concerned. In operation, to be sure, these rules are often grossly violated, but they stand as law in nearly every part of the Union and serve as a goal toward which public policy is advancing. When, and if, fully realized, railways, electric-light plants, water-works, and other public utilities— all vital branches of national economy—will be removed from the domain of capitalism as historically conceived and given a special position within the sphere of government. Moreover, judicial opinion seems inclined to widen the range of affairs "affected with public interest" to include other prime necessities of life, thus marking out new lines of development for the coming years.

On its own account the federal government manages undertakings of immense economic significance. In the field of communications it once limited itself to the transmission of mails from post-office to post-office; by gradual stages the Post Office Department has extended its functions until it now delivers mail in cities, towns, and rural regions, transmits money, carries parcels, and operates postal-savings banks. In Jackson's day, the federal government proceeded on the assumption that all land, timber, mineral, and water resources of the public domain were to be transformed as rapidly as possible into private property on terms favorable to individuals. The arable land has been granted in severalty on that theory; but under an act of 1891 an enormous forest reservation has been created and is now managed under federal auspices. By later legislation Congress has retained public ownership of oil and minerals beneath the surface of public lands, even when granted to individuals, and has declared all water-power sites under federal jurisdiction to be inalienable national property. These resources the government now undertakes to administer and develop under a system of leasing and regulation. Complying with the terms of the Newlands Act of 1902, it constructs and operates enormous irrigation plants in the arid states of the West. The Panama Canal is a monument to its enterprise. It spends millions, often wastefully, in the development of inland and coastal waterways, and through a corporation in which it owns all the stock it operates fleets of barges in the carrying trade. Once, as we have noted, President Jackson halted federal highway construction with a veto; in the administration of his Democratic successor, President Wilson, Congress began a great system of federal aid to states for highway construction, and under federal supervision a giant network of improved roads has been spread throughout the land. For the toll road of 1833 has been substituted the free and open road of our time, and the motorist can now travel freely from coast to coast, from the Great Lakes to the Gulf, over highways built under public auspices.

Within their sphere of power the states have departed even farther than the federal government from the simple postulates of Jacksonian democracy. In a flash the whole process is revealed by a comparison of the simple structure of any state government in 1833 with its present organization of departments pertaining to agriculture, industry, labor, health, highways, conservation, education, public works, and public welfare. Several states carry on enterprises of considerable magnitude. Many administer forest domains and water-power sites. North Dakota operates grain elevators. Two have public coal-mines. All engage in highway construction, especially since the federal government has come to their aid. A few have ventured tentatively into the realm of insurance, particularly with reference to compensation for workers injured in industries. All except two or three undertake to control public utilities on principles discussed above. With varying detail all regulate the conditions prevailing in mines, mills, and other branches of industry, and the most advanced have elaborate codes covering the ventilation of factories and workshops, the use of dangerous machinery and materials, the safety of boilers, elevators, and other devices, the provision of sanitary conveniences, and the methods to be tolerated in dangerous employments of every kind and class. In other words, the thesis so dominant a century ago, to the effect that industries may make their own laws and that those who work in them must accept whatever they find, including all the hazards and strains, has been definitely abandoned by state governments. They now assume the responsibility of establishing and enforcing at least minimum standards throughout the entire domain of industrial economy. As technology makes new conquests, every year marks increases in their burdens, taxing to the uttermost their ingenuity in legislation and administration.

The greatest strain on the simple institutions inherited from Jacksonian democracy has come, of course, in the cities, three of which combined exceed in population the entire country in 1833. The rise of industries, the influx of foreigners by the million, and the drift of the rural population to urban centers have made a revolution in American life beyond all calculation, and have forced municipal governments to take on responsibilities that would have seemed staggering to the men who framed the federal Constitution. They must build miles of costly pavement, provide water supplies and sewers, remove wastes by the ton, light, clean, and patrol their thoroughfares, establish or regulate gigantic utilities to furnish gas, electricity, transportation, and telephonic communication, administer schools and in some cases colleges, assure safety in the construction of buildings, fight fires and fire hazards, lay out and maintain parks and playgrounds affording recreation to teeming millions, clear slums and guarantee minimum standards in houses and tenements, plan for their future growth, wage war on disease and suffering, and, where they possess harbors, build dock and terminal facilities. If in part the story of this development is marred by folly and corruption, it is in the main a record of amazing achievements, bearing witness to an extraordinary capacity of the people for self-government. The proof of this assertion is to be found in a comparison of the sanitary and living conditions of New York or Boston in the age of Daniel Webster with the state of affairs prevailing today. And in this connection it would be well to remember that the devastating horrors of epidemics which swept through our cities in the middle period of the nineteenth century are now as obsolete as witchcraft and alchemy.

THE HUMANE SPIRIT IN GOVERNMENT AND LAW

Although many of the functions thus far mentioned may be ascribed to practical necessities, such as the growth of machine industry and the spread of natural science, that is not the whole story. No small part of the increasing burdens of government are to be explained by the rise of the humane spirit in America. Steadily through the past hundred years, crudely manifested, no doubt, it has made headway and is reflected in the development of legislation and institutions pertaining to welfare. It can be concretely illustrated in the annals of any state. Using Illinois as an example, we find the following entries made according to chronology:

    1839 School for the deaf
    1847 Hospital for the insane
    1849 School for the blind
    1854 State superintendent of public instruction
    1865 Asylum for feeble-minded children
          Soldiers', orphans' home

	Eye and ear infirmary
1867	Industrial university
	State reformatory
	State library
1869	State board of public charities
	Normal university
1872	Board of agriculture
1877	Board of health
	Humane agents
1879	Commissioners of labor
1883	Mining board and mine inspectors
1887	Industrial home for the blind
1889	Asylum for insane criminals
1893	State factory inspector
1899	Food commissioner
1903	Board of prison industries
1907	Food standards commission
	Examiners of registered nurses
1909	Library extension commission
1911	Park commission
1910	Mine rescue commission
1917	Reorganization and consolidation of state administrative agencies.

A similar record, still more impressive, could be made of humane advance in any of our great cities through the same period of years, revealing in the century so often reviled as materialistic a determination of the American people to wage war on suffering, disease, ignorance, and misfortune.

Not content with providing public facilities for the care of the defective and dependent, the states continued to improve their services by making use of the latest advances in medicine and psychiatry, thus bringing modern science to the aid of their humane purposes. A comparison of an insane asylum of 1833 with a contemporary institution reveals an immense movement of informed intelligence. Yet institutional care for the dependent was not enough. In 1908 the city of San Francisco, without special warrant by law, began to apply certain funds available to the juvenile court to the maintenance of widowed and deserted mothers in their own households, in an effort to keep families together at home. Three years later Missouri enacted a statute permitting the juvenile court of Kansas City to make direct payments to mothers instead of committing their children to institutions. These tentative experiments may be called the beginning of mothers' pension legislation which has rapidly extended until today more than three-fourths of the states have provisions of this character, grading the amount of the pension on the basis of the number of children and their age.

From this it was but a step to old-age pensions, long combated as a socialistic importation from Germany. Confronted by appalling statistics on dependency, notwithstanding our boasted prosperity, Nevada, Montana, and Pennsylvania enacted old-age pension laws in 1923, with varying qualifications for eligibility. Although the Pennsylvania statute was declared unconstitutional, other states took up the issue, and by the summer of 1931 the American Association for Old Age Security could announce that fifteen states had made provision for their indigent poor in the form of regular pensions. In twelve other states the question was ardently debated, bringing the prospect of universality within the range of the immediate future.

The spirit of humanity which is building institutions of welfare and providing security for worthy dependency has made its way into all departments of civil and criminal law. In general the process is to be observed in the fusion of law and equity. According to the traditions of England, whence our system of jurisprudence was largely derived, equity, tempered with mercy and intelligence, grew up in the King's chancery to provide remedies in the interest of justice, when the law, owing to inadequacy or technicalities, failed to afford them. It involved, however, special litigation in separate courts, and in the course of time was itself overlaid with technical restrictions. In the United States, on the other hand, save in a few jurisdictions, equity and law have been united and are administered by the same tribunals, to the benefit of litigants and the advancement of reason in jurisprudence. Even where equity and law are kept apart, the bewildering technicalities of the common-law procedure have been largely eliminated by state legislation and the process of arriving at justice liberalized. In keeping with this development, the ancient law of evidence, once beset by technicalities, has been loosened up by the extension of the area of logical relevance and the limitation of technical exclusionary rules. Although the history of American law remains unwritten, so great has been the lawyers' neglect of their science, enough is evident on the surface of things to reveal at work here the spirit of democracy and fair play so apparent in other departments of American life, in spite of all contradictions and violations.

This generalization may be illustrated by a few details. By the middle of the eighteenth century, the idea had begun to creep into American colonial law that a certain minimum amount of every debtor's property should be exempt from seizure to satisfy the claims of creditors, especially with the thought of preserving his homestead intact. Although bitterly fought as class legislation, later as a violation of the obligation-of-contract clause of the federal Constitution, this protective device, conceived in the interest of the poor, spread widely during the nineteenth century and was generally incorporated in the laws of the new commonwealths as they came into the Union. A number of Southern and Western states even permit a landowner to register a limited number of acres as a homestead, which is then safeguarded permanently against seizure and sale in the satisfaction of debts.

Akin to legislation of this character has been the abolition of imprisonment for debt. At the opening of the nineteenth century it was a common practice to throw men and women into jail for failure to pay debts, and if the sum was small in amount there was no prospect for a stay of execution. Although murderers and burglars were fed at public expense, these poor wretches were required to provide their own food, and if their friends or charitable persons failed them they died of starvation. Everywhere in the land, especially in the cities in time of depression, the jails were crowded with thousands of unfortunate debtors, many of whom languished in prison for five years or more until death or a stroke of fortune relieved them. Stirred by the horror of it, Pennsylvania began to relent in 1817 and provided that no one should be imprisoned for a debt of less than twenty-five dollars. Slowly the law relaxed everywhere until by the middle of the nineteenth century the ancient custom had been generally relegated to the museum of barbarities.

In time the liberating spirit reached domestic relations. Under the common law of England which prevailed, with modifications, in the Colonies and was continued by the states, a married woman's personal property, even her jewelry and clothes, belonged to her husband and he had the right to manage her real estate as he saw fit, without let or hindrance. In short, the personality of the married woman was merged with that of her husband, was in fact sunk in it. "So great a favorite," wrote the genial Blackstone, "is the female sex of the laws of England." Notwithstanding many changes, these principles were so completely incorporated in American law that the jurist, James Kent, writing on our jurisprudence two hundred years after the landing of the Pilgrims, had only to enumerate them and add a few variations to complete the picture.

Yet within a few years the feminist movement, gathering force with the spread of political democracy, was attacking every point in the ancient law. In 1839 Mississippi emancipated women from bondage in the matter of property; in 1848 New York, Indiana, and Pennsylvania took similar steps; two years later California and Wisconsin swung into line; and at the close of the nineteenth century women had everywhere made gains toward equality in the ownership and management of property. Their interests were also favored by the abolition of primogeniture and entailment, which once kept landed estates intact in the hands of eldest sons—a legal reform which had begun in Colonial times, received an immense impetus from Jefferson's famous Virginia statute of 1776, and spread throughout the country as the frontier advanced to the

Pacific. To the disgust of traditional lawyers, the old rule that women could not testify in court against their husbands was gradually abandoned, as women made their way forward in education and industry. And to cap the climax, the ancient practice of granting divorce by legislative act, an expensive procedure open only to the privileged, was abandoned, and the authority transferred to the courts, as the grounds for separation were extended and humanized.

In another significant relation, to cite one more example of liberality within the limits of the space here permissible, the old law has been touched with the humane spirit. Under the common law, as interpreted by the courts, every employe had to assume practically all the risks of industry, no matter how hazardous; the result was poverty for himself and his family in case of serious injury, and distress for his family in case of his death. In other words, employers were liable for damages only when they were themselves personally responsible—that is, they were not liable when an accident was due to "unpreventable causes or to the carelessness of the employe himself or one of his fellow employes." Studies of accidents revealed appalling conditions—that more persons were killed and injured in American industries in proportion to the number engaged than in any other country in the world, and that more than half of the accidents arose from occupational hazards. Moreover, when an employe had legitimate claims for damages, he could only collect by a lawsuit, which was likely to be costly and tedious. When it was suggested that industry should bear the burden of its casualties, as it insured against fire and cyclone, the reply was made that to require employers to compensate their workers for injuries was to confiscate their property for the benefit of others.

For many decades this argument prevailed, but shortly after the opening of the twentieth century there began a decided movement in the direction of shifting the burdens of accidents from the victims to industry. In 1908 Congress made all railway common carriers subject to its jurisdiction liable to action for damages in case of injury or death due to defects, insufficiency of equipment, or negligence, even if there had been some contributory neglect on the part of the employe himself. About the same time state after state began to abrogate the common law rules in this connection and to compel employers to compensate injured employes whenever accidents arose from the negligence of the former or the necessary risks and dangers of the industry itself. At the present time only a few commonwealths adhere to the outmoded doctrine of the common law.

To make the new principle still more effective, a large number of states have adopted a system of insurance and prescribed that compensation for injuries shall be automatic and simple, obviating entirely the resort to expensive lawsuits. Where such a scheme is in force the terms and conditions for making awards to injured employes or their families are set forth explicitly in the law; industrial commissions hear complaints and make grants on application, avoiding from beginning to end the technicalities of the courts, with their endless appeals. In some states the compensation laws have been extended to cover diseases arising from occupations, bringing these hazards also within the scope of insurance. Thus by legislative fiat soldiers of the forge, loom, and lathe are placed on a level with soldiers of the sword and declared worthy of protection at the hands of the state.

In the domain of criminal law the century here under review has witnessed many departures from the barbarities inherited from primitive times. The number of crimes for which the death sentence is meted out has been generally reduced and in a few states capital punishment has been entirely abolished. Federal courts, and as a rule state courts, may in certain cases extend mercy to convicted persons—putting them on probation instead of sending them to prison—and thus allow them the privilege of working out their own fate. Only when they fail to comply with the conditions stipulated are they compelled to serve terms in jail. Akin to this advance has been the widespread adoption of the indeterminate sentence, which gives a generous discretion to the courts. Under such an arrangement a condemned person is not incarcerated for a fixed period, but for an indefinite term, ranging from a minimum to a maximum, and the exact length of his punishment depends upon his conduct. In the administration of the law under parole boards prisoners may be released on their word of honor when they have served the minimum term and may remain at liberty as long as they keep faith with the prison authorities. And yet, while it is proper to enter upon the record noteworthy achievements in the field of criminal law, it would be blindness to overlook the difficulties still in the way and the numerous failures to apply the best principles evolved by the science of penology.

In recent years, perhaps the most significant tendency in the study and exposition of the law has been what may be called, for want of a better name, the growth of realism. In essence it means an emancipation from the tyranny of self-deception and dogmatic principles and a propensity to get at the facts and forces behind the making and administration of law and to inquire into the actual effects of such operations on the property, persons, and social relations involved. The rise of this tendency was signalized many years ago by Justice Oliver Wendell Holmes, who was capable of making a good working distinction between his own naïve feelings about law and the stream of events moving in the world of reality. His influence was later strengthened by Justice Brandeis, who brought to the law a passion for statistics, facts, and results which gave a broad social and economic setting for the formulation of judicial determinations. In the better law schools there is now apparent "a distrust of the received set of rules and concepts as adequate indications of what is happening in the courts," and a wide-spreading search for an exact knowledge of what is actually happening, in terms of causes, operations, and social effects.[1] It is impossible to escape the conclusion that in the near future this realistic spirit of inquiry and exposition

[1] See "Some Realism about Realism," by Karl N. Llewellyn, *Harvard Law Review*, vol. xliv, no. 8.

will exert a profound influence on the theory and practice of the law.

CENTRALIZATION

Towering above the prodigious movement in government and law here described in faint outline and associated with it, in many respects, has been an immense centralization in legislation and administration. This centripetal action appears in the states in connection with health, highways, education, industries, institutions of public welfare, and other departments of government. It is particularly striking on the national stage. Under apparently limited powers Congress has enacted whole volumes of statutes pertaining to industry, agriculture, railways, pure food and drugs, labor, public health, safety, and morals, placing under federal jurisdiction matters undreamt of by the fathers or reserved to the states in the days of primitive economy such as prevailed at the close of the eighteenth century. By grants-in-aid to assist the states in education, agricultural experimentation, highway construction, vocational rehabilitation, and other significant functions, Congress has in effect imposed national standards upon them and brought them within the jurisdiction of federal authorities in Washington. Under the broad and elusive terms of the Fourteenth Amendment, every act of every state and local legislature touching seriously the property of citizens and corporations is subject to review and annulment at the hands of the Supreme Court. Although the march of centralization is assailed at every step in the name of ancient liberties, it is inexorable, indicating that the problems of democracy and planned economy, now looming fatefully on the horizon, are to be met on a national scale, involving international relations of the first magnitude, and solved in a manner befitting the great society into which the scattered rural communities of Jackson's day have been transformed by steam, steel, and electricity.

If an American citizen of today, covering the past hundred years with the mantle of oblivion, could take his stand in 1833, penetrate the curtain of the future, and see the Union, preserved, spreading from coast to coast, chattel slavery abolished, the suffrage widened, political and religious disabilities canceled, popular education established and set on the road to universality, great functions of administration undertaken in the interest of humanity at large, government made more efficient as its burdens multiply, and the law liberalized in innumerable directions, he might well rejoice in the prospects of his country in spite of all the leaden discouragements evident in the Jacksonian scene—its crudeness, demagogy, and distressing mediocrity.

So the citizen of the present era may look upon problems still unsolved—lynching a disgrace to the nation, crime widespread and shocking, corruption breaking forth with baffling virulence, periodical industrial crises bringing poverty and misery in their train, vast areas of rural and urban slums, civil liberties so often trampled underfoot, the spirit of bigotry rampant, natural resources wasted, lowly aliens in our midst treated with heedless cruelty, religious intolerance stirred by partisan angers, incompetence still present in government and economy, preparations for wars notwithstanding the pledges of the Kellogg Peace Pact, vulgarity standardized and worshiped on a national scale, and American civilization challenged as the apotheosis of materialism. And yet seeing these things with open eyes, without extenuation or illusions, and recalling the noblest triumphs of the past, he may look forward with confidence, trusting that the nation which has carried its destiny thus far through the years will rouse itself, gird its loins, summon its powers of creative imagination, and advance inexorably upon the future, armed with all the instrumentalities of modern science, inspired by the finest traditions of the past, and make the coming century a still more imposing reach of achievement, remembering in the darkest hour of uncertainty that, as the poet has exclaimed, hope may create even from its own wreck the thing it contemplates. If not, what is the alternative?

CHAPTER NINE

THE PROCESS OF SOCIAL TRANSFORMATION

"During the same winter (1889-90), three boys from a Hull House Club were injured at one machine in a neighboring factory for lack of a guard which would have cost but a few dollars. When the injury of one of these boys resulted in his death, we felt quite sure that the owners of the factory would share our horror and remorse, and that they would do everything possible to prevent the recurrence of such a tragedy. To our surprise they did nothing whatever. . . . The visits we made in the neighborhood constantly discovered women sewing upon sweatshop work, and often they were assisted by incredibly small children. I remember a little girl of four, who pulled out basting threads hour after hour, sitting on a stool at the feet of her Bohemian mother, a little bunch of human misery. But even for that there was no legal redress."
—JANE ADDAMS, *Twenty Years at Hull House*, pp. 198 ff.

The Process of Social Transformation

By Jane Addams

THE decade beginning in 1830 has been described from the point of view of Boston as "the day of emancipation and hope, opening paths of progress in all directions," with an output in letters never equaled before—and scarcely since—upon this continent. But while the country as a whole shared with Boston the proud consciousness of having won political independence and of having established republican institutions, various stages of development were represented in the United States.

EARLY MATERIAL AND SPIRITUAL CONDITIONS

The city of Chicago, for instance, in the same decade still centered around old Fort Dearborn, which had been unoccupied during a period of peace with the Indians, but had been regarrisoned in 1832 at the time of the Black Hawk War. The earliest settlers at the foot of Lake Michigan found their resourcefulness taxed in many directions. In the first place, they were living in the middle of a swamp, upon land so low that the river flowed east into the Lake or joined its waters to the Desplaines River on the west, as the season indicated. My own father, landing from a Lake boat in the little city as late as 1844, found his horse and light vehicle so firmly mired in one of the main streets that my mother had to be carried out on the broad back of a friendly Irishman. From her safe position on the sidewalk she watched the valiant efforts of man and beast to rescue themselves from the stream of quicksand which still flows under Chicago, that they might travel further westward.

THE PROCESS OF SOCIAL TRANSFORMATION

Harsh conditions obtained within as well as without in the newly opened territories, for although philosophers had arrived in Boston to resurrect a belief in man's inherent worthiness, the soothing doctrine had not yet reached the sturdy men and women who occupied the great stretches of country to the west and south. These early settlers were constantly told by their spiritual guides that the terrors of hell awaited them as evil-doers and added this fear to the sweat and toil which broke up innumerable quarter sections of tough prairie.

The reverse of this doctrine, that rewards come only to virtue, was easily transformed into the belief that the man successful in dollars thereby established his right to all the honors due to virtue itself, and this social tenet became more widely held than the theological doctrine. This may have been because it fell in quite easily with the economic theory of *laissez-faire* which at that moment ruled the English-speaking world. Never was a doctrine better fitted to the predilections of these pioneer agriculturalists lately emerged from the covered-wagon stage and with governmental free land still available for its surplus population. This free land administered by the governmental land office was so swiftly taken up, that throughout my childhood the phrase "doing a land-office business" was descriptive of any thriving activity.

By the early 'sixties, when machine industry began to invade the Mississippi Valley, all the orthodox beliefs of church and state protected it from "interference" of either industrial or fiscal legislation. As one after another of these industries attained giant proportions in Chicago, they claimed and received the adulation of a population which by this time numbered more than a hundred thousand. Just as the pioneers gave the railroads everything they wanted because the development of the country was so dependent upon their services, so they enthusiastically cherished the first industries and considered it unpatriotic to criticize them. The fact that Chicago became the center both of the manufacturing of farm machinery and the slaughtering of livestock also influenced the situation. The individualistic farmer, made even more independent by the ownership of a harvester, was to be placated as a buyer and the Western rancher as a seller of raw material. In any case the farmer and rancher cared little for the problems of the European workman and considered both trade unions and governmental control over conditions in industry as subversive of American principles.

The United States was therefore inevitably slow in any attempt to regulate industry in the interest of the workers, although there were other reasons as well. By 1832 England had enacted the first of her successive Reform Acts covering land taxation and the enfranchisement of labor; incidentally she abolished slavery as early as 1834, and proceeded almost uninterruptedly to the legalization of trade unions and the enactment of the Factory Acts of the 'eighties and 'nineties, but during these same sixty years the moral energy of the United States was diverted. During the thirty years between 1830 and 1860 it was largely absorbed in the effort to prohibit the extension and finally to abolish the institution of chattel slavery from American soil. To great statesmen like Charles Sumner and the best of his colleagues, year after year it became the absorbing preoccupation. It is curious how children catch the glow of the moral enthusiasm of their elders, even although it touches them most remotely. I must have been less than four years old when I first saw a slave, but I have never forgotten the impression, nor the fact that the black man whom I found sitting in quiet conversation with my father, as I entered his room one sunny morning, was being helped to Canada, where he would become a free man, and I still recall my chagrin that he could not be free with us.

By the time the abolition of slavery had been accomplished public-spirited citizens were obliged to deal with another social problem which long absorbed their energy. A new type of commercial and political corruption had followed the Civil War and, worst of all, a combination of the two was represented by many of the war contractors. Again and again men were elected to office solely upon their war records, without regard to their fitness to perform their official duties. Much

THE PROCESS OF SOCIAL TRANSFORMATION

American energy inevitably went into a reform of political institutions so prolonged that at the end of the next thirty years, in 1890, Theodore Roosevelt, a civil-service commissioner in Washington, believed that such a post at that moment afforded the highest type of service to be rendered to this country by a citizen of courage and public spirit.

BARRIERS IN THE WAY OF THE SOCIAL MOVEMENT

For many reasons, therefore, the period of social reconstruction in the United States was postponed until the 'nineties, by which time the situation had grown correspondingly acute, and by 1890 we were well behind England and the Continent in all types of social legislation designed to safeguard the health of laboring-men and their standards of life.

It was quite obvious by this time that the attitude of native Americans toward the immigrants, who have always composed so large a proportion of those engaged in performing the rough labor of the nation, was a factor in our delayed manifestation of social compunction. This was discernible as early as 1830 and, in spite of many checks and exceptions, it has remained essentially unchanged to the present moment. The early attitude toward them was in marked contrast to a certain aspect of the pioneer spirit, for in spite of the somewhat dour theology held by the sturdy men "who thought as they plowed," these earliest Americans often revolted against the brutal and unremitting industry which alone could spell success. In fact it is said that the real American spirit had always displayed a great tolerance toward lack of enterprise, and was inclined to take it humorously; that the early pioneers loved to loaf and to whittle a stick, to swap stories and to crack jokes. This view of things illustrated by Mark Twain's stories of the Mississippi River and also by that typical American joker, Abraham Lincoln, has been called "the American adjustment to reality, the compromise by which was made a livable bridge between the austere ideas of theology and education and the realities of human nature." Like Conrad's heroes, these pioneer Americans were in conflict with nature but at peace with man.

This tolerant attitude, however, very largely broke down in regard to immigrants. From the very beginning of the 1830's to the opening of the World War, there had been an ever-increasing number of immigrants, until in 1913 the annual arrivals were more than a million. Because these immigrants were associated with the unskilled and undesirable labor of building railroads and opening coal-mines, there gradually developed a superior attitude toward them, resulting in a tendency to exalt the Yankee and to put the immigrants into a class by themselves. Naturally every approach to labor problems had to do with immigrants, and it is quite likely that Americans were less concerned for the well-being of aliens than they would have been for their own kinsfolk. This was perhaps inevitable, for as the immigrants increased in numbers they tended more and more to live in colonies by themselves and, separated in many cases by religion and almost always by language and customs, Americans had naturally few contacts with them.

SIGNS OF CHANGE

The imposing World's Fair in Chicago in 1893 was followed by a period of severe business depression, at which time a young man from Princeton University made what was then considered an astonishing experiment. He tramped across the continent without money or recommendations, supporting himself by the work of his hands, going hungry when that failed. Before he left a given city, however, he looked up his friends among the professional and business men, and made it quite clear that, in spite of their dictum "that a man can always find work if he wants it," it had not always been possible for him, a man of good habits, of trained mind, and of American parentage. A reluctant consent to his discovery was one of the first of many indications that the self-sufficiency of the individual citizen had begun to break down even in the flux of the economic and moral energy still pressing westward.

In the decade 1880 to 1890 the number of industrial employes throughout the nation had doubled over the previous decade and the cities had begun to exhibit the overwhelming and unprecedented problems of a new industry. These included overcrowded tenements, widespread misery in the periods of unemployment, the lowered health and vitality resulting from long hours and low wages, and many another familiar problem. There was also found the corruption of city politics by the local boss who delivered the immigrant vote. The alderman of this type most familiar to Chicago held his office, with short interruptions, for forty years. This popularity was largely founded upon an early habit of giving to his constituents in largess a goodly share of what he had received from the traction interests.

The United States at 1890 was almost a generation behind the industrial nations of the world in applying public resources to meet great collective needs, although many political attempts had been made to obtain protection for industrial workers sometimes as a concession to the labor vote. The workingman was constantly told, on the one hand, that the protective tariff was enacted solely on his behalf, and on the other, that his salvation lay in self-help. He made many efforts in this direction through various types of labor organizations which received very little general recognition until the period of the World War, when the American Federation of Labor was accepted, through its officials, as an important part of the body politic and in Washington as elsewhere has since retained this position of an "estate," as it were. During the 'nineties, however, as if to make up for lost time, the United States developed a mood somewhat similar to the impassioned feeling created in England by a study of London poverty which had taken place during the previous decade. The research there ranged from Charles Booth's huge volumes on the *Life and Labor of the People* to Walter Besant's *Palace of Delight,* all concerned with the problems of the working class and with the darker problems of the submerged tenth. The match girls' strike in East London had disclosed conditions of work and wages which were calculated to break down the stoutest virtue, and when these were supplemented by William Stead's sensational report in his *Bitter Cry of Outcast London*, which dealt with what later came to be called the white-slave traffic, the English conscience was thoroughly aroused.

Tentative remedies for the situation were presented in the House of Commons by eager young members who believed that at last representative government was performing its legitimate function, and whose zeal outstripped the reformers themselves. About the same time the theory of socialism was presented in England most reasonably through the efforts of the Fabian Society and through the beguiling organization founded by William Morris, in contrast to its presentation in the United States, where it was first widely advocated in a doctrinaire fashion by Germans and Russians and brought an odium upon the term itself which has lasted for many years.

In that remote decade the young men's movements in America in the church, in labor organizations, in philanthropies as diverse as the social settlement and the Salvation Army, were all characterized by a desire to get back to the people, to be identified with the common lot; each of them magnified the obligation inherent in human relationships as such.

Among the attempts then made by various bodies toward a social reconstruction upon a more collective basis were the settlements, which were established in the late 1880's and grew in surprising numbers throughout the next decade. These were in the main founded by college people who had revolted against the "economic man" of the class-room and, wishing to know conditions for themselves, went to live voluntarily in the industrial districts of the great cities. These districts were occupied largely by immigrants whose lives had been so strangely isolated from prosperous Americans, although the very first months in the Hull House neighborhood, at least, showed the most interesting overlapping of nationalities and customs.

Our section of Chicago had also retained much that was

THE PROCESS OF SOCIAL TRANSFORMATION

reminiscent of frontier life. For many years at Hull House each New Year's day we held an Old Settlers' Meeting. I recall one presided over by Mrs. Joseph T. Bowen, a Hull House trustee, whose mother had been born within the stockade of Fort Dearborn and whose earliest memories were identified with the growth of the young city in which she afterward became such a potent factor for good citizenship. At this meeting two old men told pioneer stories which were fairly well authenticated; one had bought an Indian pony for four quarts of whisky just outside the stockade of Fort Dearborn, and the other had shot a deer "which was making its way to the river for a drink, but which stopped and nibbled grass just about where Hull House now stands." He made his little joke, faintly reminiscent of the yarn of the pioneer, that, because the blood of the deer must have soaked into its very foundations, he, the hunter, had a right to call himself a Hull House founder. It was easy to find a basic resemblance between the spirit of the pioneers and many of the immigrants, numbers of whom had been peasant farmers in Bohemia, in Poland or Croatia. They too had cherished their independence and grew restive under the conditions that made them interchangeable units in a great industry. Between their Slavic land-hunger and their desire to be regarded as individual citizens, many of them made desperate efforts to buy "a house and lot" and often accomplished it at the expense of the nurture and education of their children.

At first the settlements were largely committed to education, which is the traditional American approach to every problem, probably because we all instinctively realize that our public-school system is our proudest contribution to popular government. From the very beginning, however, the educational process was mutual; Greek plays were given by Greek immigrants, already familiar with their own classics although they could not make the lines scan; reading parties in Dante were led by a local Italian editor; we heard harrowing tales of refugees from the Kishinef massacre of 1903. Promising young people in our earliest clubs entered the universities or professions and gradually became merged into the life of cultivated America. But we also saw many instances in which the promise of youth was frustrated by premature labor or by malnutrition in childhood. It was not only the greed of employers, but of consumers and even of parents which had to be curbed before leisure for their own education could be secured for them, and it was perhaps inevitable that efforts to secure a child-labor law should have been our first venture into the field of state legislation.

It was also natural from our affiliations that after the Pullman strike in 1894 we should urge the formation of boards of arbitration in which we hoped the most reasonable representatives of capital and labor might discuss their differences, and we ventured to predict that both groups would eventually unite in promoting legislation to correct untoward conditions. At least we saw, even in the early days of settlements, that the interest of employers, workingmen, and the consuming public are inextricably involved.

Hull House early advocated legislative control of the outrageous conditions all about us, which were involved in the sweatshop system of manufacturing clothing. The first Factory Act of Illinois went into operation in July, 1893, with a Hull House resident as chief inspector and two others on the staff. During this period a code of labor legislation was gradually built up throughout the country, receiving a fresh impulse from time to time, as when Dr. Graham Taylor of the Chicago Commons served as a member of the Illinois Mining Investigation Commission, which, in fixing the responsibility for the Cherry Mine disaster in 1909, described it as "pitifully preventable." The verdict was disquieting to the consciences of thousands of his fellow citizens who knew full well that the miners' union was the strongest in the state and had yet been unable to secure a degree of safety for its members which an employers' liability law would so easily have guaranteed. Since then various compensation Acts administered through state departments of labor have been steadily recommended by commissions composed of employers and members of univer-

THE PROCESS OF SOCIAL TRANSFORMATION 243

sity faculties, as well as of state Federations of Labor, until in two decades workmen's compensation laws have spread to all but five states in the Union.

The settlements were early obliged to consider an aspect of education having to do with the streets rather than the schools, as we found conditions responsible for disaster to the second and third generation in immigrant families: the inevitable temptations of commercialized amusements, the difficulties of a blind-alley job, meaningless conflicts with the police, the different standards of conduct required by Old World parents and by young Americans. It was the knowledge of such conditions in Chicago on the part of Mrs. Lucy M. Flower, Julia Lathrop, and others which led to the establishment of the Juvenile Court, the pioneer of its kind in the world, with a widespread probation system, followed ten years later by the first psychopathic clinic for the study of delinquent children, the Mothers' Pension Act administered in the interest of the dependent child, the addition of a vocational bureau to the public-school equipment, and such ample provisions for public recreation that the small park system of Chicago became one of its most noted achievements. There was a careful inspection and reform of the public dance-halls, the latter carried on under the auspices of the Juvenile Protective Association, whose activities have continued for thirty years. This association has succeeded in establishing various safeguards for city youth, often in spite of the inherent difficulties of philanthropic effort in the face of municipal indifference. It was during the late 'eighties that many philanthropic undertakings were begun which, later modified, are carried on as parts of our permanent equipment for social amelioration: the Visiting Nurses' Association for the care of the sick poor in their own homes and the establishment of infant-welfare stations, the careful and later scientific treatment of family poverty. It became necessary inevitably that training-schools should be established in preparation for such tasks; an experimental one in Chicago was afterwards taken over by the University of Chicago and has become a graduate professional school for "social workers," bearing the same relation, in kind although not yet in degree, to the undergraduates as do the professional schools of law and medicine.

The early settlements practically staked their future upon an identification with the alien, and considered his interpretation their main business. We stuck to this at some cost, for we believed that America could be best understood by the immigrants through a connection with their past history. We extolled free association and the discussion of common problems as the basis of self-government, and constantly instanced the New England town meeting. We especially urged upon the immigrant that he talk out his preconceived theories and untoward experiences, believing that widespread discussion might gradually rid him of old compulsions and inhibitions.

In time the settlements came to find the truth of the saying that "shared experience is the greatest of human goods" and settlement activity gradually extended from its own group to a participation in neighborhood and national undertakings. This came about not only through our search for shared responsibility, but because we found this extension was the only way to get anything done. The neighborhood playground was not secured until it became part of a system covering the whole city; better housing was as dependent upon rapid transit as upon a good tenement-house code. Social intercourse, leisure-time activities, as it was later called, became an important part of the settlement program because a certain freedom of the spirit, a mutual understanding and revelation, can come in no other way. The freedom and ease thus evoked in time came to express itself through drama and music, as well as through the graphic and plastic arts; and from the Hull House theater and studios, as from other settlements, notably the Henry Street Settlement and Greenwich House in New York, came contributions to the common American fund. Through such agencies as the University Extension Lecture Course and the evening high schools, largely supported by the immigrants, at least a few of them and their children came to consider cultural living as important as material success. As to

THE PROCESS OF SOCIAL TRANSFORMATION

the residents in the Settlement itself, the following description written by one of their own number may best estimate them:

> With the devotion of a new age upon them, domiciled within the shadow of hardships and confusion which it had brought, with their minds strangely reorientated by initiation into industrial unrest and aspiration, tinctured with alien outlook upon their native land, straitened with the sense of an urgent mission to those groups out of which they had come, they urged forward their pragmatic appeal for the more human administration of education, industry, and government; for the "nationalization of good."

Just when usages long established come into collision with new standards, resulting in a moral revolution, it is hard to say. Certainly the first decade of the new century exhibited features of such a coming revolution. The investigation of the Equitable Life Assurance Society in 1905 and the challenge of the methods of the Standard Oil Company in 1907 were examples of a fresh consideration of a business procedure which for years had been taken as a matter of course. These sensational affairs were partly the result and partly the stimulus for five years of so-called "muck-raking" in the national magazines that exposed one evil after another in business and in politics.

SOCIAL PHILOSOPHY IN PROCESS OF FORMULATION

It was following this period that the Progressive party was organized in 1912. Many people throughout the country, social workers among them, had become convinced that certain industrial evils could not be corrected unless they were considered a national responsibility and treated from a broader scope than any one state could furnish. It was also believed that only a nation-wide discussion of these social needs would arouse an enthusiasm and understanding adequate to secure national measures. In August, 1912, a platform was adopted in the Coliseum of Chicago by the newly organized Progressive party, which expressed the stirring hope of thousands of citizens, including a goodly representation of scholars and workingmen. The platform formulated in political terms remedies for the industrial situation, believing that the United States, which had been so unaccountably slow in this direction, might now attain the standards of protective legislation adopted by other great industrial countries and also make her own characteristic contribution to the social problems of our generation.

The Progressive party campaign was spirited and doubtless of educational value, but it was apparently premature. It was unfortunate, too, that its plans for the future were cut across by the volcanic eruption of the World War. But is it too much to hope that the logic of reality will at last bring men to a concerted endeavor to make government a coherent and skillfully planned activity for human betterment? A generation capable of the scientific and practical advances of our time will not long be content with political mechanisms so uncertain, so difficult to control, so ineffective, and so productive of waste as are the major political parties. These are retained largely because it is so easy to confuse democracy itself with the special development of political machinery with which we are most familiar.

Some of the measures advocated by the Progressive party and others that had been enacted into laws by the various state legislatures were declared unconstitutional by the Supreme Court of the United States and thus made inoperative. This was true of an attempt to limit the hours of work for the bakers of New York State, of certain minimum-wage laws for women, and of similar legislation. This judicial nullification on the grounds of constitutionality set a distinct limitation to the experiments through which the nation might increase its fund of social knowledge. It curtailed the opportunity for utilizing experimentation as a method for progressive government. Intelligent attempts to deal with a social evil, if made the basis of legal enactment, were less likely to receive judicial approval, than an outworn method of dealing with the same evil, which had the doubtful value of an early precedent under conditions long since obsolete. Big industry, while itself operating over interstate areas, was always afraid

THE PROCESS OF SOCIAL TRANSFORMATION 247

of an opening wedge for federal legislation on industrial matters and its arguments were generally sustained by majority decisions of the Supreme Court, although a minority opinion in each case made clear that a process of adjustment to current conditions was a necessity, if constitutional government was to endure.

Perhaps it was because mass production had so raised the standards of living that more and more people lived in much the same sort of essential material comfort, resulting in a pressure to make them take on a like similarity mentally and spiritually. It is hard to tell what actually produced such a situation—doubtless fear of Russia was an element in it—but certainly for a decade after the war there was less scope for individual self-expression within the ordered framework of the state than there had ever been before on American soil. Yet this widespread desire for conformity was accompanied by such an overwhelming admiration for our early individualism, that it was considered patriotic to oppose governmental measures for old-age security, for instance, because they would lessen the sturdy independence of the workingman.

A tendency always present in America thus intensified resulted in a dogmatic nationalism which inevitably bred new intolerance toward immigrants. This was clearly exhibited in the congressional discussion, at moments reaching the point of self-righteousness, which in 1920 accompanied quota regulation of immigration. It revealed both an unconscious contempt for immigrants and the early religious notion of making men good by fear of punishment.

But because the Simon-pure American made an exception of himself—what was good for the immigrant was not necessarily good for him—he exempted himself from laws which he would like to see enforced upon others, with the result that the individual often voted for laws that he himself had no intention of obeying. For instance, the Southern man voted for the Eighteenth Amendment because he wanted to keep drink away from the Negro, the Northern man because he wished sober immigrant labor, and so on. The result of such voting has been analyzed by an Englishman as follows:

> Because law in the past has proved capable of preventing men from committing the more obvious kinds of wickedness, Americans have assumed that it can be used to make men good. And as nearly everyone naturally supposes that he is good enough already, the law has come to be regarded as an instrument for making other people good.

If such a diagnosis is in any measure correct, can it not be traced to that sense of division which has corroded our social relations and has undermined that natural democracy characteristic of the frontiersmen, who, impressed with the essential soundness of human nature as they knew it day by day, instinctively kept a balance between their moral convictions and their human experience? When immigrants became isolated into groups by themselves and were somewhat feared because of their very strangeness, it was perhaps inevitable that the tendency should develop legislation born of abstract convictions rather than a code arising from a body of common human experience. It also resulted in that type of dogmatism which has been described by a brilliant American journalist, "as much more interested in affirming ideals than in facing the problems of applying them." It may have been an uneasy apprehension of this or an honest desire to settle a vexed moral issue, but certain it is that, as the question of slavery overshadowed all other efforts for social reform during the thirty years preceding its abolition, so after the World War the need for all other social changes tended more and more to become submerged under the discussion of prohibition. The constitutional amendment passed in the most orthodox manner was discussed from every possible approach. The doctrine of personal liberty was doubtless overworked, but it made many people uncomfortable and, as Walter Lippmann has pointed out, it has resulted in a curious dilemma:

> The very people who had enacted the program of moral reform were also great lovers of liberty. . . . They inherited a tradition of profound distrust of executive and judicial authority, which was embodied in the Bill of Rights. Nothing has ever been able to induce them to set

up a really strong executive government in America. Thus by their moral convictions they prohibited all sin, and by their liberal convictions they have kept the prohibition from being enforced.

Possibly there is no way out of the impasse save through an understanding of the tissue of objective causes and inner motives which bind people together and which has unique power to soften our antipathies and lessen our dread of one another. Whether this dilemma is insoluble, unless one ideal is sacrificed to the other, remains to be seen. An analysis ought to be easier in these later years when psychiatrists are lending aid to all sorts of public undertakings.

It was unfortunate for the development of social solidarity that the earliest outbreaks of gang violence in Chicago—more or less typical of those throughout the country—should have been associated with colonies of immigrants, although these outbreaks have long since ceased to be found only there and have spread throughout the metropolitan area and are also spilling over into questionable roadhouses in the adjacent country. The men who were bootlegging, racketeering, conducting gambling-houses, or systematically stealing automobiles had been able to buy the connivance and even the participation of the police themselves. Their successful methods of corruption often involved far-reaching political and business affiliations. It was commonly said in this later period that the police were not so much concerned to protect the community as to protect themselves from exposure, and that the criminals had the police "on the run," using an expressive phrase from the Irish Revolution. Although everyone knew that such lawlessness was the result of political corruption, the community was slow to act because, so long as the Sicilians who composed the first powerful bootlegging gang killed each other, it was considered of little consequence. Such preferential indifference is a grave symptom of a breakdown in democratic government and may be another indirect result of venturing to consider immigrants of less consequence than the rest of us. It was a wise man who said, "the essence of immorality is to make an exception of oneself."

Perhaps the differing elements throughout the country were pulled together somewhat under a sense of common disaster during the sharp period of unemployment which involved the nation in the early years of the 1930 decade. This pressure upon them, resulting from modern economic conditions and their daily experiences, gave a twist to the actual social theory held by the so-called practical man, in two directions: first because there was a reluctant apprehension that neither the tariff nor any other political device could keep America from sharing a world-wide depression, since the economic world was curiously interdependent, and secondly owing to the fact that business men inaugurated among themselves a whole line of collective effort in which the most individualistic of them might conscientiously engage. President Hoover in his term as Secretary of Commerce had often made the statement that national prosperity depended upon high wages; only through them could the manufacturer be assured of a buying public; under healthy economic conditions the producers should be the largest consumers.

This was an easy doctrine so long as business was good, but a hard saying in bad times. Nevertheless, the need for the stabilization of industry during those difficult winters following the collapse of 1929 brought forth experiments in one city after another—an elimination of seasonal work, a sinking-fund for wages similar to the funds set aside for dividends, a system of unemployment insurance inaugurated by the General Electric Company, the extension of loan funds to employes based upon the credit of their return to work, first tried out in Chicago by the International Harvester Company, but quickly followed by many other great industrial concerns.

Investigations were made in one city after another by competent economists as to the extent and inevitability of unemployment. The National Federation of Settlements contributed a poignant document dealing with the effects of long-continued unemployment upon self-respecting people throughout the United States who had never before been subjected to this strain. The loss of houses, furniture, and

THE PROCESS OF SOCIAL TRANSFORMATION

radios already partly paid for was a trifle compared to the defeat of the human spirit, especially among the young men doomed to idleness during their very first years after college or high school and the old people too shamefaced to eat the food needed by their grandchildren.

Hundreds of men standing in line for food and shelter made it obvious that it was the United States which depended upon a "dole" in such a crisis, and not those countries possessing a well-considered system of unemployment insurance. When the American Association for Labor Legislation presented to Congress a bill designed to secure uniform legislation throughout the nation, it led at least to widespread discussion of the industrial situation. Bad as it was, it tended to become obscured by the plight of the farmers, who, throughout a wide belt across the continent, had suffered from a season of drought, added to the difficulties of a transitional period in the methods of producing and selling foodstuffs.

The appointment of a Farm Board two years earlier had been a distinct departure from the American tradition of laissez-faire which had always been most pronounced in relation to agriculture. But even the most individualistic farmer saw dimly that the remedy for overproduction should not be restriction of output. When millions of people were starving in China, a loan to that distracted nation of the very money spent for the farmers' grain would have put his hard won products into hungry mouths instead of barren elevators. He too wished to bear a man's share in the readjustment during the period of unemployment and overproduction in which the world was involved. The churches everywhere throughout these hard winters stressed as never before the social obligations of religion and brought their members under conviction, if not of sin, at least of dereliction, emphasizing the fact that a nation as well as an individual is condemned if he liveth unto himself.

The widespread discussion concerning the entrance of the United States into the World Court and the signing of the Kellogg Pact brought thousands of people nearer to a realization of the inevitable relationships between nations. The newspapers were responsible for slowing down such a realization in the Middle West, but happily the radio and the news reel are constantly making us less dependent upon a single agency. It was through the radio that a leading business man in 1931, urging the United States to be an easy creditor, called attention to the fact that our economics and our politics are in conflict everywhere; the first, making nations constantly more interdependent; the second, growing more nationalistic; that the problem of "reconciling the two is the most intricate and difficult problem in the world."

The coming century may be faced with this problem in all its efforts at social reconstruction as the century closing with this World's Fair, in 1933, has persistently tried to reach an equilibrium between individual and group responsibilities.

CHAPTER TEN

THE CHANGING POSITION OF WOMEN

When in 1848 the first woman's rights convention at Seneca Falls put forth a new declaration of independence containing principles then strange and revolutionary, now generally accepted, the document was greeted by almost universal jeers. With this document for a text, says Elizabeth Cady Stanton, one of the leaders at the convention, "it seemed as if every man who could wield a pen prepared a homily on 'woman's sphere.' . . . So pronounced was the popular voice against us, in the parlor, press, and pulpit, that most of the ladies who had attended the convention and signed the declaration, one by one withdrew their names and influence, and joined our persecutors. Our friends gave us the cold shoulder, and felt themselves disgraced by the whole proceeding."—*Elizabeth Cady Stanton,* I, p. 148.

The Changing Position of Women

By Grace Abbott

WHY the progress of women, ask those impatient with the attention which women have been compelled to demand with respect to their position or lack of position because they were women. Why should the contribution of women to the progress of the century not be recorded in the history of the sciences, the arts, education, the organization of industry, and the development of government? Well, why not, is exactly what women have been asking as they found one door after another at which they knocked for admission closed to them. Because of these closed doors, the progress of women during the past century must be measured less by their contributions than by the removal of legal, social, or economic barriers which have prevented them from making their contribution to our common life.

THE OLD ORDER AND SIGNS OF CHANGE

"We do not expect so much of women; the heroic virtues as little as the vices. They have not to unfold the scroll of character," wrote Meredith in *The Egoist* as his mind played with the difficulties which made responsible decision and action so hard for Clara Middleton in the choice of a husband.

Here we have a basis for a possible measure of progress. How has the public expectation of what women should be and do changed in the hundred years which are just behind us?

The year 1833 does not mark the beginning of the change. But the century which began then includes the great leaders who set forth the injustice to women and the loss to society resulting from the artificial barriers which had been erected

THE CHANGING POSITION OF WOMEN 255

to prevent their progress. It was the period of earnest and sometimes bitter defense of the old order. And it was the century when men generously gave up great privileges and prerogatives which belonged to them exclusively because they had come to believe it was not just nor in the interest of the public welfare that they should retain the position of authority with reference to women which they occupied when the century began. But this generous yielding has always been accompanied by many reservations. Were they not, after all, the superior and women the inferior sex, and would not the losses resulting from these concessions outweigh the advantages? Moreover, in America, which was dedicated to a new social order, the opposition to any revolutionary change was for that very reason the greater. Any new application of the principle of democracy must be cautiously considered. The American experiment was great enough as it was. So that as one door was opened to American women it was felt necessary to make sure that the others were securely bolted.

This is not to say that the doors had not always been open to some few women. There had been empresses and queens, women of learning, women soldiers, women physicians, and women writers before the nineteenth century. But they were regarded as altogether unusual. Their accomplishments or the privileges which were for a brief time accorded them established no precedents for other women. Or they were women of noble or royal birth and were not to be judged by ordinary standards of what women should or should not do. Such women usually felt themselves superior to other women and did not encourage the belief that what they did constituted an argument for enlarged opportunities for women in general. On the contrary, the influence of these women, flattered into the belief that they were unlike other women, has frequently given support to the denial of opportunity to the sex in general.

But before the nineteenth century women had emerged from the darkest period of their history. Theologians were no longer discussing whether a woman had a soul and the world was ashamed of the hangings and persecutions of the witchcraft period.

Still, in 1833, when the century which we are reviewing opened, divine law as then interpreted and expounded by the clergy, the law of the land, and social custom decreed that women had few rights and no duties except those their husbands or guardians might give them. It was a period of extravagant praise of the virtues and the charm of womanhood—the kind of praise which made it clear that man must be her protector and guardian through life. The child wife—Dickens has given us her picture in Dora of *David Copperfield*—was the popular ideal woman of the time. Ignorance made her more attractive and physical weakness more appealing. While the century furnished many Doras, there were many able and intelligent women who exerted great influence in their family and community life and a few who influenced public thinking. During the first quarter of the century, Mary Lyon had appeared as the champion of educational opportunity; Elizabeth Cady Stanton, Lucy Stone and Susan B. Anthony of political equality for women; Dorothea Dix had braved public opinion in order to secure decent treatment for the insane, and Harriet Beecher Stowe had contributed *Uncle Tom's Cabin* to the anti-slavery cause.

THE REVOLUTION IN WOMEN'S LEGAL STATUS

In an analysis of the progress that has been made, consideration of the change in the legal position of women should perhaps come first. Legal rights are only legal rights. They make possible but do not create opportunities for service nor change a public opinion which makes any departure from the old standards so difficult that only those of great independence or those who suffer from a serious abuse of power will undertake it. On the other hand, a legal denial of rights raises a serious barrier and its removal lays the foundation on which independence and self-direction can be based.

In 1833 wifehood and motherhood were the only careers approved of for women, and yet it was in exactly this field that

THE CHANGING POSITION OF WOMEN

they were most circumscribed by the law. The English common law, practically unmodified in its crossing of the Atlantic, still controlled the relations of husband and wife and parent and child in most states in 1833.

Under the common law, upon marriage, a woman lost her rights, her responsibilities, and even her identity. Whether the courts in formulating the rules which governed her were acting upon the theory that the husband was the guardian or master of his wife, or that, because of the unity of married people, the wife's legal status became merged and lost in that of her husband, her position was most unequal and unjust. Speaking broadly, upon her marriage all the property a woman had or might in the future acquire became her husband's. Some distinction was made in the kinds of property. In her real estate he had a freehold right during the life of both and, if a child was born, the right to curtesy or a life interest in all her freehold estates, upon her death. With certain insignificant exceptions, all her personal property, including her earnings, passed into the absolute possession of her husband. But in his property the wife was given no corresponding share. She was entitled to what was called her *paraphernalia*, and, upon his death, under the right of dower, to a life interest in one-third of all the estates of inheritance held by him during coverture.

Since the wife could exercise no control over the property, quite logically she was not allowed to contract, except as her husband's agent, to make a will without his consent, or to sue or be sued in her own name. Of necessity the law made the husband liable for his wife's ante-nuptial debts and for her post-nuptial debts when contracted in the purchase of necessaries and for her ante- and post-nuptial torts.

Although Blackstone declared that "even the disabilities which the wife lies under are for the most part intended for her protection and benefit. . . . So great a favourite is the female sex of the laws of England," John Stuart Mill was much nearer the truth when he said, "If married life were all that it might be expected to be, looking to the laws alone, society would be hell upon earth." That it was not this in most cases Schouler thinks due to the fact that "Woman's weakness has been her strongest weapon . . . and if her life has been legally speaking at her husband's mercy, her constant study to please has kept him generally merciful." But even though the wife "studied constantly to please," her husband was not always "merciful" and the harsh rules of the common law would probably have been modified earlier if wealthy and influential fathers had not found in the "equitable separate estate" some protection of their daughters' interests against profligate sons-in-law. Through the equitable estate, the courts found a device for protecting the property interests of the wealthy woman; but the independent ownership of her property and earnings by the woman of ordinary means, the right of any mother, rich or poor, to be joint guardian with her husband of their children, and her responsibility to society for her own acts came to married women only by legislative enactment after a long struggle.

It is not possible to discuss in detail the gradual statutory change in the position of married women in the several states. They can, however, be briefly summarized. During the period from 1839 to 1860, the earliest Acts passed gave the married women the right (1) to acquire and hold free from all liability for her husband's debts, both real and personal property, (2) to an independent legal existence entitling her to contract, to devise her property and to sue and be sued in her own name, and (3) to be herself responsible for her own debts and contracts.

As American legislation on this subject preceded English by about thirty years, there were no precedents to follow, no statutes that had already been tested by experience and judicial interpretation, to serve as guides. However revolutionary the general plan a legislator desired to carry out eventually might be, there was every reason for proceeding cautiously in making these first innovations in the family life. Keeping this in mind, two theories upon which the legislatures acted either consciously or unconsciously may be worked out. The first

THE CHANGING POSITION OF WOMEN 259

was that the position of the wife under the common law was, generally speaking, what it should be; but that, under it, the wife or, more correctly, the wife's property, needed some protection against this very large authority of the husband. In other words, that the husband was and should remain the head and legal representative of the family, but that the wife's property should be protected against his extravagance. This was the theory of the "equitable doctrine" and did not contemplate the economic independence of the wife nor her responsibility to society for what she did or failed to do. By a "clause against anticipation" her property was protected not only from the extravagance of her husband, but, as though she were still a child, against her own imprudence. As a matter of fact, the equitable separate estate was apparently resorted to, not for the purpose of enlarging the powers of the wife, but to protect the interests of her heirs.

On the other hand, the early Acts of Massachusetts, Michigan, Maine, Pennsylvania, New York, and Illinois, however tentative and incomplete they were, clearly indicated that the legislatures which passed them believed that the common law was unjust to the wife and that her position should be changed by placing her on an equality with her husband.

For in these states she was given control as well as ownership of her property, made a legal person responsible for her own acts, and later given equal powers with her husband over their children. So with the New York law of 1848 and the Illinois Act of 1860 a revolution not only in the law of property but in family relationships was begun.

When the English bill of 1868 was under consideration, a number of Americans testified before the Select Committee on the Married Women's Property Bill. According to them the change from the common law was necessary on the grounds that "objection . . . was made to the state of the law on the part of persons with small fortunes and of women earning money by their own exertions"; that mothers should have the power to supply the wants of their children when the husband refused to do so; that the wife's separate estate might be preserved from liability for her husband's debts; and "the belief that the mutual affection of husbands and wives would be more promoted by their standing upon an equality than by one being made inferior to the other." The fear that the recognition of the independent legal existence of the wife would destroy the family was general among Englishmen and they were not sufficiently reassured, even by the statements of the American lawyers, merchants, and manufacturers that the change in America had not destroyed "the proper authority of the husband"; so they refused to pass the bill at that time. That the Americans had had many grave doubts as to the wisdom of the departure from the common law is shown in the testimony of Hon. E. Washburn, at one time Governor of Massachusetts, a professor of law at Harvard University, and a member of the legislature which passed the bill giving the wife the right to make a will in Massachusetts—"I voted for the bill," he said, "but with doubt and hesitation."

But doubts and hesitations have been gradually conquered and from 1860 to the present the movement toward a dual equality in marriage has continued in the United States. Intellectually the case has been won and most of the old legal restrictions have been swept away, but there still remain in the statutes of many states relics of the old ideal and an inconsistent and contradictory mingling of the old and new. In a few states the wife does not yet have the right to her earnings; in Florida a married woman may not manage her own property; in the states in which the community-property principle of joint ownership by husband and wife of the property acquired during marriage is followed, the husband is the sole administrator of such property. Dower and curtesy have been equalized in many states, but there are still about a third in which the husband and wife are not entitled to an equal share of the estate left by the other. In some the inequality reflects the common law; in others the wife has been given a larger share than the husband. The right to make binding contracts and to engage independently in trade has not yet been granted the married woman in a number of states.

THE CHANGING POSITION OF WOMEN 261

The position of the married woman as a mother is especially interesting. So much was said of the beauty and sacredness of motherhood that it might be supposed that legislatures would have been eager to make the mother the joint guardian with the father over their children. But this was not the case. In most states the wife was given property rights long before she was given a mother's rights. Illegitimate children belonged to the mother and she alone was liable for their support; but if the children were legitimate the old common law rule decreed that the father alone was entitled to the control and custody of his children and "the mother", as Blackstone puts it, "entitled to no power but only reverence and respect," and this still prevails in many states. Connecticut, the District of Columbia, Illinois, Nebraska, and New York have, however, by statutory exactment declared the authority of the father and mother equal, and Kansas has a provision to that effect in the state constitution.

A Pennsylvania statute provides that the mother's control over her children shall be equal with her husband's only when she contributes by the fruit of her own labor to their support, which puts a premium on labor performed by the wife outside of the home and clearly reckons her services in the home of little value.

The fact that so few states have passed laws placing the husband and wife on an equality in this matter may perhaps be explained on the ground that the enforcement of any such equality would be possible only in a contest over the possession of the children, and in such cases the courts have considered the welfare of the children rather than the right of the parent. The cruelty of permitting the father to separate the mother and her children, especially by the appointment of a testamentary guardian, has been very slow of modification. Here the remedy is clear. John Stuart Mill pointed out long ago the weakness of basing the case of the mother or the married woman on the abuse of the power rather than the power. Since the father usually makes every effort to enable the mother to keep the children with her after his death, the change in the old common law seems to many men to be purely academic, and most happily placed mothers are shocked to learn when some case of cruel injustice finds its way into the papers that mothers have no legal authority.

The increase in the responsibility of the married women for the support and education of her children has come even later than the legal recognition of her control over them. But early laws were passed by Pennsylvania, 1848, and Illinois, 1874, making the mother's separate estate liable for their support. At present, legal liability for necessaries for the family has been placed upon the wife by the statutes of nearly half the states. But the husband's common law obligation to support his wife still remains the law. In recent years, in an effort to reduce public dependency, desertion and non-support on the part of the husband and father have been made a penal offense.

The hard and unjust position in which the law placed the unmarried mother and her child is slowly being changed. While this movement has been primarily in behalf of the child, legal recognition that every child has two parents, who are equally obligated to contribute to the child's education and support, transfers from the shoulders of the unmarried mother a part of the social and economic burden which, under the common law, was hers alone. Statutory modification of the old rule which relieved the father of all responsibility has come with many hesitations and fear lest public morality should not be served by a more just and humane procedure. The example afforded by the Castberg law enacted in Norway in 1915 and evidence which social research has revealed as to how the common law worked in this country, have resulted in a changed public opinion and radical modification of the old law in the recently enacted statutes.

Before leaving this subject of the changed rights of married women, mention should be made of the developing tendency to put the husband and wife on an equality when they seek to end the obligations of marriage. They are not yet on an equality in all the states, but the most serious of the old discriminations have been removed. The laws which left the wife with no redress when the husband insisted on having his mistress live in the same house with her and her children and gave him the right to divorce her for any immorality on her part have been repudiated. Adultery on the part of either the husband or the wife is now a cause of divorce in every state which has legalized divorce, which every state, except South Carolina, has done. As to other causes, in more than half the states the law still favors either the wife or more frequently the husband. So far as the children are concerned, the courts increasingly recognize only the interests of the child; and this means that, other things being equal, they are more frequently awarded to the custody of the wife than of the husband. In alimony, the wife still has an advantage. But the justice of an economic claim on the husband when marriage is dissolved, and there are no children, is being questioned by many and is less frequently and less generously awarded by the courts. A woman's earning power may be adversely affected by marriage when she gives up her trade or profession to devote herself to work in the home, so that, even with the growing economic independence of women, the divorced wife may be said to have a valid claim on her husband for temporary assistance.

As these survivals of the ancient order in the relation of husband and wife and the responsibility of unmarried parents toward their children are on the way to extinction, questions as to what may be said to constitute equality between the husband and wife and what reciprocal services the law should enforce remain to be answered. When there are no questions of the care of children and the wife can support herself out of her own earnings, should support by the husband be required? And what about her obligation to contribute to his support in the event that her means or her earning capacity are greater than his? These are questions which are easy for the individual husband and wife to answer for themselves, but a legally enforced obligation raises difficulties. The logic of the movement of the last century is clear, but mere logic is usually a poor basis for a public policy.

WOMEN AND EDUCATION

In addition to these new rights, new opportunities, and new duties of married women, other changes came, sometimes so slowly that the leaders should have been much discouraged, and then so rapidly and with such complete acceptance by the public that people soon forgot that there had been any change. In the *Vindication of the Rights of Woman,* widely read and widely condemned by the clergy and by public speakers in the United States during the first part of the nineteenth century, Mary Wollstonecraft had declared that the real reason for woman's inferior position was her lack of education—or her education merely as an object of flattery, an ornament to society rather than as a rational human being preparing for a useful life. This was no new doctrine. Defoe had said, in one of his essays published in 1697, that "one of the most barbarous customs in the world, considering us a civilized and a Christian country, is that we deny the advantage of learning to women." Defoe would have erected no limitations. "To such whose genius would lead them to it, I would deny no sort of learning." But his essay aroused neither the enthusiasm nor the opposition which is necessary for a change. But with Mary Wollstonecraft's impassioned declaration of what was basically wrong with the position of women in England the battle was on—to the present day a more difficult one in England than in the United States. But here, too, the beginning was slow and timid.

All the first steps were taken with the greatest caution. The educational opportunities of boys must in no way be jeopardized or retarded. Woman's slender intellect must not be overtaxed. She should not be taught things the knowledge of which would cause any lady to blush; she should not think an education meant approval of speaking in public, serving on committees of charitable societies or other unladylike activities.

It is so short a time—a hundred years. I have heard the story from my mother and my grandmother, a member of the Society of Friends who had inherited, as a member of that fellow-

THE CHANGING POSITION OF WOMEN 265

ship, religious convictions against war and slavery and in favor of equal opportunities for men and women. Mother and grandmother remembered the serious discussions of neighbors and friends. It was not amusing to them. Physiology was, of course, not suitable, but was geography? And higher mathematics—*higher* being algebra and geometry? The female mind was unequal to that. Latin and Greek were in that day the evidence of culture and learning among men; but should any but the most exceptional women tax their minds with the syntax of any language but French? It might produce a revolution in family life and the experiment must be most cautiously undertaken.

But before the Civil War we find these forbidden subjects taught in the progressive academies for women in New England, New York, and the South, in the pioneering "female seminaries," and, most important of all, the few public high schools of the Middle West to which rich and poor sent their sons and daughters. And the victory for secondary education was complete, at least in principle.

Then the arguments and the fears were all canvassed again, when the suggestion was made that a college education should be made available for women. Was she physically equal to the strain? Would a college graduate want to marry? And, more serious, would any man want to marry her? Would she be a willing and competent mother if she were allowed to eat the intellectual food heretofore regarded, except for the very few, as suitable for masculine diet only?

Oberlin's great step was taken with a full realization of the dangers to womanhood which a coeducational college offered. The faculty wives were vigilant supervisors of manners and morals, and the courses to be opened to the girls were carefully considered. Lucy Stone's bold stand against the rule that the "young ladies" of the college might write their graduating essays, but that some faculty member must read these productions for them while the boys read their own, was much discussed. Public sympathy was with the college authorities in denying Lucy her diploma when she made an issue of it. One must stop somewhere! And the timid friends of education for women deplored her obstinacy.

But this was only an incident, soon forgotten; and, as the number of girls asking to be admitted increased, more college doors were opened to them. In the East, privately endowed women's colleges—Mt. Holyoke, and much later, Vassar, Wellesley, Smith, Bryn Mawr—then Barnard and Radcliffe, women's annexes to colleges for men, were established, while equal opportunities were assured to women at Cornell and Johns Hopkins by their founders. The greatest gains, however, came in the Middle West and West, where the public educational system from kindergarten to professional school rapidly developed. These were practically all coeducational, and the numbers of women and girls in college and high school rapidly increased—in the latter so rapidly that for a time the number of girls considerably outnumbered the boys. Thus at the beginning of the twentieth century, 58 per cent of the pupils in public secondary schools were girls and 42 per cent were boys. There are several explanations of this fact. Boys of high-school age had a wide choice of occupations and for those who were to learn a skilled trade, an early start was considered more valuable than the training received in high school. On the other hand, except in domestic service, there were few work opportunities for a girl who had no education. A high-school education paid in the case of girls. Most parents hoped that their daughters would not have to work outside their own homes. That was still the approved social standard, and many families, by great sacrifice and at the cost of their happiness and independence, kept their daughters in school. At the same time they made it quite clear that economic dependence on fathers and brothers, in the event they failed to achieve matrimony, was a necessary corollary to being a lady. A high-school education was considered an asset in marriage, and if the daughter should be compelled to accept the social disgrace of self-support, a high-school education would help her to secure clerical or office work, or to become a teacher.

There have been great changes since that time—none more

revolutionary than the changes in high schools and high-school attendance. The reports of the United States Office of Education show a half-million attending the public high schools in 1900, nearly a million in 1910, more than two million in 1920 and nearly four million in 1928.

With this increased attendance came changes and expansions in the curriculum—mathematics and the classics had become relatively less important, and new courses in business and in practical trade training, never before offered in secondary schools, had been added. At the same time the age of employment and of compulsory school attendance was raised. Employers increased the educational requirements for the better paying jobs, so that some of the temptations which led boys to leave school were removed. At the same time, public opinion about the employment of girls was likewise undergoing a change. Gainful employment before marriage became much more general and was no longer regretted by most parents. And then some well-to-do women continued to work after marriage. As a result, the reorganized high schools have met the developing needs of the girls as well as of the boys, so that their attendance has also greatly increased but proportionately a little less than the boys. In 1928, although the girls still constituted a majority in the public high schools, the proportion had shrunk from 58 per cent in 1900 to 52 per cent in 1928.

In the colleges, too, the last quarter of a century has brought great changes in curriculum and an almost overwhelming increase in numbers. Prior to this, the number of women in college had been steadily growing. Each year many fathers and mothers were converted to a college education as desirable for their daughters and many more girls asked to continue their education after graduation from high school. In addition to the earnest students who went in the earlier days, women now went for the same general and widely different reasons that influence men. But the old ideal of educating women as ornaments to society has lingered—a finishing-school and a grand tour for the girls and college for the boys is still considered the proper thing by a decreasing number of well-to-do families. In families able to send only one to college, it has been the son and not the daughter who was sent, while in high school, if only one could go, the daughter was quite apt to be the chosen one. Nevertheless, in 1900, sixty-three years after the first college in America opened its doors to women, there were 36,051 women in its institutions of higher learning—not including normal schools and teachers colleges, where the women greatly outnumbered the men, nor the professional schools where the reverse condition obtained. This meant that by 1900, 35 per cent of the total number of college students were women. Since 1900 the returns show an increase. There are now 292,977 women in college, which is 42 per cent of the total attendance.

It is in point to ask how the women have met the test of college education. Has it proved too difficult for their less robust physical equipment or their weaker minds, as so many feared at the beginning of the century which we are reviewing? Happily the prophets of disaster and failure were wrong. College has helped to bring a new standard of health for women. And long hours of intellectual work are no more fatiguing to the girl than to her brother—seemingly less so—but he has probably felt freer to complain than she who has been regarded as more or less on trial in intellectual pursuits, down even to the present day.

And has she passed the test in scholarship as successfully as the men? In coeducational colleges the lists of those elected to the honor society of Phi Beta Kappa usually reveal a larger number of women, although they have constituted a minority in attendance. If the record had been the other way, it would have established the greater intellectual capacity of men and the greater profit to society in their advanced education. But when the tendency is for women to excel in scholarship, some other explanation must be found. More than the men, women have been "grinds," have worked for marks, have sought to please their instructors rather than strike out on original and independent lines, which may not have the approval of an

THE CHANGING POSITION OF WOMEN 269

unprogressive faculty. This is the explanation that often has been given; and it has some basis in fact. Women have been trained to conform from childhood—which does not mean that they are by nature more timid in their convictions, in their selection of interest, or in their pursuit of an objective than men. For many years one objective of college women has been to prove that they were not unequal mentally and physically to performing the tasks that were set for college men. And in their eagerness they have perhaps more than proved it.

Another explanation of the large number of women who have won scholastic honors in college has been made. The men have "gone in" for other campus activities, one hears, at the sacrifice of many hours which would otherwise have been spent on class assignments. It is true that except for the purely social activities, women have devoted themselves more exclusively to class work. In the past they have had no other alternative in the coeducational college. The faculty have been almost exclusively men, and this has created a campus public opinion which has led to the preferment of men as class officers, as editors of college papers and magazines, and other student undertakings. Moreover, men students have been in the majority, so that the election of men students was to be expected. But a change here too is in progress. Men are yielding the advantage which they have in numbers and tradition. The first woman student to hold this or that college post has come and gone in most coeducational colleges and we may expect a college public opinion which offers an equal chance for leadership to men and women.

THE PROFESSIONS AND SOCIAL SERVICE

But what about women in professional schools? With their first early efforts to secure a college education, women sought opportunities to prepare for the learned professions also. Their ambitions extended to theology, law, and medicine. There the opposition was much greater. The question of the practice of the profession was involved. A college education might lead women down new and, to the conservative-minded, dangerous pathways; but this did not necessarily follow. The question of what she was to do next could be postponed until after college. Fathers trusted that, with a little pressure and an appeal to affection and duty, daughters would yield to their mothers' insistence that they remain at home as companions in social and homemaking activities. Moreover, any other course presented great difficulties. Work, except teaching in the elementary or secondary schools, was difficult to obtain. This was the vicious circle of the prevailing viewpoint that the daughters of the well-to-do and of those who tried to make the public believe they were well-to-do, should be spent in comparative idleness while the sons were encouraged to hard work as the basis of a future career.

College education was becoming for many women the usual sequel to secondary education and, with college completed, they often took up the same life as the girl who had gone to finishing-school. But there is evidence of a great change. Society no longer offers a career even for the rich girl as it was supposed to offer a hundred years ago. Society, as it was known in Europe and transplanted to America, required leisure-class men as well as leisure-class women, and it could be really successful only when an aristocracy could raise barriers difficult to surmount. Some of the well-to-do college-trained girls have been quick to discover that society as a profession offered only the pursuit of happiness, not happiness, and they are turning to a business or professional career to test their training and their capacity. Thus gradually the social barriers to employment are being removed and the girl who undertakes to earn her own livelihood does not have to drop down into a lower social level. Those with little money are taking advantage of this fact, and their parents are no longer raising the objections which they did a generation ago. How large a percentage of college women are looking forward to work outside their own homes is illustrated by the report which newspapers have carried, to the effect, that every graduate from Smith College in 1931 sought the help of the college placement bureau in ob-

THE CHANGING POSITION OF WOMEN 271

taining a position. But this great change has come very gradually.

Attendance at a professional school created a head-on collision with the old traditions. A request for admission meant in most cases a determination to follow the profession after the education was secured. Prejudices were not to be gradually worn away, but ignored or defied. And there were other hurdles beside public prejudices. The professional schools said no. The professions said no. Women doctors and lawyers and ministers—the mere suggestion struck at the dignity of many whose claim to a place of honor in the community rested on their profession rather than on the contribution to public welfare which it made possible. Elizabeth Blackwell's admission to the Medical School at Geneva, New York, constituted no general precedent for other women or other medical schools. After its establishment, some women attended the Medical School of Howard University in Washington, taking advantage of the fact that although dedicated to the higher education of Negro men and women, it had never excluded white students. For a time the organization of special professional schools and women's hospitals seemed the only road to professional education for women. A few were founded, but the increasing cost of professional education, especially medical education, made an expansion of such a plan impossible. Improved educational standards required fewer and better schools.

As medical and law schools were established in connection with state universities, these were opened to women. But privately endowed schools were very slow to change their policy. As recently as fifteen years ago, the only important privately endowed law schools open to women were found at the University of Chicago, Stanford, and Northwestern. Johns Hopkins was the first to make its fine medical school available for a small number of women students. Today women are admitted to every important law or medical school in the country save only Harvard. In spite of a tradition of consideration for changing social and economic conditions in the organization of its professional schools, Harvard lives in the past. Dedicated to the selection of students not on the basis of right or privilege but expected professional capacity, and with nearly half of all the college students of the country women, the Harvard professional schools refuse even consideration of the capacity that any woman might possess.

In the professional schools opened to women the road has not been an easy one. The difficulty of admission was only the first of the hurdles. The men students, reflecting the general attitude of the faculty and the profession, often made the professional school an unhappy place for the women students. Moreover, the opposition which delayed the opening of the professional schools to women made it difficult and sometimes all but impossible for women to secure a practice, once they were equipped for it.

While in many parts of the country, particularly in the far West, opposition is rapidly disappearing, it still flourishes in others. The woman physician usually finds herself in a vicious circle. Whatever her standing may have been in medical school, it has been difficult for her to secure an internship in the best hospitals. This means that the best of the women must begin less well trained than the best of the men. The leaders of the profession do not assist her to a practice as they do promising men doctors. With less experience, she develops less skill than the men she may have excelled in medical school.

But the change is coming. Women lawyers who have suffered greater handicaps in establishing a practice than women doctors—apparently we are more cautious about our money than about our health—have made great headway in the last ten years. The sense of fairness and the interest of the leaders in their profession will finally lead to the discouragement of no one who gives promise of advancing it. The fact that up to 1910 there were no women teachers in our professional schools, and that in 1928 there were 542, indicates the capacity of the women and the willingness of the professional schools to recognize and use it.

The history of women in the graduate schools which prepare for college teaching and research has been similar to their history in the professional schools. At first the best were not open to women; their attendance was not encouraged by fellowships and scholarships. At the present time, although very few women are employed as instructors or professors in the coeducational colleges, the pioneering has been done.

A century ago the church was the most articulate part of the opposition to the changing positions of the wife and mother, to education and political rights for women. The old order was the divine order, and entrance into the ministry was described as blasphemous by leading clergymen when the first woman minister in the United States, the Rev. Antoinette Brown, was not only ordained, but secured a pastorate. While here, too, opinions have greatly changed, opposition or prejudice to women in theology has continued. Out of 127,270 clergymen, enumerated in the census of 1920, only 1,787 were women. Nor have women made great efforts to break down the barriers to service in this field. This may seem strange in view of the fact that, more than men, women are deeply religious and interested in church activities. They have shown gifts as speakers and are not less self-sacrificing in the service of others than men. Perhaps the fact that many and generous scholarships are available in the theological schools for men, but not to women, may explain the small number of women students in the divinity schools which are open to them; and an eagerness to hold men to the church has made women willing to accord them the leadership, while the women have carried a heavy burden of church service.

In general it may be said, then, that the road to professional education for women has been improved. Gravel roads have taken the place of the dirt roads of the earlier period. The fellowships made available to women by the Association of University Women have been increased and, as more advanced scholarships are made available by foundation assistance, more are being awarded to women. The professional associations are gradually admitting them to participating membership.

Professional women are today in a better position to devote their time and intelligence to their professional work instead of to the controversy as to whether they should be given an opportunity to work. Numbers show that they have chosen the professions in which there were the fewest obstacles to professional opportunity. More than 600,000 were teachers in 1920, more than 13,000 librarians, more than 10,000 college presidents or college professors, more than 7,000 physicians and surgeons, nearly 6,000 editors and reporters.

Concern over the unjust and needless suffering of others has during the entire century been the motive power which compelled many timid women who hated the publicity to insist on their right to write and speak and, finally, to vote their convictions as to how social wrongs should be righted. The exclusion of women delegates at the London anti-slavery meeting nearly a century ago resulted in the first plans for an organized "woman's rights" movement. It is, therefore, not surprising that women have played a major part in the development of the new profession of social service. Here they have worked under handicaps similar to those women have met as teachers. If a well-paid administrative position is to be filled, the few men in the profession are usually carefully canvassed before the qualified women are considered. In 1909 the National Conference of Social Work, which had met annually for more than thirty years, elected as its first woman president, Jane Addams, who had been for many years the conference's most distinguished member. This was accomplished without the struggle that was necessary to elect Ella Flagg Young as president of the National Education Association, at about the same time. But social workers had some fears. Not as to Miss Addams, who was beloved and respected by both the men and the women of the conference; but what about the precedent?

As nurses, women may be said to have had no competition. Florence Nightingale, having demonstrated what a woman of ability, determination, and industry can do when confronted with stupidity and incompetence in high places, gave courage to others to undertake important administrative positions and

THE CHANGING POSITION OF WOMEN

to change our whole conception of adequate care for the sick. Led by Lillian Wald as the pioneer public-health nurse in America, they have played an important part in the initiation and development of public-health services.

LETTERS AND ARTS

But what about creative work in literature and art? At the present time women are writing some of the best, the most popular, and some of the poorest of the novels that appear each year in this and other countries. So great and general has been their success in this field, that it might almost be said that when their workmanship is found to be different from that of men novelists, that difference is judged on its merits.

Was this not true a century ago? In England, at least, some good novels were written by women. But for women who wanted to make the attempt there were inhibiting conventions—they and their families might be objects of ridicule. Without much social position they would lose what little they had. They might take the risk for themselves, but there were affectionate mothers and fathers who knew what was best for their daughters. And if they were not economically independent, how could they make the experiment? Virginia Woolf, quoting from the *Memoir of Jane Austen*, by her nephew, James Edward Austen-Leigh (the hyphenated name is significant), gives us a picture of the difficulties under which she wrote *Pride and Prejudice*, some hundred years ago. The general sitting-room was the only study she ever had and "she was careful that her occupation should not be suspected by servants or visitors or any persons beyond her own family." It was not womanly to write novels, and "Jane Austen hid her manuscripts or covered them with a piece of blotting-paper and was glad when a hinge creaked" and she could avoid being discovered working at a manuscript. Women were, so to speak, in at the birth of the novel and have had their part in its development. But George Eliot and George Sand felt safer under the *nom de plume* of a man.

Even with greater embarrassment they essayed poetry a

hundred years ago. In the development of what is today called modern poetry women have played as significant a rôle as men.

On the stage, so important a part do women now play as actors and producers that it is hard to believe that in Shakespeare's time it was generally accepted that women should not be permitted to attempt a rôle because, of course, they could not succeed.

In the fine arts, painting, sculpture, and architecture, women have had some success which, like the work of the women novelists of the last century, should make real achievement easier for women artists in the coming century. As for the woman composer, Mrs. Woolf says, "She stands where the actress stood in the time of Shakespeare."

WOMEN IN INDUSTRY

We have been discussing the most favored or the most gifted women—those who come in the main from families able to give them, or to help them to secure for themselves, the best that the nation has to offer in education and professional training. But these women constitute a relatively small group. What about the daughters and wives of workingmen? What has a "century of progress" meant to them? The problems which the well-to-do, the educated or professionally trained, women have had to meet have been quite different from those which the untrained and unskilled women have encountered. While they have, as women, much in common, the economic independence of a more favored group can be achieved with the position of the workingwoman unchanged.

In a large measure this is what happened during the first three-quarters of the century we are considering. In her history of *Women in Industry* Edith Abbott wrote in 1909:

> The efforts of the professional woman to realize a new ideal of pecuniary independence, which have taken her out of the home and into new and varied occupations, belong to recent, if not contemporary, history. But this history for her covers a social revolution, and the world she faces is a new one. The woman of the working classes finds

THE CHANGING POSITION OF WOMEN

it, so far as her measure of opportunity goes, very much as her great-grandmother left it.

This is not to say that there have not been profound changes which have affected the working life of both men and women. But from the beginning, women were an important factor in the development of manufacturing in America. The Colonial attitude toward their work is found in court orders, laws, and public subscriptions which were intended to help women to be self-supporting. The Puritan condemnation of the sin of idleness applied not only to men and women, but to children also, and during the Colonial time women were gainfully employed outside their own homes and, even more important, they were employed in craft industries and in manufacturing in their own homes. The United States was an agricultural country, and agriculture was in the hands of men, while it might be said that before the industrial revolution of the nineteenth century, manufacturing was in the hands of women. Thus, when Alexander Hamilton sought government help for the development of the factory system in America, he argued that it would give profitable employment to large numbers of women.

It was at the end of the transition period, when the factory system was becoming established in this country, that the century we are reviewing began. From a series of "Documents Relative to Manufactures" collected by the Treasury in 1832 and from the industrial census of Massachusetts of 1836-37 a list of more than a hundred occupations in which women were then employed can be compiled. But in a large per cent of these occupations only a very few women were actually employed, which is, of course, also true of the much longer list of occupations in which the census enumerators find women employed today—carpenters, masons, and blacksmiths, for example.

But at any rate women are not a new element in industry in the United States. New industries and new occupations have come in which they have found employment; the conditions under which they work have greatly changed; the number employed in manufacturing has greatly increased, but the per cent of increase in trade and transportation is much greater. In the cotton industry, one of the earliest to develop, according to the census of occupations, women constituted 64 per cent of the employes in 1850, 42 per cent in 1900, and 42 per cent in 1920. In all industries they constituted 24 per cent of the employes in 1850, 19 per cent in 1900, and 20.5 in 1920. But these percentages do not tell the whole story. The number of industrial women had increased from less than a quarter of a million in 1850 to more than a million in 1900 and more than eight million in 1920.

Throughout the century, the women who sought industrial employment have been easy to exploit and much exploited. Their strength in bargaining did not increase as their numbers grew larger. On the contrary, at the beginning of the century, when professional and clerical occupations were closed to women, women of more education and with more influence in the community sought employment in the factories that were then being opened. These women, who had taken a leading part in the early efforts to improve working conditions, later became teachers and clerks, and the women factory workers were drawn increasingly from the immigrants who began to arrive in great numbers at this time or were daughters of the unskilled workmen who belonged to what was, in each generation, called the "new immigration." The wages paid the men were low, but the wages of the women were lower. Skilled work and opportunities to acquire skill were not open to women. Their working life in the factory usually ended with marriage, and those who remained were apt to be women specially handicapped by home conditions or in other ways. There was, therefore, among the women workers, no experienced group to organize and lead in the struggle for better wages and working conditions. As the labor movement developed they were left behind rather than included. It was the aristocracy of labor, the skilled workmen, who were first successful in organization and collective bargaining.

Workingmen had fewer prejudices against the employment of women than professional men, but they were not without prejudices. Influenced by the prevailing public opinion, they hoped by raising their own wages to keep their wives and daughters from the necessity of gainful employment outside their homes. They were influenced by the current talk of "women and the home" which the struggle for political rights and professional opportunities for women had provoked. But they knew that women who sought opportunities to work twelve or fourteen hours a day in factory, laundry, or as scrubwomen were driven by the same economic necessity by which they were driven. However, it was, they thought, a temporary necessity and during this temporary period, they regarded the women as underbidders and poor trade-union material.

Employers' theories about womanhood, and the chivalric part men were to play in their protection, applied only to women of their own class who were confined to the home, sometimes against their protests. Workingwomen, on the other hand, were locked out of the home by economic necessity, and employers saw in their youth and helplessness an opportunity to make money. Business was highly competitive. The employer was in business for profit, and women became cheaper and more helpless under the mass pressure which the new organization of industry made possible.

In spite of the apparent hopelessness of their position, leaders among the workingwomen began to emerge and to interest other women and the general public in the improvement of their condition. It was clear then, as it still is, that as yet, neither as individual bargainers nor by trade-union organization, could they hope to meet the economic pressure which the employer can invoke against them. And so here, as in England and on the Continent, they sought, with the aid of other women and of the organized labor movement, legislation as a means of limiting the powers of the employer. Forty-four states now regulate the hours of work for women, but the legal working-day is long in all but a very few of these states. Twelve states have passed minimum-wage laws for women. Twenty-five states have prohibited the employment of women in occupations presumed to be hazardous to them. In four of these states the prohibition is so general as to make its construction a possible hindrance instead of a protection to women. In general, trade-union women now advocate that no occupations should be closed to them except those which are peculiarly hazardous to women and after the facts as to these special hazards have been established by scientific investigation. As industries change, as protective devices are developed, what is peculiarly hazardous today may not be hazardous tomorrow, so that it is generally held that occupations should not be placed in the prohibited class by law but by the ruling of an industrial commission or board.

Labor legislation must always run the gamut of a court test of its constitutionality. The United States Supreme Court has held the regulation of the hours of work for women and prohibitions on their employment which are in the interest of their health to be constitutional, but it has held unconstitutional certain mandatory minimum-wage laws.

In general, labor legislation has been less regarded as a method for improving labor conditions in this country than in Europe. The constitutional difficulties, the non-political tradition of the organized labor movement, the fact that the leaders among the men have felt that in the long run the greatest gains were to be secured by organization rather than legislation are offered as the explanations of this state of affairs. However, the most powerful of the men's organizations found themselves too powerful to resort to the strike, and the reduction of the hours of work for the trainmen on interstate railways was secured by statute. Legislation for women has been sought not because of their strength, but of their weakness in their relation to their employers, because of their youth and their consequent lack of appreciation of the hazards of industry. It has been upheld because, as potential mothers, society was especially interested in their protection. Workingwomen are not embarrassed by the apparent inequalities which such legislation creates. They know its effects and they know that legal equality does not always mean equality in fact.

Trade-union women and their friends and supporters maintain that the woman problem is not all of a piece, that the needs of business and professional women are not at the present time, at least, identical with the needs of wage-earning women. Hence they are asking for protective legislation for women employed in factories, stores, restaurants, and laundries, and not for business and professional women. Perhaps the homeless, unskilled laboring-men who build the railroads and dams and work in the lumber camps are a weaker group. But they begin with an advantage. They are at least men and the road up is easier for them.

In general women are at every level the weakest group in the industrial market. The improved conditions which legislation has brought have made organization easier, not more difficult. The increased class consciousness which workingwomen are beginning to develop, appreciation on the part of organized labor of their importance in the struggle for group justice rather than individual gains, and a changing viewpoint in regard to women and their employment outside the home will all be helpful to them. The tendency toward increased employment among married women, which began before the depression, will be accelerated by the diminishing immigration. The Negro in the South and women in the North will be the source of supply which industry will draw on in the new expansions to come with the next era of prosperity. It will be approved by capital because it will mean increased consumption, and by the family unit because it will mean a higher standard of living.

It seems safe to prophesy that, building on the foundation already laid, organization among women will be more easily accomplished, and men and women trade-unionists will work together more effectively in the future than in the past. Will they be able to solve the problems of industrial relationships, of stability of employment and changing production processes? These are the greatest problems of the century. Whether in the future these industrial questions will become our major political problems, it is impossible to say. But whether or not workingwomen in the future resort to legislation as a necessary tool in their effort to secure social justice will depend on many factors, and it may be assumed that their decision will be determined by experience and by a careful study of the changing factors in the situation rather than by the adoption of some theoretical formula.

WOMEN IN POLITICS

In the movement for what was first called "woman's rights," political equality was regarded as the touchstone in both the United States and Great Britain. In Germany, where political rights for men were longer delayed and where the public had been taught by Bismarck that good government rather than self-government was the ideal, the vote did not seem so important and was less discussed by feminist leaders. To the American and the British leaders it was in itself of great importance as well as useful in effecting other reforms in which they were interested.

The demand or the request that women should be granted the vote came before 1833. It had been temporarily secured for the women of New Jersey by the efforts of a Quaker preacher in 1776. But those who asked for the right asked as individuals until 1848, when Elizabeth Cady Stanton, Lucretia Mott and a number of other Quaker women called the first women's convention ever held in the United States. That Seneca Falls Convention was the beginning not only of the woman-suffrage movement, but of organization among women which was later to assume such vast proportions in the woman's club movement. These early American feminists favored the abolition of slavery, temperance reform, extension of property rights to married women, and many other social reforms; they wanted the right then generally denied them of holding meetings, of speaking and writing in behalf of these causes, and they wanted also the right to vote for them. The right to speak in public, to write and to organize in order to

THE CHANGING POSITION OF WOMEN 283

promote social and political reforms came to women in spite of opposition and with relative rapidity, but the right to vote was delayed for nearly seventy years. Mrs. Stanton, Lucretia Mott, Lucy Stone, and Susan B. Anthony did not live to share in the victory. Mrs. Catt and Anna Howard Shaw saw the women of the Scandinavian countries, of Germany, Austria, and the new republics created at the close of the Great War, given the vote before the English and American women who had borne the burden of the fight for women in every place had won it for themselves.

While it was greatly elaborated and restated in terms which the average man and woman could understand and apply to their experience, nothing was added to the argument which John Stuart Mill gave us in his essay on the *Subjection of Women* published in 1869. Intellectually the case was won at that time. But the long, slow, and painful triumph over prejudice and indifference was yet to come.

The early leaders thought that it would come at the end of the Civil War when the Negro was emancipated and gratitude for the war services of women was general. But the men who had worked side by side with the women for votes for women and the abolition of Negro slavery said that this was the hour of the Negro and the claims of women should not be pushed to his detriment. The Fourteenth and Fifteenth Amendments were passed and were tested in the courts by women in many states, Miss Anthony's case attracting most attention. The Supreme Court decided that these amendments neither gave to, nor withheld from, women the right to vote; it was still a question for the states. So the women took up the burden of educating an electorate scattered from California to Maine and from Canada to Mexico. The first gain came in the Territory of Wyoming in 1869, and although heralded as a precedent for victory everywhere, Wyoming was the only suffrage state until in the 'nineties, when Colorado, Utah, and Idaho granted suffrage to women.

In this interval there had been constant activity. Constitutional amendments were submitted and defeated in many states; Utah's long, legal struggle was complicated by the question of polygamy; Washington Territory granted suffrage to women in 1883, and then denied it when it adopted its state constitution in 1889. School suffrage was gained in many states, and municipal suffrage in a few. And then for nearly two decades there were practically no gains. Again the movement began to take on new life. Victory in California, statutory suffrage in Illinois, and finally the winning of the referendum in New York State in 1917 convinced politicians everywhere that ultimate victory was assured. Further delay and postponement in the changing post-war period became politically dangerous. The Nineteenth Amendment was adopted by Congress in 1918 and ratified by the last of the required thirty-six states in 1920. The right to vote had at last been granted to American women.

What has this brought them in the way of political recognition and political power? It is impossible in the allotted space to summarize the gains women have made or have failed to make in forty-eight states and hundreds of cities. Thousands of women are elected to local offices each year, usually as county superintendents of schools, as city, town, or county treasurers, as county clerks, and occasionally, particularly in the smaller towns, as mayors. In Michigan alone in 1929, 590 women were elected to local offices. These local elections are of the greatest importance, because it is the nearest government that most influences the daily lives of its citizens. It is in the state that public policies with reference to education, labor, and social welfare are adopted and administered. It is these issues that have especially interested women as relating to the welfare of children and home life. But the gains in the federal government may be used as a convenient yardstick for the purposes of this discussion of the progress that women have made in government in the past ten, twenty, or fifty years.

One of her biographers tells us that Clara Barton was the first woman clerk to be employed by the federal government some seventy-five years ago. If true, this achievement may be said, in a sense, to detract from the later triumphs of Miss

Barton which are so much more widely known. It is not surprising that, having succeeded in her determination to use pen and ink as a copy clerk in the service of the government, she had no hesitancy in asking for passes to the front line for herself and her kit of jellies, bandages, and medicines. "I shall give battle-front service," she said when she arranged for her clerical tasks to be performed by her fellow clerks who remained in Washington. To some of us it seems as though she must have given "battle-front service" if in the eighteen-fifties she succeeded in breaking down the prejudice which had up to that time prevented any woman from entering the government service. Forced by the exigencies of the Civil War, the numbers of women in the government service increased greatly during the 'sixties just as they did during the World War. Doubtless this admission of women to the lowest ranks of officialdom was regarded by many men in Washington and, particularly by those who spoke in the most exalted terms of the sacred calling of women, as one of the horrors that the Civil War brought.

The progress of women in government from 1860 to the present has not been rapid. An analysis, made by the Women's Bureau of the civil-service examinations that were open and those that were not open to women during a six months' period in 1919, revealed that women were excluded from 60 per cent of all the examinations held, from 64 per cent of the examinations for scientific and professional positions, and from 87 per cent in the mechanical and manufacturing services. This report was submitted to the Civil Service Commission before publication, and ten days after receiving the report the commission passed a ruling opening all examinations to both women and men, which left women free to compete in the examinations and the appointing officers free to appoint men, if they preferred men. What the preference of the appointing officers was the Women's Bureau discovered six years later in a second survey of the status of women in the government service. Most of the government services were included in this second survey. It showed that only 16 per cent of the women as compared with 51.8 per cent of the men received salaries of $1,860 or more per annum. No more recent survey has been made, but those having an opportunity for close observation see only a slight movement in those percentages.

An editorial in the New York *Times*, commenting on the fact that the Director of the Census was appointing 24 women as supervisors in the decennial enumeration in 1930, when only 5 women were so honored in 1920, made no mention of the fact that this was 24 out of 546 already appointed, and out of a total of 574 supervisors to be appointed. The closing sentences of the editorial were as follows: "In addition to this valuable work there will be many jobs of less prominence for women. They will do much of the house-to-house canvassing and a large share of the clerical work will fall to them," which describes the opportunities available to women in most government bureaus. In the same period the number of women appointed postmasters has increased from 255 out of a total of 2,487 in 1919 to 940 out of a total of 5,338 in 1930. The increase has been, it is probably unnecessary to point out, in the smaller towns, where the salary offered produces few qualified men competitors.

The story of women in major executive positions is a very short one. Julia Lathrop was appointed chief of the Children's Bureau in 1912 by President Taft, and a woman was continued in this position when she resigned in 1921. Between 1917 and 1920, when there was a lively expectation of immediate political power for women, women were first appointed to the following positions by President Wilson: chief of the Women's Bureau, member of the Civil Service Commission, assistant attorney-general which carried the status of an assistant secretary in the Department of Justice, and commissioner for the District of Columbia. Secretaries and bureau chiefs also opened the doors to a number of less important offices carrying administrative responsibility during that period.

When the Republicans returned to power in 1921 women were reappointed to all these positions except that of district commissioner. Some additional first appointments were made at that time. A woman was appointed for the first time as chairman of the Federal Employees Compensation Commission; the Home Economics Bureau was created in the Department of Agriculture, and a woman under Civil-Service appointment was promoted by the Secretary to be director of the bureau; women were admitted for the first time to the diplomatic service and a few minor appointments of women to this service were made; two or three women were made assistant trade commissioners in the Bureau of Foreign and Interstate Commerce; women were appointed to serve as collectors of internal revenue for the first time; and a few were appointed by bureau chiefs to minor executive positions in the scientific services which never had been previously held by women.

At the present time there is no woman assistant attorney-general and no woman has the rank of an assistant secretary. A woman has recently been made a judge of the Customs Court, and another a member of the Board of Tax Review in the Treasury Department; a woman has been appointed to the newly created position of assistant commissioner of the Office of Education and another as head of the Passport Division of the State Department. The head of the Army Nursing Service has the rank of major, and her assistant is a captain in the army. If we try to evaluate the present position of women in the executive branch of the government not by total numbers employed, but by the number holding important administrative or scientific and professional positions, we must say that women have had up to this time very limited recognition in the administrative branch of our federal government.

There are more women in Congress now than in 1921, eight in 1931 as compared with one then. Although still only a little band, this increase is encouraging and it is safe to say that the members can do their work more easily now than the same women could ten years ago. Except for a courtesy appointment, the Senate still awaits the coming of a woman, but it would not use the word "await" in the sense in which women use it.

The modern political Diogenes, we are told, is going about

with a lantern in his hand, seeking, not an honest man, but the perfect woman to appoint to office. She must, of course, be charming, well dressed, but not a leader of fashion; able, yet in the habit of constantly concealing that fact; she must have had much political experience so that the politician will say, "She deserved the job"; on the other hand, she must bring such technical training and professional experience to the position that the newspapers and the independent voter will say, "This was not a political appointment, but evidence of a policy of selecting only women with special qualifications for the work." It is easy to predict that such women will not often be found.

A cynic has suggested that, if women did not have the vote, the easy answer to all their difficulties would be the old slogan, "Votes for women." Yes, that would be the answer just as, if farmers did not have the right to vote, we could solve at once the problem of farm relief by votes for farmers.

"Votes for women" has supplied the general ticket of admission to the political fair-grounds, but not to the side shows. But general admission *is* general admission and, in spite of all that might be said that is depressing about women in government, their position might be worse. In her clever book, *A Room of One's Own*, Virginia Woolf discusses women as novelists and poets, but what she has said applies equally well to women in government, or in business, or in the professions —to any effort of women toward freedom in the effective expression of their convictions and the utilization of their capacities. Mrs. Woolf points out how unpleasant it is to be locked out, but how much worse to be locked in. Women, are, politically, no longer locked in the four walls of the home; and they are not yet locked in the citadels of the political parties, and by that token they are still free to negotiate the terms on which they shall be admitted, what they will surrender for what they are to receive.

Those who are discouraged at the political progress that women have made should remember that the woman voter, in most parts of the United States, has only a decade behind her.

She is still very young in political experience and her future usefulness in government depends upon her independence in political action, whether she will be prepared to risk failure, to retain her confidence in her judgment, and to keep up the fight when the decision of others goes against her.

"Life at best," Mrs. Woolf reminds us, "is arduous, difficult, a perpetual struggle. It calls for gigantic courage and strength. More than anything, perhaps, creatures of illusion as we are, it calls for confidence in oneself. Without self-confidence we are as babes in the cradle." Self-confidence is of slow growth. In his book entitled *Understanding Human Nature*, Alfred Adler says: "It is a frequently overlooked fact that a girl comes into the world with a prejudice sounding in her ears which is designed to rob her of her belief in her value, to shatter her self-confidence and destroy her hope of ever doing anything worth while." Less than in the past, and less in the United States than in other countries, this sense of inferiority on the part of girls and of superiority on the part of boys is being deliberately cultivated.

Some women and some men wonder why change in the position of woman has been desired. Repeatedly they ask: Why should anyone choose the "strenuous life," seek a part in the struggle to end the injustice and ugliness of our modern life? They are the lotus-eaters who prefer to live in a gray twilight in which there is neither victory nor defeat. It is impossible for them to understand, that to have had a part in the struggle, to have done what one could, is in itself the reward of effort and the comfort in defeat.

There is, then, we may conclude, evidence of great progress in the removal of restrictions which a hundred years ago denied to women the opportunity for education and for effective use of such talents and capacities as they possessed; which left them, as married women, without economic independence and, as married mothers, without authority over their children, while fathers of children born out of wedlock had no legal responsibilities; which nourished prejudices that induced the well-to-do woman to shun all useful work and placed the woman wage-earner in a peculiarly disadvantageous position.

Prejudice may be said to have nine lives. The social physician, who feels the pulse of the prejudice responsible for the position of women a century ago, reports that several of its lives are ended, the fatal illness of others, and a discouraging life expectancy in the remainder. But the physician's diagnosis may be wrong. The death of the ninth may be near at hand or those now reputed to be expiring may make a miraculous recovery.

CHAPTER ELEVEN

THE ADVANCEMENT OF NATURAL SCIENCE

This story opens in the scientific dark ages of Andrew Jackson's administration. In spite of the labors of centuries, the heavens above are still a vast mystery. There is no huge telescope with which to explore its deeps. Chemistry and physics have no appliances for discovering the composition of the stars or measuring accurately the velocity of light. Distances are still partly a matter of guesswork. The substances of the earth at our feet, of the hills and forests about us, and of the food we eat are even more mysterious. Electricity is in the toy stage. The history of the earth—its geological formation—is little more than a crude primer, mainly wrong in its surmises. Large sections of the North American continent, its early civilizations, its natural resources, and its primitive peoples await scientific exploration. Knowledge of the human mind is a confusion of speculations and superstitions. In the colleges there are only a few chairs in science, and these are relatively unimportant. In this chapter are sketched in outline the researches by which great departments of science have been developed in the United States. It is a record of many microscopic advances and of few spectacular achievements. It is a survey of patient labors by hundreds of workers, of martyrdom, and of triumphs—the history of a conquering army, captains and privates, ever pushing onward the frontiers of knowledge respecting the nature of man and his universe.

The Advancement of Natural Science

By Watson Davis

CIVILIZATION'S true immortals are the scientific pioneers who discover and inject new truths into the turbid, eddying thought stream of human intelligence. Great kings, statesmen, warriors, and industrialists, even poets and philosophers, may flicker out in the memory of the future; the scientific innovators will place their impress upon the world so long as our evolving civilization endures.

A hundred years is a short time in the growth of man's present estate, but the current tempo of science is so accelerated that vast gains in organized knowledge have been made within the memory of those now living. America's contributions to science have achieved world-commanding importance within the last fifty years, almost since the turn of the century. In the early days of the Republic, when the pioneer fringe was creeping slowly westward, only an exceptional few were able to pursue the secrets of nature. Scientific advances are not made on the extreme advancing frontier of civilization unless it be by exploratory parties whose roots rest in the more affluent cultures of the further east. Even a hundred years ago, when America had been settled along the Atlantic seaboard for over two hundred years, this continent was largely a frontier.

SCIENTIFIC BEGINNINGS

A firm foundation of scientific achievement and tradition had been laid, however, by such men as Franklin, Count Rumford, Audubon, Alexander Wilson, and a few others. Benjamin Franklin had won the scientific applause as well as the admiration of the Old World before the American Revolution. The unique individual who was born Benjamin Thompson, a poor New England boy, and who became Sir Benjamin and Under Secretary of State in England, Count Rumford of the Holy Roman Empire, Privy Councilor and Minister of War to the Elector Palatine, etc., may be numbered among America's early scientists even though he fought for England in the Revolution, founded the Royal Institution of Great Britain, and lived abroad most of his career as intellectual soldier of fortune. In a sense America may claim also Joseph Priestley, the discoverer of oxygen, whose scientific conservatism in strange contrast to his religious liberalism caused him to cling to the phlogiston theory which his own experiments disproved. In the case of Priestley, as in many others, religious intolerance in England contributed to young America's intellectual growth. Audubon and Wilson, with their artistic recording of America's bird life, were the spiritual ancestors of our colorful biological and geological explorers of today who record in photographs, monographs, and collections the flora and fauna of the least unknowns of the earth.

Philadelphia was perhaps the leading center of science in America in Colonial days and for a few decades following the Revolution. There labored Dr. Benjamin Rush, called the "Sydenham of America," who somewhat ahead of his times urged that the insane be treated as diseased rather than inhabited by devils. There also lectured Dr. James Woodhouse, who ridiculed the idea that the visitation of yellow fever was a punishment by God for the sins of his people. Robert Hare invented the oxyhydrogen blowpipe and presented the instrument to the Chemical Society of Philadelphia.

The colleges then in existence in some cases had professorships of mathematics and natural philosophy, which then included the whole field of science. John Winthrop, called the first American astronomer, held such a chair at Harvard. A narrowing of academic fields began soon after the Revolution. For instance, Benjamin Silliman occupied, in the first decade of the nineteenth century, the first professorship of chemistry and natural history at Yale.

Impetus to science in the early days of the United States came from high political quarters. Thomas Jefferson has his niche in the hall of scientific fame as well as the annals of democracy. At his bidding Lewis and Clarke had ventured upon a journey of exploration into the vast area from the Mississippi to the Pacific which he had acquired for the United States by purchase from France. His inquiring mind concerned itself with new plant varieties that might prove useful when introduced into this country from the Old World. He was a pioneer in paleontology when he discovered and reported the tracks of strange animals in the sandstone quarried for his home at Monticello.

More than a hundred years ago, in the more urban portions of the rapidly expanding nation, nuclei of scientific organization had formed. The American Philosophical Society of Philadelphia had a long record of usefulness before the Revolution, since it was the outgrowth of the Junto founded by Benjamin Franklin in 1727. Its transactions date from 1771. During the Revolution in 1780, the American Academy of Sciences at Boston was chartered. Local organizations had been formed and served as additional channels for the distribution of scientific intelligence. Early American scientists used the pamphlet method of making known their researches, and in some cases they sent their reports and letters to European societies for their *Proceedings*. Benjamin Silliman's *American Journal of Science*, established in 1818 at Yale, was the first successful chronicle of original American research outside the *Proceedings* of a society.

A century ago the scientific ferment had been firmly inoculated into America's fresh intellectual soil and it had begun to work. What was to be the result? Thousands of scientists engaged upon research problems of pure science; hundreds of competent scientists emerging each year from graduate courses of universities that dot every portion of the nation. Institutions spending millions of dollars a year on pure science alone. Industries built on scientific discoveries that were announced with the boast that they "could serve no useful purpose." A

THE ADVANCEMENT OF NATURAL SCIENCE

plethora of scientific literature that nearly chokes the assimilating power of the scientific world and the storage facilities of the libraries. A happier, healthier, better world that still seeks avidly to solve the unrevealed mysteries of the universe.

No branch of human knowledge is more universal and less provincial than science. New findings produce their reactions sooner or later at all points on the globe. Scientific progress is successful only to the extent that new facts and meanings, wherever discovered, pass into the world's general store of recorded and utilized knowledge. Because the advance of science is an interplay between all centers of research it is difficult to stamp a forward step indelibly and unequivocally as "Made in U. S. A." Nevertheless, the American scientific scene for the past hundred years can be viewed briefly for its summits and its trends.

Over a vast range of extremes in size the fertile genius of American science has left its imprint in the past century. Astronomers with the aid of their brother physicists have pushed outward the limits of the universe. Experimental biologists have discovered the extremely minute mechanisms of heredity. These two fields of research stand out in bold relief in the historical panorama.

ASTRONOMY

The stars have always enticed the imagination and curiosity of human beings. Thus it was natural that simple telescopes, which were part of every eighteenth-century gentleman's library furnishings, came to America early in its settlement. In those early days, also, laymen and seafaring men made far greater use of the stars in their wanderings, looking to them for guidance and orientation in time. Although the 1700's saw the beginnings of astronomy followed by a rising interest, particularly in New England where John Winthrop earned the reputation that caused him to be called the "first American astronomer," there were no astronomical observatories worthy of the name until the third decade of the nineteenth century. And not until 1873, when the Naval Observatory at Washington acquired its famous 26-inch refractor, did America have a truly notable telescope. That telescope, now dwarfed by a score larger, was in its prime a world's wonder.

In the years since that time America has acquired telescopes and equipment that have aided it in becoming the leading astronomical country of the world. Lick Observatory, on Mt. Hamilton, California, with its 36-inch telescopic twins, one a reflector and the other a refractor, has been in the astronomical vanguard since its founding through the will of James Lick, who died in 1876. Yerkes Observatory, inspired by George E. Hale, funded by its Chicago millionaire namesake and mothered by the University of Chicago, grew around its telescope with a 40-inch object glass. Mt. Wilson in California with its world's largest 100-inch mirror, its 60-inch telescope also a reflector, and lesser solar telescopes, the result of Hale's enthusiasm backed by the support of the Carnegie Institution of Washington, is now the most powerful astronomical concentration in the world. Soon the new 200-inch telescope now projected will rise from some southern California mountain peak. At Harvard College Observatory (Cambridge, Massachusetts), Lowell Observatory (Flagstaff, Arizona), Detroit Observatory (Ann Arbor, Michigan), Seward Observatory (Tucson, Arizona), Ohio Wesleyan University (Delaware, Ohio), Allegheny Observatory (Pittsburgh, Pennsylvania), there are seeing-posts of the heavens equipped with lens or mirrors of 30 inches or greater diameter. Throughout the country in universities and elsewhere there are lesser telescopes, many used for fruitful research. Nor must there be forgotten the essentially American observatories of Harvard, Yale and Michigan in South Africa for studying the southern skies.

Gigantic telescopes do not alone make successful observatories. The light gathered from the outposts of the heavens must be recorded, and these records must be pondered. Photography greatly increased the capabilities of telescopes, for the eye cannot add up the light falling upon it while the emulsion of the photographic negative can. The first celestial photograph ever made was a daguerreotype of the moon taken in 1840 by John William Draper, professor of chemistry at New York University. Ten years later Whipple, under the direction of W. C. Bond, first director of the Harvard College Observatory, took the first photograph of a star. Thus began Harvard's leadership in photographic recording of the sky. At Harvard also Henry Draper in 1872 photographed the first stellar spectrum in which characteristic lines were visible. Logically, Harvard astronomers concentrated on this field of work, setting standards for the relative brightness of stars, and finally under the famous E. C. Pickering established the great and growing collection of thousands of photographs, the Henry Draper Memorial photographic library of the sky, which recorded the brightness, spectrum, and places of millions of stars. This great collection is playing a major rôle in the exploration of the universe. The gyrations of comets and minor planets, the inconstancy of stars, and all the changes of the heavens are recorded there for future generations of astronomers. There discoveries are made with microscopes instead of telescopes, for variations in the sky are found by comparing minute dots of light on many photographs.

Working over the Harvard photographs, Miss Henrietta S. Leavitt discovered the important relationship of the Cepheid variable-star periods of variation and their true brightness, a measuring-stick that has proved invaluable in the hands of Harlow Shapley, present Harvard director, and other astronomers in scaling the expanse of space now known to be dotted by "island universes" like the Milky Way, of which our sun is a rather mediocre star.

The pushing of the limits of known space outward, which has occurred in the last decade since the Mt. Wilson 100-inch telescope has been swung upon the heavens, is one of the most thrilling advances of astronomy, stirring philosophic and religious bents innate in astronomers as well as other men. The observations of Edwin P. Hubble and M. L. Humason, which have shown spiral nebulæ and other great star systems scattered throughout space as far as the telescopes can reach, furnish a factual basis for the mathematical physical theories of Einstein and others of the new school of physics. Important also have been the tests applied by Americans, notably Albert A. Michelson, Charles E. St. John, W. W. Campbell, Robert A. Millikan, Walter S. Adams, Richard C. Tolman, and Edwin P. Hubble, to the Einstein theories that have ushered in the new physics. This chapter of astronomy recalls the days of Galileo and Newton when the common conceptions of the material world were revolutionized.

It took years to discover that not only is the earth merely one satellite of the sun, that the sun is one of many thousands of stars, but that our galaxy is merely one of many thousands of star systems, each a gigantic "universe" of its own.

American observations plowed out many of the heavenly facts. Hundreds of astronomers gave their lives to explorations of the universe which, like all pioneer work, had its many hours of routine drudgery and its momentary rises to the heights of understanding. Stars were found to move like flocks of birds through space. Lewis Boss of Albany pioneered in discovering star-streaming, and his star catalogue is a monumental work. W. W. Campbell, then director of Lick Observatory, by observing the reddening and purpling of stellar spectra, measured star movements. Double stars were discovered and watched for years. Sherburne Wesley Burnham, stenographer by profession and astronomer by interest, catalogued 13,665 of them, and R. G. Aitken and W. J. Hussey of Lick captured additional thousands. The spectroscope in the hands of Walter S. Adams of Mt. Wilson yielded the distances of stars. The immensity of the stars is apparent from the interferometer measurements of star diameters made by Albert A. Michelson and Francis G. Pease at Mt. Wilson. The star Betelgeuse is 300 times the diameter of the sun, and the orbit of the earth and moon could be sunk, with many thousands of miles to spare, within this star's bulk. The dark companion of the bright star Sirius is known to be not only three times the earth's diameter, but 200 times as dense as platinum. Alvan Clark found it while testing one of his famous telescope mirrors in 1860. The life history of stars has been discovered and classifications have been devised, notably by Henry Norris Russell of Princeton.

The solar system itself has not been neglected. The European discovery of the ninth planet, Neptune, in 1846 stirred popular and scientific imaginations and spurred on American astronomy. Most of the positional astronomy of today rests firmly on the labors of Simon Newcomb, under whose genius and tedious labor the American nautical almanac set standards still relied upon. Whenever the orbit of a comet or asteroid is computed, Newcomb's fundamental constants are used. In fact, only in recent years has E. W. Brown added to Newcomb's work with his researches on the motion of the moon. From the time of Asaph Hall's discovery of the moons of Mars in 1877 at the Naval Observatory to the exciting discovery of Pluto, the tenth planet, by a Kansas farm boy, who, untouched by college, had left his home-made telescope to participate in the Lowell Observatory's searchings for the transneptunian planet, there were many important discoveries to the credit of American astronomy. The finding of Pluto was the result of Percival Lowell's conviction that continued to inspire his associates after his death. Lowell will be remembered for his connection with Pluto even after his colorful observations on ruddy Mars are forgotten. To the question of the origin of the earth and the rest of the solar system, two University of Chicago scientists, Thomas C. Chamberlin and F. R. Moulton, brought the planetesimal hypothesis which fits the facts better than the Laplace nebular hypothesis.

The sun, most important of stars to earth dwellers, has been studied because it is the source of earthly light and heat as well as a typical star. Whenever the moon eclipses the sun, astronomers go to the narrow band of darkness, though they must travel halfway around the earth. Then the corona of the sun can be observed for a few fleeting minutes. Then the inconstancy of the moon's motion can be checked. Since Colonial times American astronomers have been observing solar eclipses. The Princeton scientist, Charles A. Young, first observed the flash spectrum at the 1870 eclipse and at another eclipse he identified the coronal spectrum lines. Samuel Pierpont Langley, secretary of the Smithsonian Institution, is remembered for his study of the sun as well as his pioneer aëronautical experiments. With his bolometer, he showed that the sun's energy extends into the infra-red and that its output of energy varies. Charles Greely Abbot, his successor in sun study at the Smithsonian, has refined the observations of the sun's inconstancy, erected observatories on cloudless peaks of this country, South America and Africa, and daily he announces the sun's heat which can be correlated with earthly manifestations, including the weather. George E. Hale, father of both Yerkes and Mt. Wilson Observatories, investigated sunspots, which he found to be fields of localized magnetic influence. Strangely, his studies showed that with a change of the eleven-and-one-half-year cycle, the sunspots reverse their polarity.

While observatories push forward the frontiers of space and astronomical knowledge, imported planetaria recreate in light the changing processions of the visible heavens for the edification of the public of Chicago, Philadelphia, and Hollywood.

PHYSICS

Since light, heat, and other physical manifestations are the same whether they originate upon the earth or in the heavens, astronomy and physics have close kinship, particularly in these days when physicists go to telescopes to view experiments not possible on earth and astronomers go to physics laboratories with the hope of duplicating here on earth some conditions in outer space.

The history of physics in recent years is largely a story of the spectrum of ether vibrations, so called, that begin with the slow pulsations of electrical current, and continue with increasing shortness of wave length through radio waves, heat, light, ultraviolet, X-rays, radium and cosmic rays.

In the investigation of light Albert A. Michelson ranks with the other pioneers of precise measurement. His application of the interferometer to problems of physics, the Michelson-Morley experiment on ether drift that laid a physical foundation for Einstein's special theory of relativity, and his repeated measurements of light's velocity are permanent accomplishments. He was the first of America's Nobel-prize men in science, and under his guidance and inspiration the Ryerson Physical Laboratory of the University of Chicago became a producer of eminent physicists. It is significant that Michelson's development of the interferometer was contemporaneous with the invention of the linotype.

The production of concave diffraction gratings for splitting up white light into its colors was just one of the many achievements of Henry Augustus Rowland, long associated with Johns Hopkins University. So perfect were Rowland's diffraction gratings that his spectroscopic measurements were a hundred times greater in accuracy than those of Ångström, and Rowland's photographic atlas of the solar spectrum is a monumental work. Rowland is also remembered for his Berlin experiment of 1876, showing that an electrostatic charge put in motion produces a magnetic field similar to that of a voltaic current. Lewis Morris Rutherford, lawyer by profession and scientist by inclination, ruled the best gratings before those of Rowland, and a physician, Dr. David Alter of Freeport, Pennsylvania, in 1854-55 used the electric spark to map the spectra of eleven different metals.

The last decade has seen the demolition of the partition separating radiation and matter, and American physicists have been prominent members of this constructive wrecking crew. Robert Andrews Millikan won international recognition and the Nobel prize for his measurement of the charge on the electron, one of the most fundamental of the constants of physics. Arthur H. Compton applied the quantum theory to the problem of scattering of X-rays, predicted the effect that bears his name, and incidentally won a Nobel prize. C. J. Davisson and L. H. Germer showed that streams of electrons have many of the characteristics of radiation and can be deflected and concentrated much like light. Other physicists have worked in these fields which are almost daily yielding fruitful results.

The continuity of the radiation spectrum was emphasized when E. F. Nichols and J. D. Tear closed the gap between the optical and the Hertzian waves. Robert Andrews Millikan, in addition to his building up of the California Institute of Technology and his participation in national and international science organization, showed through his cosmic-ray researches that there is radiation far shorter than the most intense gamma rays of radium. American physicists have applied X-rays and radium to research with vigor, and W. D. Coolidge of the General Electric Company developed the X-ray tube of usual form, pioneered in the production of tubes of high voltage and developed the cathode-ray tube.

Joseph Henry was throughout his long career one of America's most outstanding scientists. His first secretaryship of the newly-founded Smithsonian Institution, beginning in 1842, opened new fields of research and stimulated coöperation between scientists and the government. His organization and operation for thirty years of a nation-wide weather service with volunteer observers was the forerunner of present government weather-forecasting. But Henry's greatest achievements lay in his work in electricity. In 1831 he anticipated in discovery, but not in publication, Faraday's fundamental finding that electric currents have the power of inducing other currents, a discovery that is utilized in every motor and dynamo. In 1829 Henry devised the first electromagnetic motor, the forerunner of all electric motors, and the "henry" as a unit of inductive resistance memorializes his discovery of the phenomenon of self-induction in 1832. His early electrical experiments were made at Albany Academy, and later at Princeton, among other achievements, he recorded the transmission through walls and floors of electric waves of the type used in radio and telephony today.

Benjamin Franklin and Joseph Henry were the forerunners of a long line of electrical experimenters and inventors in America. From these researches arose the telegraph invented by Samuel F. B. Morse; the telephone invented by Alexander Graham Bell; the electric light by Thomas A. Edison; the electric dynamo, motor, and transmission devices by a host of inventors, among them C. F. Brush, William Stanley, and Elihu Thomson.

From these electrical industries, born of science, have sprung in recent years research laboratories, created largely for their immediate utilitarian aid to the industry. With foresight some on the technical staff are allowed to indulge in "pure science," and their work has proved in repeated instances to be more profitable than the direct attacks on the more immediate problems. Talking motion pictures arose full-blown out of research laboratories to catch a self-satisfied industry unawares. And so to a large extent with television.

To radio, America has made vital contributions, although the long scientific lineage of this great industry was largely in Europe in its early years. Fundamental was Edison's discovery of the "Edison effect," the unidirectional flow of current from a heated filament to a plate within a vacuum tube. Lee de Forest in 1907 added the grid, the third element, and made the electron tube that finds such wide use today.

The art of communication has called upon the physics of the universities to aid it, as for instance when Michael I. Pupin eliminated distortion and attenuation in long-distance telephone circuits by loading coils devised by the application of electromagnetic theory. Or when the Ryerson Laboratory physicists contributed so largely to the development of electronic repeaters for long-distance telephone lines and then applied these amplifiers to the radio telephone and talked across continents and oceans without wires.

To another great industry, aviation, physical research has brought its gifts. The wind tunnel, in which the craft stands still and the air moves, aided design from aviation's early years, and now the Langley Memorial Aëronautical Laboratory of the National Advisory Committee for Aëronautics, with the world's largest wind tunnel and the world's largest towing basin for seaplane testing, promises design data for better airplanes and airships.

To industry in general the National Bureau of Standards at Washington in its three decades of activity has given fundamental scientific aid of incalculable value, particularly in the field of measurement.

In the formulation of the principles of energy, the name of one American, Josiah Willard Gibbs, mathematical physicist of Yale, stands out prominently. He is the parent of much of the thermodynamical thought of today as well as of physical chemistry. His paper, "On the Equilibrium of Heterogeneous Substances," published in 1878, laid a foundation for the interpretation of chemical phenomena which was not utilized for many years afterwards in this country. The industrial value of the phase rule, a by-product of Gibbs' thermodynamic discussion of chemical equilibrium, would annually support an army of scientists, while his work on the laws of thermodynamics, particularly his *Elementary Principles in Statistical Mechanics*, laid foundations built upon by later work in the field of energy. To this realm of heat and energy other American mathematicians-physicists-chemists, among them G. N. Lewis and R. C. Tolman, have brought reconciliation or compromise between the views of classical physics and the new formulations of the era of relativity.

When scientists ponder on what they have observed and begin to apply logic to discover the ways of nature, mathematics becomes both handmaiden and master. Benjamin Peirce of Harvard was one of the most noted of American mathematicians. His mathematical demonstration that the theory of solid rings of Saturn is untenable dealt a mortal blow to the Laplace idea of the origin of the solar system. In recent years, many other mathematicians, among them E. H. Moore of Chicago, working in the field of general analysis; Oswald Veblen of Princeton and his group researching on analysis situs, or how to put colors on a map; and George David Birkhoff of Harvard, developing boundary value problems—have added to mathematical knowledge.

THE ADVANCEMENT OF NATURAL SCIENCE

CHEMISTRY

A century ago the era of organic chemistry had just been ushered in by Wöhler's synthesis of urea in Germany. The basic facts of some of the fundamentals of chemistry, such as combustion, the composition of water, the general laws of chemical combination had been recognized for several decades. America followed Europe in most chemical endeavor. But in a relatively new country the desire to discover the composition of things is strong, and for this reason, perhaps, chemical analysis developed early.

To mineral analysis Henry Seybert of Philadelphia, J. Lawrence Smith, Oliver Wolcott Gibbs, F. W. Clarke and W. F. Hillebrand and others brought new methods. Wolcott Gibbs of Harvard, not to be confused with J. Willard Gibbs of Yale, also pioneered in the field of double and complex salts and electrolytic analysis.

Edward W. Morley, the Morley of the Michelson-Morley ether drift experiment, determined the ratio of hydrogen to oxygen in water which resulted in the choice of O equals 16 in atomic weight values. Because of his work on the photochemistry of silver salts, Carey Lea was a pioneer leading to the development of photography. Largely American in its development is the work on liquid ammonia solutions which is associated with the name of Edward C. Franklin. In the modern development of qualitative analysis A. A. Noyes and his comprehensive system is notable, as is his work on electrolytic dissociation. W. A. Noyes is known for his work on positive and negative valences.

Ira Remsen of Johns Hopkins was one of the great teachers and developers of chemical research, and his making of saccharine in 1879 was a notable synthesis.

In more recent years the chemical workers have greatly increased in number until the American Chemical Society, founded a little over fifty years ago, is one of the largest of scientific societies. It is not possible even to name many of the more recent workers in this field.

In the theory of organic structure Julius Stieglitz did notable work, Roger Adams has made important progress in the chemistry of catalysts, anesthetics and other organic compounds, P. A. Levene has pioneered in the field of protein chemistry, Treat B. Johnson worked on the organic chemistry of nitrogen, Moses Gomberg opened up a new chapter in organic chemistry by his discovery of triphenylmethyl, C. S. Hudson gained recognition for his studies of sugars, and Samuel C. Lind studied the chemical effects of radiation.

To the discovery of the properties of the 92 chemical elements America has made notable contributions. Theodore W. Richards, America's only Nobel-prize winner in chemistry, made fundamental determinations of the weight relationships of the elements, and Frank W. Clarke participated over a long period of years in the international establishment of constants of the chemical elements. The last three of the 92 chemical elements promise to be American discoveries. B. S. Hopkins found number 61 and named it illinium after his university. Fred Allison and associates of Alabama Polytechnic Institute reported the detection of 87, closely related to radium, sodium, and potassium in the periodic table, and then 85, which is the "eka-iodine" of Mendeleev, while Jacob Papish at Cornell also claimed the discovery of these elements.

Many of the world's new industries have arisen out of chemical discoveries, inventions, and researches in America. Some of these have come out of chemical laboratories where the pursuit of knowledge is the principal objective; some of them have come out of the great urge to innovation that has made our country a land of inventors.

INDUSTRIAL APPLICATIONS

The whole vast petroleum industry began when Benjamin Silliman, Jr., chemist and son of a chemist, in his Yale laboratory in 1854 analyzed petroleum and showed that it could be used as illuminating oil, lubricating oil, and oils for making gas. This was five years before the first oil-well was drilled. Since then chemistry has aided the oil industry continually;

THE ADVANCEMENT OF NATURAL SCIENCE 307

for instance, when William M. Burton developed the cracking of petroleum and when Thomas Midgley, Jr., devised chemical additions to gasoline to prevent "knocking."

Out of applied chemistry came such important industrial processes as the vulcanization of rubber by Charles Goodyear in 1839, the electrolytic production of aluminum by Charles Martin Hall in 1886, the making of collodion by J. Parkers Maynard in 1848, the making of celluloid by L. S. and J. W. Hyatt in 1874, the invention of carborundum by Edward Goodrich Acheson in 1891, the production of special alloyed "high speed" steels by Frederick Winslow Taylor and Maunsel White in 1898. It is possible to mention only a fraction of the important industrially revolutionary chemical processes made in America.

In photography George Eastman applied the flexible celluloid film as the support for the emulsion, a step that allowed amateurs to use the camera. Nearly as important was the example set by the Eastman Kodak laboratories under the direction of C. E. K. Mees in demonstrating that an industry can keep itself evolving by conducting research without too much commercial pointedness.

The General Electric laboratories at Schenectady, under the guidance of Willis R. Whitney, have made major contributions, among them Irving Langmuir's gas-filled electric lamp and atomic hydrogen welding. The chemists of the many and diverse du Pont interests, engaged in group applications of science to industry, have raised to fruitfulness numerous processes, among them those for production of explosives of various sorts and the pyroxylin lacquers.

Leo H. Baekeland created a new class of materials when he made the carbolic-acid-formaldehyde synthetic plastic which is known as bakelite. R. B. Moore pioneered in the production of radium and he was one of the group who, during the World War, following H. P. Cady's analyses of natural gas which showed helium content, produced helium gas on a large scale for use in airships and balloons. F. G. Cottrell gave industry the process of electrical precipitation.

The World War vigorously demonstrated the need of American chemical independence, and the growth of the American chemical industry was sustained by public interest created in large measure by the interpretations of Edwin E. Slosson.

BIOLOGICAL CHEMISTRY

To the knowledge of the mechanism of the human and animal body America has made important contributions. The discovery, isolation, and application of internal secretions or hormones in the past three decades opens a new field of physiology and medicine. John J. Abel has been a leader in this field with his pioneer work on the isolation of epinephrine, or adrenalin, the secretion of the adrenal glands, his researches upon the posterior portion of the pituitary or hypophysis, and his crystallization of insulin. Jokichi Takamine, working in America, shares the credit for producing adrenalin. The secretion of the thyroid gland was isolated by Edward C. Kendall and later was synthesized abroad. Spectacular and important was the isolation of the secretion of the pancreas, by the University of Toronto group, consisting of F. G. Banting, J. J. R. Macleod, C. H. Best, and J. B. Collip. This secretion, insulin, was found to be effective in the treatment of diabetes. These three hormones have saved many human lives. So also has the liver extract for pernicious anemia which resulted from the physiological work of George H. Whipple, the clinical work of George R. Minot and the chemical work of E. J. Cohn. The pituitary hormones have been investigated, those of the anterior lobe by T. Brailsford Robertson and Herbert M. Evans among others, and those of the posterior lobe by John J. Abel and Oliver Kamm. J. B. Collip of the Canadian insulin group has also studied the parathyroid glands. The hormones of sex, both male and female, are now being investigated intensively by many physiologists and biochemists, among them E. A. Doisy and E. Allen.

The fuel of the human body, food, has been considered chemically and physiologically only in the last sixty years. It

THE ADVANCEMENT OF NATURAL SCIENCE

is true that just a hundred years ago, William Beaumont, American army surgeon, was studying the digestive process, using Alexis St. Martin, "the man with a lid on his stomach," as his subject. Beaumont's experiments are classic. But not until 1869 did W. O. Atwater, then just out of college, make America's first proximate analyses of a food, corn. Atwater continued to do fundamental work in nutrition and the diet, extending his extensive listings of what people eat to the determination of how the body uses the foods. For this purpose he and E. B. Rosa made a respiration calorimeter which measured the oxygen consumed and the heat produced by human-life processes. In the hands of Henry P. Armsby the calorimeter was adapted for farm animals, and thus the nutritional economics of livestock feeding could be studied. Francis G. Benedict and his associates of the Carnegie Institution nutrition laboratory have studied extensively the metabolism of matter and energy in the human body. Not so directly applicable to human nutrition, but agriculturally important, are the experiment-station investigations which in dozens of fields are conducted at centers in every state. These began with fertilizer analysis work by Samuel W. Johnson in 1853 in Connecticut.

Out of the usefulness of chemists in assuring farmers of the quality and value of fertilizers, feedstuffs, etc., rose the movement for pure food and the food and drug control which was led by Harvey W. Wiley.

Early investigators in nutrition recognized only the chemically obvious fat, carbohydrate, and protein constituents of foods. About the turn of the century came the realization that that was not the whole story. Thomas B. Osborne at Yale recognized that all protein is not the same, and his work on the amino-acids that make up proteins is classic. It was shown that many of these protein building-blocks are essential to life. With Lafayette B. Mendel, Osborne showed that life cannot be maintained on a diet of purified food and thus took steps toward the discovery of the vitamins. At the University of Wisconsin, under the guidance of Stephen M. Babcock, famous for his butter-fat test, there was a group working on long-continued animal-feeding experiments. Among them were Edwin B. Hart, Elmer V. McCollum, and Harry Steenbock, names well known in modern work on nutrition. These tests showed that unknown substances not contained in some foods are necessary to life, and with similar research abroad they ushered in the era of vitamins. Casimir Funk, a European biochemist who worked in America, coined the name "vitamine." The list of vitamins has grown until now the first six letters of the alphabet are used to designate them and more letters will undoubtedly be needed in the future. The more the vitamins are studied the more complex they are found to be, and the finding of two substances in material previously considered to contain but a single vitamin may be expected to happen again. Elmer V. McCollum and his associates at Johns Hopkins pioneered in many phases of work on vitamins A, B, and D, and also in the significance of manganese and magnesium in the diet. Important to health is the fact that various vitamins are protection against deficiency diseases; A is necessary to normal growth and its lack causes an eye disease, B prevents beriberi, C prevents scurvy, D prevents rickets, and lack of Herbert M. Evans' vitamin E causes sterility. G, or P-P, as Joseph Goldberger called it in his work on pellagra, is a protection against that disease. Of industrial application to food production and of clinical use was the work of Harry Steenbock and Alfred F. Hess on the relation of ultraviolet rays to vitamin-D production and the possibility of activating fats with this sort of light. Other American scientists working on vitamins include H. C. Sherman, Walter H. Eddy, R. A. Dutcher, Atherton Seidell, Robert R. Williams.

A century ago pain was an unavoidable accompaniment of any surgical operation. Anesthesia, which spelled the "death of pain," was given to the world by America. Independently, Crawford W. Long of Alabama in 1842 and W. T. G. Morton of Boston in 1846 used ether to cause surgical patients to lapse into blessed unconsciousness while the knife was applied. Long used ether first, but did not publish his results, and it

THE ADVANCEMENT OF NATURAL SCIENCE

was from the Boston focus that the use of anesthetics spread to the rest of the world. Horace Wells in 1844 used nitrous oxide for teeth extraction and thus inspired Morton's use of ether. Since those epochal days other anesthetics have been discovered or synthesized, notably many drugs for local use. The one major addition to general anesthetics in recent years was made by Arno B. Luckhardt, who recognized the anesthetic properties of ethylene gas, now used extensively.

Even before the days of Pasteur and his new foundation for medicine, Americans were building portions of the structure of scientific medicine. In 1843 Oliver Wendell Holmes pointed out the contagiousness of puerperal fever. Joseph Leidy, anatomist and great descriptive naturalist, discovered in 1846 the cysts of trachina in pork and in 1849 he placed the bacteria among the plants rather than with animals. Two years later he transplanted malignant tumors and in 1879 he was the first to separate out the parasitic amœbæ. One of the earliest discoveries of the bacterial cause of a disease was made by Thomas J. Burrill, who found the bacillus causing pear blight in 1877.

Soon after the initial successes of Pasteur, the science of bacteriology had its American beginnings in the work of George M. Sternberg, army surgeon, and Theobald Smith, pioneer in the theory of infectious diseases. Sternberg in 1880 discovered the pneumococcus simultaneously with Pasteur, and his later work on microörganisms of the air cleared the way for the army's yellow fever commission and its proof of the mosquito transmission of this disease. The world's first experiment in immunization was made by Theobald Smith in 1886 when he demonstrated that immunity to hog cholera could be secured by injection of the filtered products of the specific organisms. His demonstration with F. L. Kilborne in 1893 of the parasite of Texas fever, and its transmission by the cattle tick, was a great step in the science of protozoan disease. Theobald Smith, among his many other researches, differentiated between the human and bovine tubercle bacilli and demonstrated anaphylaxis.

William Henry Welch greatly advanced bacteriology and pathology, and from his laboratories at Johns Hopkins University came many eminent followers. With George H. F. Nuttall, the biologist, Welch in 1892 described the gas bacillus. Nuttall previously in 1888 had discovered the bactericidal powers of the blood.

The proof of the mosquito transmission of yellow fever given by the yellow fever commission's experiments in 1900 is an epic of medicine and wrote the names of Walter Reed, James Carroll, Jesse Lazear, Aristides Agramonte, members of the commission, and the memory of their volunteer subjects for the experiment, indelibly in the history of civilization. Thus was proved the suggestion made twelve years before by Carlos Finlay, Havana physician, that mosquitoes are guilty of spreading the disease. Hideyo Noguchi of the Rockefeller Institute nearly three decades later found the parasite causing yellow fever and, probing further in Africa, died a martyr to science.

Another martyr to medical research, Howard Taylor Ricketts, who died while studying Mexican typhus fever, showed that Rocky Mountain spotted fever is transmitted by ticks. Edward K. Francis' work on tularemia revealed this disease of rabbits often transmitted to hunters. The discovery by C. W. Stiles of hookworm disease, widespread in southern United States, and his instigation of the Rockefeller campaign have begun the conquest of this disease of backwardness. Scarlet fever has been added to the list of diseases brought under control by the work of George F. Dick and Gladys H. Dick, bacteriologists and husband and wife.

Alexis Carrel won the 1912 Nobel prize in medicine for his researches in physiology and physiological surgery, and his tissue cultures outside the living body have shed new light on biological processes. The Carrel-Dakin solution was used during the World War for infected wounds.

Simon Flexner, as head of the Rockefeller Institute for Medical Research, and George W. McCoy, as head of the United States Public Health Service Hygienic Laboratory,

THE ADVANCEMENT OF NATURAL SCIENCE

now the National Institute of Health, have directed important medical researches in addition to their own experimental work. Victor C. Vaughan, Frederick George Novy, and Ludwig Hektoen are also great names in American bacteriology and pathology. Notable also are the investigations of W. B. Cannon on emotion in disease, and those of William H. Howell on the coagulation of the blood. To the continued story of the life of germs there is now being added a new chapter on their invisible stages that have been discovered through the researches of Philip B. Hadley and Arthur I. Kendall.

As important as these medical discoveries is their application, and this is the function of the public-health movement that has gained momentum in the last fifty years. The conversion of typhoid fever from a common to an infrequent disease, as the result of improved sanitation and the cleaning up of milk and water supplies, is typical.

BIOLOGY AND ENTOMOLOGY

Early biology in America was largely a matter of looking at living things, describing them and putting them in their places in the scheme of classifications. The early students of animal and plant life were naturalists rather than experimental biologists. They were exploring collectors. Out of this pioneer phase, which will never be entirely completed, there emerged interest in how and why living things have their particular forms, how they grow and how they live. With the crystallization of evolution by Darwin there came a period of intense investigation of the relationships in time and ancestry. The rediscovery of Mendel and the influence of experimental medicine put living things in the laboratory as living things rather than as museum specimens.

Louis Agassiz and Asa Gray belong to that early era of the great naturalists. Agassiz, who had won fame in Europe before he came to America to rejoice in a new and nearly untouched collecting field, gave an immense impetus to zoölogy. Gray's researches in botany remain fundamental even today. Spencer Fullerton Baird, first commissioner of fisheries and second secretary of the Smithsonian Institution, enlisted the energy of thousands of government officials at home and abroad in building up the collections of the United States National Museum, established famous Woods Hole as a marine biological station, and did significant work on birds, fishes, and lesser creatures. Another great naturalist was George Brown Goode, Baird's associate. George Engelmann was the St. Louis botanist-physician who inspired the Shaw Botanical Garden and worked on the flora of the Mississippi Valley when it was frontier country. A century ago John Torrey of New York was the best-known botanist in America. The most important of early anatomists was Jeffries Wyman of Harvard, who in 1847 described the gorilla as a new species of ape, using specimens sent him by a New York medical missionary, Dr. Thomas S. Savage, who brought them from Africa. William Keith Brooks, a follower of Agassiz, and Alexander Agassiz, the son of Louis, were among the marine naturalists. Versatile Joseph Leidy ranged over whole fields of science, including botany, zoölogy, minerology, paleontology, and anatomy, as a great descriptive naturalist. Of more recent years was the zoölogical work of David Starr Jordan.

The stimulating idea of evolution did not at first find complete acceptance in America, for Louis Agassiz opposed it with scientific candor but earnest vigor. In Asa Gray, however, Darwin had a vigorous and stanch adherent to his theory. In fact, Darwin outlined his theory in a letter to Gray over a year before the publication of the *Origin of Species*. It did not take many years for the evidence given by nature, much of it literally unearthed by the American paleontologists, to win practically unanimous adherence to the basic idea of evolution both in America and abroad.

To the grand idea of the rise of animate nature up the stair steps of life, America has contributed its researches in experimental evolution. Beginning in the studies in embryology and cytology, particularly those of E. L. Mark and Edmund B. Wilson, these researches flowered in the first three decades of the present century following the rediscovery of the Mendel theory of heredity. Mendel published obscurely his fundamental law of heredity in 1856, but not until the turn of the century did it gain currency and appreciation.

Then the search began for the mechanism by which each parent hands on to the offspring unit traits that are not influenced by any other hereditary units. Thomas Hunt Morgan and his associates, especially A. H. Sturtevant, H. J. Muller, and C. B. Bridges, demonstrated in long series of experiments on the fruit fly, *Drosophila melanogaster,* that the genes are the bearers of heredity. The chromosomes of the fertilized egg-cell, contributed by both parents, are each linear aggregates of these genes, and the character of the offspring is determined by the manner in which these genes arrange themselves as the cell divides. This explained the mechanism of Mendelian heredity. Growing many thousands of generations of these banana-fed fruit flies, reared in milk-bottles, Morgan and his associates found hundreds of new characters that appeared by mutation—that is, they appeared suddenly as "sports." This suggested the mechanism of evolution as well as heredity. C. E. McClung in 1902 showed that sex is determined by one of the chromosomes. The X-ray was used on fruit flies by H. J. Muller to induce mutations that became hereditary, and this offered hope of speeding up and perhaps controlling the processes of evolution by artificial means.

To the chemistry of living things Jacques Loeb made many contributions, most startling, perhaps, his production of fatherless frogs by pricking unfertilized eggs with a needle. Loeb's first experiments on artificial parthenogenesis were in 1899, when he activated unfertilized sea-urchin eggs with concentrated sea water, and in 1916 he raised year-old fatherless frogs. His work on tropisms and his considerations of the smallest particle that can show all the phenomena of life were also important.

Experimental methods of biology have been used by many modern investigators, notably F. R. Lillie, E. G. Conklin, G. H. Parker, H. S. Jennings, L. L. Woodruff, William E. Ritter, and many others.

Charles B. Davenport's studies on the heredity of man and the researches in experimental evolution at Cold Spring Harbor laboratory have laid the foundation in this field. Notable also are the population studies of Raymond Pearl and the genetics researches of E. M. East, W. E. Castle, and C. C. Little.

To the farm, biology has brought many gifts. The experiment stations in each state have both spread knowledge and created it. In soil researches America has made notable contributions which have saved farmers much money and added to the record of science. Liberty Hyde Bailey's monumental work on horticulture and cultivated plants, Mark Alfred Carleton's introduction of Russian wheat, George H. Shull's breeding of superior corn, are high lights in agricultural botany. L. R. Jones and his fellow specialists in plant disease made phytopathology a major division of biology. Plants and animals were considered in relation to their surroundings when ecology arose as a mingling of botany, zoölogy, and geography.

Man's war against the insects has been waged vigorously in America. Entomology is given a strong economic urge when insects invade from other lands or native insects change their food habits to ravage the crops. Hand in hand with the so-called pure science of entomology, there has arisen applied or economic entomology. The necessity of defense against insect depredations has encouraged complete knowledge of insects and from out of museums and laboratories have come many vital suggestions to the entomological field forces. Thomas Say is perhaps the best remembered of early American entomologists, while Thaddeus William Harris is considered the founder of applied entomology. Townsend Glover was the United States Department of Agriculture's first entomologist, and Charles Valentine Riley began the federal government's large-scale participation in insect control. Asa Fitch in 1854 became the first entomologist of a state, New York. John Henry Comstock and his wife, Anna B. Comstock, established a focus of entomology at Cornell that has had wide influence. To L. O. Howard, long the leader of the Government's insect-fighters, economic entomology in America probably owes more than it does to any one other person. Howard has been in the thick of every conflict with insects in the last fifty years. His influence has extended to all countries of the world, and his work on medical entomology alone would assure him a place in history. The rôle that insects play in the transmission of disease has been discovered in the past few decades, and the public now coöperates in controlling disease-carrying insects, of which the house fly is not the most important but the most obvious. There will be other uprisings or invasions like those of the Hessian fly, the gypsy moth, the cotton-boll weevil, the corn-borer, and the Mediterranean fruit fly. With poison, with man's insect allies, the parasites, the entomologists will continue to wage their defensive battles.

GEOLOGY

To the understanding of the earth's crust and its behavior America has made major contributions. Geologists in the New World had a type continent to study. In its layered earth, its mountains, its valleys, and its plateaus the forces of nature had demonstrated their capabilities. William Maclure was the first important serious geological student of the American landscape in the early days. His earliest attempt at an American geological map published in 1809 is classic.

The nineteenth century saw a violent struggle between two ideas about the earth, and this was a sort of preliminary to the furor raised by Darwin's theory of evolution. The effect of religious orthodoxy was strong upon geology. Believing literally the biblical story of the Deluge, scientists, who otherwise made clever and correct deductions from the evidence of their investigations, credited the configurations of the earth's face to this legendary inundation. The catastrophic school of geology died hard. Hutton and Lyell in Great Britain contended that the forces acting upon the earth in the past were much as they are today. American geologists, notably James Dwight Dana, through their field work and deductions developed the natural conception of earth forces into a consistent theory of mountain-building and erosion that found universal acceptance. John Wesley Powell, the organizer of the United States Geological Survey, and J. S. Newberry in their surveys of the West piled up evidence. Newberry recognized that the Grand Canyon of the Colorado was the work of the river within it, Powell showed how erosion could level land down to a nearly featureless plain. Clarence E. Dutton, the brothers H. D. and W. G. Rogers, and Joseph LeConte also worked on the physics and structure of mountain-building, while William Morris Davis demonstrated the methods of formation of land forms. G. K. Gilbert's studies in the Henry Mountains of Utah showed the intrusion of igneous matter in the formation of mountains. Recognition of the part that the Ice Ages and their great continental glaciers played came to America with Louis Agassiz, and T. C. Chamberlin of earth-origin fame initiated the recognition of more than one Ice Age.

The mapping of the geography, geology, and mineral resources of this wide-flung country began with the expeditions into the new lands of the West which metamorphosed into the Geological Survey. The mapping of the seacoast and the furnishing of a precise geodetic skeleton for the other surveys has been the work of the Coast and Geodetic Survey of which Alexander Dallas Bache was first superintendent.

Dana was America's pioneer investigator of volcanoes when he studied the Hawaiian group as geologist of the Wilkes expedition, and volcanology has continued to interest scientists as well as laymen. The science of seismology in America began with the Charleston earthquake.

The fossil record of the rocks is as much a part of biology as of geology, and Darwin's theory gave an impetus to paleontology in America as in Europe. Footprints and bones of prehistoric monsters had been found in America's early days. Thomas Jefferson studied fossil bones in the White House. But the important work in American vertebrate paleontology began about 1870 when Joseph Leidy, E. D. Cope, and Othniel Charles Marsh unearthed thousands of animals, birds, and reptiles from Tertiary, Cretaceous, and Jurassic layers of

THE ADVANCEMENT OF NATURAL SCIENCE

the earth and clinched the story of evolution. Cope showed the evolution of the camels and Marsh traced the history of the horse. In their footsteps have traveled more recent students of ancient life, including W. B. Scott, Henry Fairfield Osborn, John C. Merriam, and O. P. Hay. Less spectacular but as important has been the work of the invertebrate paleontologists in building up the record of the continuity of life through the geological ages. All parts of the world have been hunting-grounds for fossils. American geologists in the last decade have uncovered the past life of Asia and linked it with America.

EXPLORATION AND EXCAVATION

Americans have always been explorers and many have gone to strange places not for excitement alone, but for science. The discovery of the Antarctic continent in 1840 was the high point of the United States Exploring Expedition under Lieutenant Charles Wilkes, U. S. N. Lieutenant Matthew Fontaine Maury's oceanographic exploring and his *Physical Geography of the Sea* gave foundations for ocean meteorology in the middle of the last century. Other oceanographic ventures, largely biological, took place on the *Blake, Albatross, Fish Hawk,* and other lesser ships. The non-magnetic *Carnegie* sailed all seas, charting the electrical inconstants of the earth. Admiral Robert E. Peary's discovery of the North Pole climaxed the many American ventures into the Arctic, and aërial exploration as practiced by Admiral Richard E. Byrd and others begins a new period of polar exploration. To Africa, South America, Asia, and the islands of many oceans American scientists of many interests have traveled adventurously to add new knowledge to the store of the world.

American exploits in archeology are just a little over a hundred years old. It was one of the signers of the Declaration of Independence who became the first American archeologist, John I. Middleton of South Carolina. His book, *Grecian Remains in Italy,* appeared in 1812. Since that time Americans have taken their part in the rediscovery of the ancient classical world. They have excavated at the famous scenes of Cretan civilization, at Corinth, at Pisidian Antioch, Olynthus, at Eretria, the Argive Heræum, Carthage, and many other sites. An American archeologist found the sarcophagus of Egypt's feminist Queen Hatshepsut. The tombs of the kings of Ethiopia were discovered by a Harvard expedition. In the lands of biblical association, American expeditions have helped to unearth remains of the biblical world. They found Kirjath-Sepher, and explored its eleven stages of habitation. They worked at Samaria, and at Ur of the Chaldees, at Kish and Beth-Shan, all names famous in biblical events. Going across the world to Mongolia, American Museum expeditions have sought the cradle of the human race in the Gobi Desert and have found artifacts and other indications of some chapter of man's very early history.

America has done its share toward helping the Old World to find its ancient history. But the most concentrated and consistent efforts of American archeologists in general have been directed toward the rediscovery of America. From seeking the oldest Eskimos of Alaska—and finding some that are certainly very old—to hunting lost cities in the tropical jungles of Yucatan and digging up the mummified bodies of prehistoric Peruvians, American expeditions have scanned the New World with thoroughness, though their leaders would still tell you that they have only begun working.

The rediscovery of ancient America, in the scientific spirit, is sometimes dated from the travels of J. L. Stephens. When he visited Yucatan, about 1840, Stephens found there broken sculptures which he deemed amazingly beautiful, and worthy to be compared with the art of Greece. Even before that the Indian mounds of the Mississippi Valley had been given some attention as ancient monuments. It is hard to say exactly when American archeology began, for there were many early explorers and travelers and some who examined relics in a critical spirit, but the real science of constructing ancient history from the discovered objects was still scarcely imagined.

The first systematic effort to conduct archeological researches in America is accredited to A. F. Bandelier, who began his work in 1880. He was the first of the long line of archeologists who have discovered the Southwest, setting the prehistoric inhabitants in order in a racial line: Basket Maker I, II, and III, Pueblo I, II, III, IV, and V, which covers the course of civilization in the Southwest from perhaps one or two thousand years before Christ down to historic times. The construction of a calendar by means of tree rings, so that the prehistoric Pueblos might be dated exactly is one of the triumphs of American archeology.

In the Mayan and Mexican country, both Mexican and United States archeologists and British archeologists, too, have made important "rediscoveries." The ancient history of the Mayas, from their archaic beginnings down through their Old and New Empires, will surely be understood eventually. Already we know how their cities looked and can say with some certainty when different cities and tribes flourished. Interpreting the Mayan hieroglyphics and studying their system of recording time have been valuable aids to understanding the remarkable Mayas.

One of the chief puzzles of America's past has been to identify her first inhabitants. How long ago those first Americans came, and what they were like, are questions still not satisfactorily answered. But M. R. Harrington brought science a long step nearer a solution when he explored Gypsum Cave in Nevada and found traces of camp fires and human artifacts associated with remains of the old ground sloths. These sloths were among the animals of the American Pleistocene. They may have lingered on earth somewhat longer than has been supposed. But, even so, to find that man lived in America in the days of the ground sloths indicates a measure of antiquity that has not been accorded him by some scientists heretofore.

The latest feats in the rediscovery of America are being achieved from the air. A reconnaissance flight over the Mayan country and the Southwest by Charles A. Lindbergh for the Carnegie Institution of Washington demonstrated the usefulness of planes to American explorations. Since then, Peruvian cities have been mapped from the air, and new ruins discovered; Pueblo canal systems, almost obliterated, have been spied out by the eye of the airplane camera.

ANTHROPOLOGY AND PSYCHOLOGY

The past of mankind can be studied in the primitive peoples of today as effectively as in the ruins of the houses and temples of their ancestors. And the minds and physiques of peoples other than our own help us to understand ourselves. Ethnology, anthropology, and their related sociological categories have developed in the past few decades. John Wesley Powell, the geologist, spent the last years of his life pioneering in American Indian studies. W. H. Holmes and J. Walter Fewkes were other leaders in the government's ethnological work. Franz Boas, Ales Hrdlicka, and George Grant MacCurdy have made a reputation for physical anthropology in America.

The sciences of human behavior, psychology, and psychiatry, have their roots in religion, philosophy, and medicine. Only in the past few decades have they assumed their separate entities. It is but natural that the abnormal should receive preferred attention. Dr. Benjamin Rush of Philadelphia, just after the American Revolution, led the way toward the sane treatment of the insane, and the twentieth century saw the mental hygiene movement rise to the same purpose largely through the inspiration of Clifford W. Beers, "the mind that found itself." Psychiatry is now an established branch of medicine. Psychology did not secure its divorce from philosophy until about 1890, when the experimental method successfully invaded the field and chairs and laboratories of psychology were established in universities. Before that time psychology was in charge of philosophers, like the great William James, G. Stanley Hall, and George Trumbull Ladd, American pioneers in psychology.

The trend of American workers in psychology has been toward objective rather than subjective methods. J. McKeen Cattell, first professor of psychology, was the first to "publish psychological measurements of individual differences without

THE ADVANCEMENT OF NATURAL SCIENCE

regard to introspection." Physiology was blended with psychology when animal behavior, long of interest to biologists, was introduced into the psychological laboratory largely through the mazes or "puzzle boxes" of E. L. Thorndike. S. I. Franz and K. S. Lashley went further and combined testing of animals with physiological experiments on their brains. From testing the intelligence of animals it was but a step to making similar tests for human beings, and Robert M. Yerkes' army mental tests met the emergency need of personnel selection during the World War and began an era of tests for school children. Lewis M. Terman's studies of genius have further developed these techniques. In these applications of psychology to practical problems of industry and education there is much hope for the future, and the science of management, such as the Taylor system, will probably be absorbed eventually into psychology.

Schools of psychology have arisen in the past few years, all protests against the "association psychology" of the previous centuries, and all at theories' points with one another. There is the introspective psychology of Edward Bradford Titchener, the behavioristic school of John B. Watson, the purposivism of William McDougall, while foreign-born psychoanalysis and *Gestalt* psychology have their American disciples.

Morton Prince with his studies of hypnosis and multiple personality was a forerunner of Freud, and the psychoanalytic technique has been used effectively in treatment of the mentally ill by a group that has had the leadership of William A. White.

ORGANIZATION OF SCIENCE

The growth of organized research has been an outstanding phenomenon in the last hundred years of science in America. National Research Council statistics show that there are at least 2,000 research laboratories in this country at the present time, and that there are some 800 societies of national character promoting research. The scientific literature continues to grow increasingly, and even the admirable *Chemical Ab-*

stracts, Biological Abstracts and Engineering Index hardly can keep pace with the outflow of published research.

The Smithsonian Institution is the pioneer general institution for the diffusion and creation of knowledge, and the thirty years of existence of the wide-flung Carnegie Institution of Washington has demonstrated the possibilities of research adequately supported. In special fields there are such great laboratories as the Rockefeller Institute for Medical Research and numerous and important institutes connected with universities. Great museums, such as the American Museum of Natural History and the Field Museum of Natural History, are no mere storehouses or exhibition halls for specimens, but their curators are in active service in natural-history research and their expeditions dot the globe.

The American Association for the Advancement of Science, with its 20,000 members, was born in 1848 of more specialized societies, and the American Chemical Society, except for the American Medical Association, largest of the specialized societies, rivals the A. A. A. S. in membership. President Lincoln during the Civil War called the National Academy of Sciences into existence and it has functioned as the supreme conference of the nation's science. Its National Research Council grew out of the World War and now acts to promote peace-time research.

As important as creative research itself is the growing public appreciation of the value and methods of science. Obscurantism continues to flourish in dark corners, occasionally coming into the light that is deadly to it. But newspapers, magazines, books, and radio are delivering the facts, philosophy, and possibilities of science to the people. With adequate information and education the torch of knowledge will continue to burn.

The fruits of creative science are bountiful and its practitioners are relatively few. There are many more milkmen than scientists. As the years pass, may there be more scientists in our population and more of the scientist in all of us.

CHAPTER TWELVE

MEDICINE

"In Shakespeare's time the 50's were venerable: 'Old John of Gaunt, time-honored Lancaster,' was 58 when supposedly so addressed; and Admiral Coligny, murdered at 53, is described by his contemporary biographer as a very old man. Now, when we hear of a death in the 60's, we instinctively feel it an untimely cutting off, in what should be still fresh and vigorous age, and even at 80 it seems but just fair ripeness for the sickle. The three factors which have wrought this change are advanced physical comfort, medicine and its handmaid hygiene and surgery." Sir William Osler's *A Concise History of Medicine*, Baltimore, 1919, page 7.

Medicine

By Fielding H. Garrison

IN 1836, five years after the organization of Cook County, Fort Dearborn was abandoned by the army and Chicago, a tiny village in 1831, had become a thriving city. Just about this time, American medicine had begun to make itself known in Europe through the achievements of Beaumont, McDowell, and other enterprising physicians in the backwoods. When Daniel Brainard secured a charter for the Rush Medical College in 1837, medicine in Chicago was in a fair way to become a going concern.

THE BEGINNINGS OF AMERICAN MEDICINE

Long before this event, the two main directives in the evolution of American medicine had already come into play, namely, the necessity for self-development out of their own resources among the pioneer settlers of the West and the more conservative tendency of physicians on the Atlantic seaboard to go to school to Europe. During the century following the settlement of Jamestown (1609), our solitary contribution to medical literature was a sheet of paper or broadside by Thomas Thacher (1677), advising the people of New England as to the best way to handle themselves during epidemic smallpox, which played havoc among the settlers in these early days.

The medicine of the pre- and post-Revolutionary periods was the medicine of the eighteenth century, and apart from vaccination, eighteenth-century medicine was mainly descriptive science, just as the painting, music, poetry, and polite literature of the period was mainly pattern work. Our initial impetus toward experimentation in medicine and surgery,

MEDICINE

which has done so much to advance the well-being and civilization of our country and of mankind, was to come from the spirit of the Western pioneers.

Whether on the Atlantic coast or further inland or in the isolated clearings of the ever-receding West, the thin ranks of the early settlers were continually depleted by epidemics of fevers, bowel complaints, diseases of the air passages, and infections at that time indefinable. Three months after the landing of the Pilgrim Fathers at Plymouth (1620), their mortality was 50 per cent. In 1633, smallpox was wiping out the Indian tribes in New England. Syphilis appeared in Boston in 1647, yellow fever in Boston Harbor in 1647, and diphtheria in Roxbury, Massachusetts, in 1659. The European physicians of the seventeenth and eighteenth centuries were keen to delimit undecipherable diseases by accurate observation and careful description, so that they might be readily recognized as such and properly handled. In this important matter, the colonial physicians were by no means backward. We know of some seven judicial autopsies (1674-90) performed to ascertain the cause of death from disease, and our colonial accounts of diphtheria (1736-38), of pleurisy (1736-50), and yellow fever (1745-94) are important historical records. Benjamin Rush, sometime surgeon-general in Washington's army, described cholera infantum (1773) and dengue (1780), and was perhaps one of the first to notice the effect of arsenic on cancer and the cure of diseases by the extraction of decayed teeth. Samuel Bard's essay on diphtheria (1771) was translated by Bretonneau and reprinted in his classic of 1826. Human inoculation against smallpox was practiced in the Colonies by Zabdiel Boylston, two months after its introduction in England (1721). Vaccination was introduced by Benjamin Waterhouse in 1800. The medicinal plants of the Colonies had all been investigated and classified during 1736-1820, culminating in the U. S. Dispensatory (1833). No less than thirty native plants were named after American botanists, of whom eighteen were physicians.

Benjamin Franklin, albeit a layman, employed electricity

in the treatment of paralysis (Franklinism, 1757), invented bifocal spectacle lenses (1784), and wrote on gout (1732), smallpox inoculation (1759), and lead poisoning (1786). In 1751, he founded the Pennsylvania Hospital. Fifteen years later, our first medical school, the Medical Department of the University of Pennsylvania, was founded (1765) by John Morgan and William Shippen, both surgeons-general in the American army. Prior to the Revolution, there were over 3,500 practitioners of medicine in the Colonies, but not more than 400 had acquired medical degrees, and of these, 63 had graduated from Edinburgh during 1758-88. Our first medical book was John Jones' treatise on wounds and fractures (New York, 1775), incidentally a manual of military medicine, which appeared, as if providentially, for the use of Washington's army surgeons. Through the first quarter of the nineteenth century, the medical profession on the Atlantic seaboard, like that of England, continued in the respectable eighteenth-century tradition. There were early, withal classical, accounts of spontaneous bleeding (hæmophilia) by John C. Otto (1803), of cerebro-spinal fever by Danielsson and Mann (1806), of delirium tremens by John Ware (1831) and of neurotic spinal affections by John K. Mitchell (1831).

Long before the introduction of anesthesia (1847) and antisepsis (1867-71), isolated American surgeons had established a record for bold and successful operating, not surpassed by that of any other nation in the same period of time. Successful operations for extra-uterine pregnancy were performed by John Bard (1759). Successful Cæsarean sections were done by Jesse Bennett (1774) and François Prevost (1832). John King of South Carolina again operated for extra-uterine pregnancy, with success, in 1816, and published the first book on the subject in 1818. Ephraim McDowell (1771-1830) of Virginia, in a small Kentucky village, then one of the outposts of civilization, first cut out the diseased ovaries in December, 1809, his patient living for thirty-one years thereafter. By 1820 he had done seven ovariotomies, six of them successful. He made ovariotomy a standardized procedure in surgery and performed twenty-two successful operations for stone. Physick (Philadelphia), the father of American surgery, and Dudley (Kentucky) cut for stone hundreds of times with hardly a failure. John Warren, first professor of anatomy and surgery at Harvard, amputated at the shoulder joint in 1781. His son, John Collins Warren, who founded the Massachusetts General Hospital (1811), excised the hyoid bone (1804) and introduced the plastic operation for fissured palate (1828).

Equally remarkable was the enterprise, skill, and success of these earlier American surgeons in tying the larger arteries and in operations upon the bones and joints, often in advance of similar feats by European surgeons. The pioneers were Wright Post (New York), who tied the femoral, primitive carotid and subclavian arteries with success (1796-1817), and Valentine Mott (New York), who performed no less than 138 ligations of the greater blood-vessels, some of them as thick as a clothes-line or an ocean cable. Walter Brashear (Maryland) was the first American to amputate at the hip joint (1806); Horatio G. James (Pennsylvania) removed an upper jaw in 1820, and David L. Rodgers (New York) both upper jaws in 1824. The interest of these early operations, which make a long list, is not merely in the skill and enterprise of the surgeons concerned, but in the essential fact that most of their patients recovered. The year 1831 was remarkable for the passage of an Act by the legislature of Massachusetts permitting the use of bodies of unclaimed persons for dissecting, one year in advance of the Warburton Anatomy Act of England (1832); and for the discovery of chloroform (prior to Liebig and Soubeiran) by Samuel Guthrie, of Sacketts Harbor, New York.

At the beginning of the century we are about to consider, our first books on surgery (John Jones, 1775), materia medica (J. D. Schoepf, 1787), dentistry (Richard C. Skinner, 1801), obstetrics (Samuel Bard, 1807), anatomy (Caspar Wistar, 1811-14), insanity (Benjamin Rush, 1812), gynecology (John King, 1818), diseases of the eye (George Frick, 1824), and medical jurisprudence (T. R. Beck, 1823) had already appeared.

ADVANCES IN PHYSIOLOGY

The three fundamental disciplines of medicine are anatomy, or the structure of the body; physiology, or "animated anatomy," which shows how the body functions; and pathology, or altered physiology, which considers the changes brought about in the structure and functioning of the body by injury and disease. Our start in physiology was made by John R. Young (Maryland), who, by experiments on the frog, showed for the first time that digestion in the stomach is effected by an acid in its secretions (1803).

But an investigation of far greater moment, which gave American physiology its place in European text-books, was to be made at an isolated army post in the wilds of Michigan. On June 6, 1822, William Beaumont, an army surgeon stationed on the frontier at Fort Mackinac, Michigan, was called to attend Alexis St. Martin, a young Canadian *voyageur*, who had sustained an extensive gunshot wound in the abdomen, penetrating the stomach. In spite of Beaumont's efforts, the wound in the stomach could not be made to close, and as the patient recovered, a permanent gastric fistula developed.

Seeing the possibilities of this unintentional bit of physiological surgery, Beaumont took his pauperized patient into his home for an indefinite period of treatment, and kept in touch with him for twelve years running (1822-34).

In 1825, Beaumont published his first observations and experiments on the process of gastric digestion, as seen with the naked eye through St. Martin's fistula, which were summarized in his famous book (Plattsburg, N. Y., 1833), our first important medical classic. Beaumont saw that digestion in the stomach is essentially a chemical process; that the food in the stomach is dissolved by what Young assumed to be phosphoric acid, and what Prout, in 1824, had shown to be hydrochloric acid; that the gastric juice is poured out only in the presence of food, accompanied by rhythmic contractions of

MEDICINE

the walls of the stomach; and that, contrary to the view expressed by Magendie, the flow is neither continual nor continuous, but profoundly affected by mental or emotional disturbance and determined as to quantity by the amount and nature of the food ingested.

Beaumont's experiments upon the effect of the gastric juice upon different kinds of food and his estimates of the digestive and nutritive values of such foods are the basis of our modern dietetic scales and tables. His chemical analysis of the gastric juice showed it to be made up of hydrochloric acid plus some unknown substance which Schwann, in 1835, proved to be pepsin. Thus, Beaumont not only obliterated the fantastic notions of the older physiologists, who regarded the stomach variously as "a mill, a fermenting vat, or a stew-pan," but really said all that was to be said about gastric digestion. He even anticipated the psychic secretions of Pavloff, discovered over a half a century later. As Dr. Joseph Collins observed of Edgar Poe, Beaumont "put us on the map" in the eyes of Europe. But no other American work of consequence was to be done in this field until after the Civil War.

In 1834, a start in pathology was made by William E. Horner, a Virginian anatomist attached to the University of Pennsylvania, who showed that the rice-water discharges in Asiatic cholera consist of epithelium stripped from the small intestine, and was author of the first text-book on pathology (1829). This was followed ten years later by the larger pioneer treatise of Samuel D. Gross (1839). But after these modest beginnings there was again long silence in the records. American endeavor in the three basic disciplines of medicine had to wait upon the development of worth-while medical schools and of laboratory facilities attached to universities.

ORIGINS OF MEDICAL EDUCATION

Medical education in the United States started from certain important centers, each of them associated with the names of prominent or eminent physicians who were prime-movers in these foundations. Minot has, in fact, likened our earlier developments in science to insignificant seed-plots or nurseries laid out by the thrifty husbandman, destined in time to produce very respectable orchards, sometimes scrawny sycamores, sometimes mighty oaks. Minerva medica, in Osler's view, used Philadelphia (1765) as her jumping-off place to flit variously and capriciously to New York (King's College, 1768), Boston (Harvard, 1787), New Haven (Yale, 1810), Fairfield, New York (1812), Lexington, Ky. (Transylvania, 1817), Cincinnati (1819), Charleston, S. C. (1824), Chicago (1837), Ann Arbor (1880), and Baltimore (Johns Hopkins, 1893). President Charles W. Eliot initiated the epoch-making reforms at Harvard (1871), and President Daniel C. Gilman, Billings, Newell Martin, Welch, and Osler the new departure at the Johns Hopkins Hospital (1889-93). Nathan Smith was not only a prime mover at Dartmouth (1798), Yale (1810), Transylvania (1817), and Bowdoin (1820), but, at Dartmouth, after the fashion of the day, held down "a whole settee of professorships," *viz.*, anatomy, surgery, chemistry, and practice of medicine.

Medical students migrated from city to city to sit at the feet of remarkable men, and as stated, there was an established tendency in the East to seek the great European centers and to utilize talented incoming Europeans, while Western medicine was built up originally by Eastern pioneer immigrants, with a coastward migration of native sons to complete their medical education at Eastern schools. The College of Physicians and Surgeons at Fairfield, New York (1812-40), in its prime the second largest medical school in the country, had a total roster of 3,123 students and 589 graduates, some of whom were founders of the five leading medical schools of Chicago.

The most dramatic and outstanding figure in these developments was Daniel Drake (1785-1852) of New Jersey, who was the greatest physician of the West in his period. The son of an illiterate pioneer in Kentucky, schooled to privation and hardship, Drake got his medical education from William Goforth, who introduced vaccination in the West. The only diploma Drake had was a simple certificate of his competency to practice, made out in his preceptor's handwriting, the first document of the kind to be issued west of the Alleghenies. In 1815 Drake completed his medical education with an M. D. from the University of Pennsylvania. He was a man of restless, dissatisfied, itinerant nature, who taught at seven different places in his life and created medical education in Cincinnati by his foundations of the Medical College of Ohio (1821) and the Medical Department of Cincinnati College (1835). He started the first medical periodical of consequence in the West (1827-32), was author of an important study of medical education (1832), described the cholera epidemic of 1832, and was one of the first to describe the local disease known as the trembles or milk sickness (1841).

CONTRIBUTIONS TO MEDICAL KNOWLEDGE

For thirty years Drake journeyed all over the Middle West to assemble materials for his greatest work, his two-volume monograph on *The Diseases of the Interior Valley of North America* (1850-54). This was the first attempt ever made to triangulate such a vast area as to topography, climate, meteorology, natural history, population, and the diseases peculiar to it. As an exhaustive study of the anomalous fevers infesting the Mississippi Valley, it is in the tradition of Hippocrates and Sydenham, and preluded the social surveys and sanitary devices whereby the ravages of dangerous communicable diseases are now aborted. Highly commended by European authorities, it was, in effect, our most important contribution to medical literature after Beaumont's study of gastric digestion (1833).

Our most important contribution to clinical medicine before the Civil War was unquestionably the definite establishment of the diagnostic differences between typhus and typhoid fevers by William W. Gerhard (1809-72) of Philadelphia, in 1837. For centuries, these two diseases had been confused, even as leprosy, syphilis, and certain skin eruptions had been confused in antiquity and the Middle Ages. Gerhard's Parisian preceptor, Louis, had already described and named typhoid fever (1829), and the characteristic intestinal lesions had been seen and understood by Bretonneau as early as 1820; but English and American physicians saw them as equally characteristic of typhus fever, which was not prevalent in Paris in the days of the great French bedside teachers. In the Philadelphia typhus epidemic of 1836, Gerhard was able to delimit its distinctive features and was thus the first to distinguish it clearly from typhoid. With the aid of his able assistant, C. W. Pennock, Gerhard made careful studies of smallpox (1832), tuberculous meningitis (1833-34), and pneumonia (1834) in children, the most important American contributions to pediatrics before the time of Smith and Jacobi.

As colonial American medicine was formed largely on the Edinburgh tradition, so, now, medical graduates in the East usually got their training at Paris, in the clinics of Chomel, Louis, and Laënnec. Next to Gerhard, the most important pupil of Louis was Oliver Wendell Holmes (1809-84), professor of anatomy at the Harvard Medical School, who made his mark and stirred up violent opposition by pointing out the danger of transference of puerperal fever to women in childbed by attending physicians, who had become carriers of the disease by hands unwashed after dissections, post-mortem sections, or conduct of infected labor cases (1843-55). Here Holmes was a forerunner of Semmelweis, who, in 1847-49, established the infectious nature of this species of blood-poisoning for all time. To Louis came also the two James Jacksons, the elder of whom first described alcoholic neuritis (1822), and the younger the prolonged sound of expiration in beginning pulmonary consumption (1833). Henry Ingersoll Bowditch (Boston), another pupil of Louis, standardized the practice of draining off the effusion in pleurisy by suction, which he learned from Morrill Wyman (who perfected the technique and tapped his first patients on February 23-April 17, 1850). The administration of opium in peritonitis (1855) was introduced by Alonzo Clark of the Fairfield, New York, College.

MEDICINE

ANESTHESIA AND SURGERY

The most important achievement in American surgery before the Civil War was the introduction of ether anesthesia (1842-47). In 1800, Sir Humphry Davy, in experimenting upon himself with laughing gas (nitrous oxide), noted the possibility of its use in minor surgery. The same conviction dawned upon Dr. Crawford W. Long of Georgia, in watching the frolics of young people inhaling ether, which Faraday had already shown to produce loss of consciousness and abolition of pain (1818). On March 30, 1842, Long removed a small tumor from the neck of a patient under ether, on June 6th, another tumor from the same patient, and on July 3rd, a toe from a negro boy, in each case with complete success. He continued to use ether in his surgical practice, and even employed it in labor cases, but unfortunately did not publish his results, and so could exert no decisive effect upon its introduction to the world at large.

This task fell to William T. G. Morton, a Boston dentist, who had experimented with the drug in the extraction of teeth and persuaded Dr. John Collins Warren to employ it in the removal of an extensive tumor around the jaw and neck at the Massachusetts General Hospital on October 16, 1846. The operation was a complete success, and at its close Warren was moved to say, "Gentlemen this is no humbug." Another trial, equally successful, was made the next day, and the results were published to the world by Henry J. Bigelow on November 18, 1846.

The high character and reputation of these surgeons was sufficient to vouch for the new method, which was taken up all over Europe with astonishing rapidity. On December 19, 1846, Bigelow operated under ether in London, where Liston took it up on December 21st. Syme followed in Edinburgh in 1847, and in the same year, Pirogoff (Russia) made ether anesthesia possible for the coming Crimean War. Meanwhile, Simpson (Edinburgh) had tried ether in labor cases, but substituted chloroform on November 4, 1847. Thus, within a year's time, the two major anesthetics employed in surgical and obstetrical practice were already in general use. The days of slap-dash, sleight-of-hand surgery and dread of maternity were over, the days of painless labor, painless experimentation on laboratory animals, and careful deliberate surgery were at hand. Ether and chloroform signalized, in Weir Mitchell's phrase, "the death of pain."

The effects of anesthesia upon American surgery were immediately apparent. The bold operating of the past, now slower and less spectacular, was carried into regions seldom, if ever, approached by the surgeon. In 1850, William Detmold (New York) opened the brain for abscess. Bigelow (1852) and Sayre (1855) cut into the hip joint. In 1858, Joseph Pancoast operated upon the bladder and Daniel Ayres upon the female bladder. Erastus B. Walcott cut out the kidney for the first time in 1861, and John S. Bobbs was the first surgeon to open the gall-bladder for gallstones (1868). Hunter McGuire tied the aorta (1868), and in 1864 Andrew Woods Smyth (New Orleans) did the astonishing feat of ligating the innominate, common carotid, and right vertebral arteries for aneurism and exhibited his patient alive in 1869. Without the boon of anesthesia, the sufferings of thousands of soldiers operated on in the Civil War would have been manifold, but its greatest immediate service was to be in the development of safe operating upon the organs of the female body.

This branch of surgery began in the South, and was to be developed almost entirely by Southern surgeons, sometimes as a set-off to the bungling of backwoods obstetrics. One of the worst sequels of faulty delivery was vesico-vaginal fistula, which entailed enormous suffering in child-bearing women and had baffled surgeons for centuries. In 1845, James Marion Sims (1813-83) of South Carolina had on his hands several hopeless cases of this kind, which he finally relieved, after many bitter disappointments, by operating on his patient in the knee-chest position, dilating the parts with a bent spoon handle, and using silver wire sutures. His first success came in the thirtieth operation on his patient in May-June, 1847. He published his results in 1852, moved to New York (1853), and from that time on was recognized as the true founder of operative gynecology, performing operations on all parts of the female generative tract. A decade spent in Europe (1862-72) was an uninterrupted series of triumphs. His book on uterine surgery (1865) was translated into German. A statue in his memory was erected in Bryant Park, New York, in 1894.

Sims engendered the remarkable school of Southern gynecologists who were so active through the nineteenth century, in particular Emmet (Virginia), who excelled in plastic operations; Bozeman (Alabama); Gaillard Thomas (South Carolina), who wrote the best book on diseases of women in his time (1868); and Robert Battey (Georgia), who introduced the idea of cutting out the appendages of the uterus for neurotic conditions (1872). These were the palmy days of gynecology, in which surgeons vied with one another in removing ovaries and uteri by the thousands; but as women came to be more robust from athletic habits, the craze for wholesale operating died away and operative gynecology is now often merged into general surgical practice.

Since 1842, efforts to improve surgical anesthesia have become an established tradition in American surgery. Nitrous oxide, first used in dentistry by Wells and Morton (1844), was revived after the suicide of Wells, and, in 1868, was first employed in combination with oxygen by Edmund Andrews (Chicago) to prevent asphyxia in major operations. Chloroform remained the preferred anesthetic in the West up to 1900, when gas-oxygen mixtures were revived, particularly by Crile (Cleveland), who introduced the new idea of aborting shock by a preliminary twilight-sleep injection of morphine and scopolamine. Conduction anesthesia, or blocking the nerves by cocaine injections, was discovered by Halsted (1885), who was already familiar with infiltration anesthesia and spinal anesthesia (Corning, 1885). About 1897-1900, nerve-blocking was introduced independently by Crile, Cushing, and Matas and became established procedure after the discovery of novocaine (1905). Gwathmey substituted anesthesia by rectal and subcutaneous injections for inhalation (1913-22) and Crile has since brought shockless surgery to the limit of perfection.

NEW ADVANCES IN MEDICAL EDUCATION

Shortly after the middle of the nineteenth century, German medicine gained an unquestioned ascendancy, and medical students, who had formerly sought the Parisian clinics, flocked to German laboratories. In the year of President Eliot's famous reforms at Harvard, Henry Pickering Bowditch, a pupil of Ludwig, became professor of physiology and founded the first labratory of experimental physiology in the United States (1871). A little later, William H. Welch, a pupil of Cohnheim and Koch, started work in experimental pathology at Bellevue Hospital (1879) and the Johns Hopkins University (1884), at which institutions, William H. Halsted, also German-trained, became the prime mover in experimental surgery.

To the Johns Hopkins University came, in 1876, one of Sir Michael Foster's pupils, Henry Newell Martin, who was destined to train many physiologists of the East in the Cambridge traditions; and to the Johns Hopkins Hospital (opened 1889), William Osler, who, as physician-in-chief, instituted the English plan of bedside training of students as clinical clerks and surgical dressers, with the coöperation of laboratory and clinic, as in Germany. All this made for an entirely new kind of advanced medical training in the East. The Johns Hopkins Hospital was planned by John S. Billings (Indiana), a medical officer of the Army, who blocked out the administrative program of its medical school, particularly as to the interlocking of bedside training with laboratory investigations on the nature and cause of disease.

Billings was also the up-builder of the Surgeon General's Library, now the largest and most important institution of its kind in the world, containing nearly a million medical items and to some extent the seed plant of the 212 medical libraries in the United States. The Index Catalogue of this Library, started by Billings in 1880, contains, in forty-six quarto volumes, the medical literature of the world, arranged in alphabetical order by authors and subjects, from the earliest times down to the date of the last volume. For nearly fifty years, the *Index Medicus*, a periodical started by Billings and Robert Fletcher in 1879, gave the medical profession a classified arrangement of the current literature of medicine, from month to month and year to year. In 1927 this journal was merged with the *Quarterly Cumulative Index* of the American Medical Association, a periodical of similar scope.

The American Medical Association, organized in 1847, became, after its reorganization in 1901, a very powerful agency in the improvement of medical education by the establishment of standards tending to suppress inferior medical schools and worthless medical periodicals, by wholesale warfare upon quackery, patent medicines, diploma mills, and similar frauds, in the improvement of the status of medical societies by reorganization, and in general instruction of the mass of the people through public lectures and progressive publications. The *Journal of the American Medical Association* (Chicago) is probably the most important and influential medical periodical in the world.

During 1878-1900, Billings took charge of the vital statistics of the United States Census for (1880-1900), before which period they were pronounced by the London *Times* to be "worse than worthless." Along with the qualified, well-educated physicians who looked after the well-being of families and communities, there were vast numbers of quacks, irregular practitioners and badly trained graduates of inferior schools, none of whom were competent to diagnose and treat disease in the living or to ascertain the causes of death in the dead. Furthermore, the general health and physical stamina of the people, as evidenced by such documents as Lemuel Shattuck's report to the Massachusetts Sanitary Commission (1850), were so poor and precarious as to be matter for the caricaturist. But now there was to be a trend toward better and more substantial things. In 1876, Billings published a searching critique of the medical literature and institutions of the country, in which American medicine was stripped down to its bare essentials. Nothing was so much needed at the time, nothing else could have been so productive of good in the long run. In the words of O. W. Holmes, the speakers and writers of the period "chewed the juice out of all the superlatives in the language in Fourth of July orations." Windbag oratory, spread-eagle proclamations, and "the passion for the beneficent edict" were the order of the day. Billings' cool douche of common sense dispensed with either boasting or cringing and exerted a lasting effect upon the quality of our medical literature by his insistence upon the fact that medical reasoning is sometimes the reverse of scientific.

BACTERIOLOGY AND CONTROL OF COMMUNICABLE DISEASES

The most telling advances, in fact, were now to be made in surgery and public hygiene, neither of which could ever have amounted to much but for our present knowledge of micro-organisms. In the 'fifties and 'sixties, Pasteur established the essential facts about fermentation and putrefaction, and out of all this came the creation of antiseptic surgery by Lord Lister (1867). In the 'seventies, Koch developed the technique of bacteriology, the science of communicable diseases and wound infections, and steam sterilization, which changed antiseptic into aseptic surgery (1881). Surgeons and sanitarians could now attack their problems with ease, surety, and security. Surgery, in competent hands, is now as reliable as book-keeping. Communicable diseases are controllable.

Laboratory methods of bacteriology were brought to America from Germany by Welch and were developed independently in the army by Surgeon-General George M. Sternberg (1838-1915), who, simultaneously with Pasteur, isolated the germ of croupous pneumonia (1880), who founded our first bacteriological laboratory (1876) and whose books on bacteriology (1892) and disinfection (1900) exerted a profound effect upon progressive sanitation in this country. Koch's determination of the bacilli causing anthrax (1877), tuberculosis (1882), and cholera (1884) created a new order of things in preventive medicine.

In 1873, Edward Livingston Trudeau, a talented physician of New York, became infected with pulmonary tuberculosis, and knowing what Brehmer and Dettweiler had accomplished with open-air treatment in Germany, he spent the summer at Saranac Lake, New York. A winter in Minneapolis left him worse than ever. At the suggestion of Loomis, he spent the next winter in the Adirondacks. Koch's tubercle bacillus (1882) gave him a new idea, that of treating patients at Saranac Lake and checking on the progress of recovery by the microscope. On February 1, 1885, Trudeau opened a Cottage Sanitarium for consumptives. In 1894, the Saranac Laboratory was built. Here he experimented on immunization against tuberculosis in guinea pigs and demonstrated the beneficial effects of fresh air upon tuberculous rabbits. Public sympathy brought in contributions of over $600,000. At his death in 1915, Saranac Lake had become a thriving village and his sanitarium had expanded to 36 buildings, accommodating 150 patients. The success of the Saranac experiment led to the establishment of similar mountain and winter sanitaria at Denver, Asheville (North Carolina) and elsewhere. Trudeau was undoubtedly the initiator and activator of organized warfare against the Great White Plague in this country.

The clinicians of Trudeau's time, such men as the elder Flint, the elder Davis, Loomis and Tyson, wrote very useful text-books on the practice of medicine and diagnosis, but, with the exception of such great diagnosticians as the elder Janeway (New York) and Da Costa (Philadelphia), who first described the irritable heart in soldiers (1862-71) so prominent in the World War, none of them accomplished anything equal to the findings of Gerhard or those of such scientific clinicians as Osler, Barker, Libman, or Herrick.

More important things came from the pathologists and bacteriologists. Horner and Gross were the precursors, Welch, Prudden and Delafield the founders of pathological research in this country. In 1872-99, Welch investigated acute œdema of the lungs, embolism and thrombosis. In 1892, he discovered the micro-organism on the skin which infects the rim of open wounds and the gas bacillus (*bacillus Welchii*), with the group of diseases allocated to it. In 1875, William Pepper, of Philadelphia, described the changes in the bone-marrow resulting from pernicious anaemia. Appendicitis was first delimited as an isolated, perforating inflammation of the vermiform appendix of the intestine by Reginald H. Fitz (1886), of Boston, who was also the first to describe haemorrhagic inflammation of the pancreas, associated with fat necrosis (1889). Before Fitz localized appendicitis in the appendix, it had been confused by European pathologists with an inflammation around the blind alley of the gut from which the appendix protrudes. In 1889, Charles McBurney of New York located the painful point above the inflamed appendix which delimits the disease, and devised the scheme of operative treatment. It is no exaggeration to say that operative control of this otherwise dangerous disease was made possible by an American pathologist and an American surgeon.

The science of children's diseases (pediatrics) was developed in the United States by one of its greatest masters, Abraham Jacobi (1830-1919), who came to New York, an exile from Germany (1853), established a pediatric clinic (1860), founded the *American Journal of Obstetrics* (1868), and gave his subject the scientific directives followed by Abt (Chicago), Koplik (New York), and the rest of the younger men. The text-book of Job Lewis Smith (New York), based upon his own experience, passed through ten editions (1869-96).

NEUROLOGY, SURGERY, AND DISEASES OF THE EYE

American neurology was cradled in the Army during the Civil War. In 1864, Weir Mitchell, Moorehouse, and Keen published a study of the effects of gunshot injuries on nerves, based upon a wartime experience at Turner's Lane Hospital, Philadelphia. This was the starting-point of Mitchell's work on ascending neuritis, neurasthenia, the effect of the weather upon amputation stumps, and other phases of nervous disorder. He described several new diseases, devised the rest-cure (1877), and became the leading neurologist in the United States. Before the late 'nineties, the American county insane asylum was too often a cross between a jail and a poorhouse, and sometimes a hotbed of political corruption and vice. The sane and humane treatment of the insane, as sick patients in hospital, was made possible through the reforms initiated by Dorothea Lynde Dix and Weir Mitchell in the East and by S. V. Clevenger and Governor Altgeld in Chicago. William A. Hammond, Surgeon-General of the Army in 1862-64, wrote the first American text-book on nervous diseases (1871). George M. Beard, who first described neurasthenia (1869), John Call Dalton, Thomas G. Morton, and other American neurologists, were also medical officers during 1861-65.

Wartime experience went to the making of many an American surgeon, notably W. W. Keen of Philadelphia, who was the ablest operator on the brain before the time of Cushing. Meanwhile, experimental surgery of the intestines and viscera, started by Samuel D. Gross (Philadelphia) in 1843, was progressing in Chicago through the work of such men as Nicholas Senn (Switzerland), Christian Fenger (Denmark), and John B. Murphy. The American tradition of skillful surgery of the blood vessels was maintained by Halsted, Murphy, Matas (New Orleans), and by Carrel (New York), who substituted end-to-end suturing for ligation in continuity. The reduction of dislocations by manipulation was improved by William W. Reid (1851-55) and Moses Gunn (1851-58) and the treatment of fractures by such devices as the splints of John T. Hodgen and Nathan R. Smith or Buck's extension (1861), and the books of Frank Hamilton (1860) and Louis R. Stimpson (1899). The plaster-of-Paris jacket of Louis A. Sayre (1877) made an epoch in the treatment of spinal deformities, which has been further advanced by Bradford, Lovett, Goldthwait, and Albee (bone-grafting). George R. Fowler did the first plastic operation on the chest (1893), and George M. Edebohls first stripped off the capsule of the kidneys in Bright's disease (1901).

American books on diseases of the eye had appeared as early as 1824 (George Frick) and 1837 (Squier Littell), and there were hospitals in Philadelphia as early as 1822 and 1834, but Weir Mitchell recalled doctors who "refused to recognize socially a man who devoted himself to the eye alone." This prejudice was soon obliterated by Helmholtz's invention of the ophthalmoscope (1851) and ophthalmometer (1852), which made ophthalmology an exact science. Weir Mitchell (1864), William Thomson (1879), and later George F. Stevens and George M. Gould, showed the effects of refractive errors and muscular imbalance upon the development of nervous symptoms. John E. Weeks discovered the (Koch-Weeks) bacillus of pink-eye (1886). George E. de Schweinitz investigated toxic blindness (1896) and the eye as an index of internal diseases. Hermann Knapp founded the *Archives of Ophthalmology* (1869).

Laryngology in America started with the colonial pamphlets on diphtheria, Physick's invention of the tonsillotome (1828), George Catlin's book on mouth-breathing (1832), Horace Green's innovation of treating diseases of the throat by local applications (1838), and Gross's book on foreign bodies in the air-passages (1854). The high point was the introduction of intubation of the larynx in diphtheria (1885-88) by the self-sacrificing Joseph P. O'Dwyer of New York.

Joseph Leidy was the first to find the trichina parasite (1846), discovered the bacteria in the intestines (1849), made the first experimental transplantation of malignant tumors (1851), noticed that flies may carry wound infection (1861-65), first separated out the parasitic amebæ that cause disease (1879) and found the hookworm in the cat (1886). His textbook on anatomy (1861) was illustrated by his own drawings. Of native-born physiologists, Weir Mitchell (a pupil of Claude Bernard) and John Call Dalton, investigated the functions of the cerebellum (1861-69), Austin Flint, Jr., discovered cholesterin (1862), William H. Howell, a pupil of Newell Martin, investigated the chemistry of coagulation of the blood (1890-1928), and Walter B. Cannon (Harvard) the internal

MEDICINE

secretions as activators of emotion (1914-16). Henry Pickering Bowditch was the first to revive a stopped heart in a chemical fluid (1871); he showed (with Kronecker) that heart-muscle will contract to the fullest extent or not at all (1871), and that nerve cannot be tired out by artificial stimulation (1885), which is the rationale of nerve-blocking in surgery.

THE WAR ON FEVERS AND INFECTIONS

The Spanish-American War (1898) was fought mainly upon the sea, and while the battle casualties were slight, there was a mortality of 3,450 from disease out of a total strength of 235,631. This was due to a break-down in the management of epidemic typhoid fever in the concentration camps and to the incidence of malarial and yellow fevers in Cuba. The subsequent findings of the Dodge Commission put the Army Medical Department in a condition of scientific preparedness for the World War.

For American medicine, the most important outcome was the annihilation of typhoid and yellow fevers as epidemic menaces. Fly transmission of typhoid fever was demonstrated by an Army board (Walter Reed, Victor C. Vaughan, and Edward O. Shakespeare) in 1899. Ten years later, Col. F. F. Russell successfully vaccinated an entire mobilized division of 20,000 men against typhoid on the Texan border (1909), the largest outdoor laboratory experiment of the kind ever made. By 1912 there were only twelve cases and one fatality in the army. Typhoid is now a negligible factor in civil life.

In 1900, an Army board, consisting of Major Walter Reed, James Carroll, Jesse W. Lazear, and Aristide Agramonte, was detailed to study yellow fever in Cuba. That the disease is transmitted by mosquitoes was the view of John C. Nott, of Alabama (1848), and the particular mosquito (*Stegomyia fasciata*) was known to the Cuban physician Carlos Finlay (1881), but he was unable to prove his case. In 1889, Theobald Smith had discovered the parasite causing Texas cattle fever and, with F. L. Kilborne, its development and transmission in a tick (1893). In 1900, Henry R. Carter (Public

Health Service) deduced from the incubation period of yellow fever the necessity of an intermediate host. That yellow fever is conveyed by infected bedding and clothing (fomites) was the commonly accepted view.

At Pinar del Rio (1900) Reed noticed, however, that enlisted men could handle such clothing and bedding with impunity and concluded that there must be a living intermediary. On August 27, 1900, Carroll voluntarily exposed himself to the bite of an infected mosquito, came down with the fever, and recovered. Lazear was accidentally bitten on September 13th and died on September 28th. Reed then established a station at Camp Lazear, near Quemados, for experimentation on yellow fever. In all, twenty-two experimental cases were produced, fourteen by infected mosquito bites, six by injection of infected blood, and two by injection of filtered blood serum, while three enlisted men remained for twenty days in a mosquito-proof hut, in contact with fomites soiled with the excretions of yellow-fever patients, without contracting the disease. Mosquito transmission was thus established by a rigorous scientific proof, fore and aft (1901), and the supposition of a filterable virus was confirmed by Rosenau at Vera Cruz (1903).

The economic consequences of this great discovery were speedily demonstrated by an immense saving of life and money through eradication of yellow fever wherever it existed. In 1901, William C. Gorgas began to destroy mosquitoes and screen yellow-fever patients in Havana, and within three months the city was freed from yellow fever for the first time in 150 years. The subsequent triumphs of screening and mosquito reduction in Havana made the completion of the Canal possible. The New Orleans yellow-fever epidemic of 1905 was promptly checked by the United States Public Health Service and infected areas in Mexico, South America, and Africa are now potentially under sanitary control. The discovery of hookworm infection in Porto Rico by Col. Bailey K. Ashford (1900), its eradication (1901-04), and his subsequent work on the parasitic origin of tropical sprue (1908), threw into relief two other carriers of communicable diseases, *viz.*, the animal parasites and the minute fungi.

Hygiene and sanitation now began to take on a new aspect. But for our present knowledge of bacteria, parasites, insects, and fungi, thousands of us would not be among the ranks of the living today. The first grand-scale demonstrations of this fact were made by the United States Army. In 1898-99, Hoff stamped out smallpox in Porto Rico by wholesale vaccination, and leprosy by isolation of carriers. In 1902-05 Craig proved the existence of human malaria-carriers in Manila, which is now absolutely free from the disease. In 1907 Craig and Ashburn showed that tropical dengue is caused by a filterable virus. Its transmission by the *Aedes* mosquito was confirmed by Siler, Hall, and Hitchens in Manila in 1925. In 1910-11 Vedder proved that emetine is a specific in the treatment of amebic dysentery and was instrumental in the prevention of beri-beri by interdiction of non-husked rice as a steady diet (Phalen, 1909). Transmission of Malta fever by goats was demonstrated in Texas by Gentry and Ferenbaugh (1911), and the relations of its micro-organism (*Brucella*) to that of contagious abortion by Alice Evans of the Public Health Service (1918). That tropical tuberculosis either immunizes the native in infancy or kills him in non-infected areas was shown by George E. Bushnell (1920).

In the United States Public Health Service, Charles W. Stiles discovered the American parasite of hookworm infection (*Necator americanus*, 1902). Its subsequent eradication in our Southern States was largely the work of the International Health Board of the Rockefeller Commission (1909-15). Tick transmission of Rocky Mountain spotted fever was demonstrated by Howard Taylor Ricketts in 1907 and louse-transmission of Mexican typhus (*tabardillo*) by Ricketts and Wilder (1907). In 1909, Anderson and Goldberger (Public Health Service) inoculated typhus from man to monkey and showed that Mexican typhus is identical with the disease discovered by Nathan E. Brill in New York in 1910. Pellagra was shown by Goldberger to be preventable by proper diet (1915-

20) and producible by faulty diet (1925-27). The bacillus causing tularemia was discovered by McCoy and Chapin (1912-14); its transmission to man by insects and rodents was demonstrated by Edward Francis (1919-21). A vaccine against Rocky Mountain spotted fever was prepared by R. R. Spencer (1925). Its affinity with typhus fever is indicated by its wide prevalence and the probability of typhus transmission by fleas (1931). The first rat-proofing of a city against bubonic plague was also accomplished in San Francisco (1907-08) by the Public Health Service, which saved the city, and perhaps the nation, by destroying the rats and ground squirrels carrying the plague-bearing fleas.

The first thorough investigation of the pollution of streams was begun on the Ohio River by Wade H. Frost (1913). In the suppression of water-borne and insect-borne diseases the engineer has ever played a prominent part. The great system of sewage canals of Chicago, authorized in 1821, begun in 1836, and eventually necessitating reversal of the flow of two rivers, will not be completed until 1940. For seventy years or more the people were exposed to typhoid fever by drinking their own sewage. The same thing was true of the water-supply of Washington and Philadelphia before the installation of sand-filters. Manhattan is still to some extent "an island surrounded by sewage." Its water-supply has expanded from the two Croton aqueducts (1837-93) and the Bronx River Conduit (1880-85) to the Catskills (1905-17). Mosquito transmission of malarial fever in the District of Columbia was suppressed largely through obliteration of the marshes of the Potomac Flats by an artificial peninsula of made ground (Hains' Point). There are fewer flies in cities since the automobile put the stable out of business.

INSTITUTIONS OF RESEARCH AND EDUCATION

At the beginning of the twentieth century there were established certain independent scientific institutes destined to render powerful aid to experimental and preventive medicine, notably the Rockefeller Institute for Medical Research, New

York (1901), the John McCormick Institute for Infectious Diseases, Chicago (1902), the Carnegie Institution of Washington (1902). Others, such as the Phipps Institute for Tuberculosis (Philadelphia, 1903), the Phipps Psychiatric Clinic (Baltimore, 1913), and the William H. Wilmer Institute of Ophthalmology (Baltimore, 1930), were attached to universities. The American multimillionaire has proved himself a thorough sportsman in the advancement of medicine.

Under the wise guidance of Simon Flexner, the laboratory experts of the Rockefeller Institute attacked in succession the problems engendered by the epidemic cerebro-spinal meningitis of 1905, with experimental production of the disease in monkeys and the perfection of a serum treatment (Flexner and Jobling); the pandemic infantile paralysis of 1907-08, with control by Flexner's convalescent serum; the Spanish influenza of 1917-23 and the epidemic (lethargic) encephalitis of 1918-19. These findings led to an experimental investigation of epidemics artificially produced in villages of mice by Flexner, Amoss, and Webster (1918-24), which have thrown much light upon the behavior and spread of communicable diseases in mass. Here also Noguchi made his investigations of syphilis, yellow fever, Oroya fever, and trachoma (1904-28). Here Peyton Rous started a new line of cancer research by experimental production of malignant tumors from a filterable virus (1910). Here Jacques Loeb produced fatherless frogs (1916) and made his remarkable experiments in comparative physiology. Here Rufus Cole and his associates typed the pneumonia bacilli (1911-18) and showed that their virulence is due to poisonous sugars within them. Here Swift investigated rheumatic fever and Cohn heart disease, while Van Slyke and Levene devised many important improvements in the technique of biochemistry. In its relation to the control of disease in the future, the Rockefeller Institute goes far toward realizing Lord Bacon's dream of a House of Science. Important work in experimental medicine has also been done by Hektoen and his associates at the John McCormick Institute (Chicago).

In the larger universities, medical teaching and investigation went forward hand in hand up to the European War. Where formerly Western men had come East to complete their medical education, younger men from the West now began to occupy prominent chairs in Eastern medical schools. Through the policies of the American Medical Association, stimulated by Abraham Flexner's stirring reports on medical education (1910-11), higher standards prevailed and inferior schools began to vanish apace. At Baltimore, Osler taught bedside medicine, Mall anatomy, Howard Kelly gynecology, Whitridge Williams obstetrics, and Halsted surgery as never before. Welch gave up his own scientific work to activate pupils, launched the *Journal of Experimental Medicine* (1896), and organized the School of Hygiene (1916) and the Institute of History of Medicine (1929). Experimentation in all branches of medicine came to be the work of ambition everywhere.

A good example of the inevitable transition from observation to experiment is afforded in American work on the development of the embryo. The prime movers in the descriptive phase were Alexander Agassiz and Minot at Harvard, Brooks (Johns Hopkins), and Whitman (University of Chicago). In the 'nineties, under the inspiration of Whitman (1878), description attained its saturation point in cell lineage or the attempt to trace the developed organ back to the particular cell predestined to produce it, or *vice versa*.

Deliberate experimentation on the frog's egg (experimental embryology) was then cultivated with great brilliancy by one of Brooks' pupils, T. H. Morgan (1894-97), who proceeded to experiment on regulation of form and regeneration in tissues (1901), heredity (1913-19), and genetics (1919-26) or the mechanisms of heredity, within the cell, in which he again became a past master. Starting from Mendel's law of the spontaneous or discontinuous origin of species (mutation), Morgan and his associates produced over 200 varieties of fruit fly and discovered that the mechanism of heredity turns upon the behavior of certain small chromatin particles paired off from the dividing cell nucleus, called chromosomes.

The unpaired or accessory chromosomes, discovered by Henking (1890) and Montgomery (1898) were identified as the determinants of sex by C. E. McClung (1902). Through genetics, the origins of such congenital conditions as hemophilia, hay fever, bronchial asthma, and feeble-mindedness became clearer. The random observations and experiments of Luther Burbank and the stockbreeders acquired a mathematical basis which may now be applied to the human stock.

Davenport and the experts of the Cold Spring Harbor station made extensive sociological surveys of the ascendants and descendants of families, particularly of isolated pockets of dissolute people, such as the Jukes, the Kallikaks, and the Dacks. Advanced statistical methods (biometrics) were applied to such problems by Davenport, and with marked ability by Raymond Pearl, who exploded several fallacies inherent in social medicine.

In 1910, R. G. Harrison, a pupil of Brooks, first demonstrated the mode of outgrowth of nerve fibers from the nerve cell in an extra vital culture. Experiments on cell and tissue cultivation were now carried forward apace. The starting-point of cell differentiation and regulation of form in the embryo was discovered by Warren Lewis (1907), while new light was thrown on the conditions of fertilization, rejuvenation and intersexuality by the studies of Corner, Stockard, and others on the sexual cycle in animals.

Another group (Mall, Teacher, Streeter, and others) showed that the stages of development in man and other animals are no longer reducible to the base-line embryology of the frog, the chick, or the rabbit, as taught in the schools. At the Rockefeller Institute, Alexis Carrel, a surgeon by training, applied Harrison's method of tissue cultivation to the preservation of arteries and other parts of the animal body employed in transplantation experiments. Thus he has transplanted the kidney successfully from cat to cat and has even kept the entire visceral tract of an animal alive and functioning in a culture medium. Carrel, Burrows, Fischer, and Drew have applied these methods to the study of cancer. During

1913-28, Maude Slye (Chicago) has illustrated the laws of genetics by successful inbreeding or outbreeding of predisposition to cancer in successive generations of mice. Thus a few modest beginnings in descriptive embryology have expanded to findings of the utmost importance in stock-raising, grafting of plants, eugenics, criminal anthropology, birth control, operative surgery, and the problem of cancer.

RECENT SURGERY

Recent surgery owes most to Röntgen's discovery of the X-rays in 1895. Here again Americans have made a number of innovations of practical importance, notably in such technical improvements as the Coolidge tube (1913) and the Potter-Bucky diaphragm (1913). As if in the footsteps of Beaumont, William B. Cannon (Harvard) first used the X-rays to study the movements of the stomach (1898) and intestines (1902) in digestion, by previous injection of bismuth paste. The latter wrinkle was carried over into surgical diagnosis by Emil Beck of Chicago (1908). The brain was first examined with the X-rays by Walter E. Dandy (1918), and the gall bladder by E. A. Graham and W. H. Cole (1923-24). The acknowledged leader in surgery of the brain since Keen is Harvey Cushing, whose fellows at Harvard have recently celebrated the two thousandth operation performed in his clinic. Cushing, an expert in technical improvements and new modes of operating, has excelled in the experimental production of such conditions as gallstones, valvular diseases of the heart, and sexual infantilism, has elucidated the pathology and surgery of the pituitary gland and has done more than any other to revise the pathology of brain tumors. Halsted improved the operations for cancer of the breast and hernia, did experimental surgery on the intestines, blood vessels, and parathyroid gland, and was the first to evolve a complete technique for the perfect healing of surgical wounds by avoidance of damage to the tissues.

George W. Crile (Cleveland) has been remarkable for minute blood-dissections in operating for cancer, has improved the technique of transfusion of blood in hemorrhage, and through his method of blocking off the brain and nervous system by general and conduction anesthesia, is the creator of shockless surgery. Frazier (Philadelphia) has had unrivaled success in operating for facial neuralgia, Elsberg (New York) has specialized in surgery of the spinal cord, and Charles and James Mayo (Rochester, Minnesota) in surgery of the stomach and intestines. The Mayo Clinic, which started from modest beginnings in 1889, is now the greatest center of organized surgery and post-graduate medical instruction in the world. Through the altruistic financial arrangements, the institution has been made self-perpetuating.

The chances of failure in operating are reduced to a minimum by careful testing of the capacity of the heart, lungs, kidney, and other organs beforehand. Many of these so-called functional tests were introduced by American workers in biochemistry and pharmacology, notably those of Folin and Van Slyke for the blood and the urine, Rowntree, Geraghty, and others for the kidneys and the total eliminative capacity of the liver, and the Meltzer-Lynn method of non-surgical drainage of the bile. At the Mayo Clinic, Mann and Whipple demonstrated that bile may be produced outside the liver and all phases of digestion and its disorders have been intensively investigated.

Elsewhere, purposeful dietetics or nutritional therapy has been advanced by the work of E. B. McCollum, Goldberger, Hess, Evans, and others on the accessory food factors, or vitamins, deficiency of which causes such diseases as scurvy, pellagra and rickets. Experimental pharmacology was promoted at the Johns Hopkins by John J. Abel (Ohio), who first isolated epinephrin, crystallized insulin and amino acids from the blood, and whose methods and discoveries have been carried forward by Rowntree, Reid Hunt, Macht, Loevenhart and other talented pupils. The beneficial effects of iodine in goiter were demonstrated by David Marine and C. H. Lenhart in 1909-13. Insulin fattening of emaciated infants was introduced by McKim Marriott (1924). The regeneration of the blood in anemia by a diet of raw beef liver was demonstrated experimentally by Whipple and Robscheit Robbins (1924) and applied to the treatment of pernicious anemia with great success by Minot and Murphy (1925).

PREVENTIVE MEDICINE

The European War demonstrated the importance of preventive medicine to the world at large by grand-scale control of communicable diseases and salvage of the wounded, through organization and teamwork. Medicine now became socialized through the necessity of caring for the peoples of doctorless rural areas and modern cities and the newer duty of the practitioner to prevent as well as to treat disease. The old family doctor began to give way to the medical center, to the public-health nurse, to the decrease in patients bedridden by communicable diseases, to the practitioner or specialist who looks after the rest in his office, and to industrial and social organization against disease and injury. Patients now gravitate to the towering Medical Center of New York as to the university medical centers of Baltimore, Boston, Philadelphia, or those of Chicago, Rochester, Madison, Ann Arbor, and Iowa City, which gather abundant clinical material from the poor and in which students are taught the ethics, the etiquette, and the fine art of office practice in innumerable little rooms.

The intense interest of the people in social medicine is manifest in the huge gatherings which convene annually to discuss the prevention of cancer, tuberculosis, venereal diseases, alcoholism, insanity, blindness, infant mortality, industrial diseases and accidents, and other phases of community welfare. The great increase in local medical periodicals and medical societies is conditioned by the extent of the country and the vast distances otherwise to be covered. Yet the actual number of physicians is decreasing and the difficulty of adjustment is indicated by the five-year program of the Committee on the Cost of Medical Care (1928). As in present-day Europe and Russia, our gravest problems are to be economic.

MEDICINE 355

As part of the general scheme of social medicine, let us consider the enormous extension of public health administration since the middle of the nineteenth century. The beginnings of all this may be traced to the Marine Hospital Service in the Treasury Department for the care of merchant seamen, established in 1798, reorganized along military lines in 1869-74, vested by Congress with the control of national quarantine in 1893, and transformed into the United States Public Health Service in 1912, now virtually our National Health Department. In 1876, only eleven states and the District of Columbia had boards of health, and by 1901 only ten states had satisfactory systems of registering vital statistics. Now every state in the Union makes some organized efforts to safeguard public health, and the more advanced commonwealths have developed agencies of unfailing efficiency. From year to year there has been improvement in the registration of vital statistics. In the larger municipalities, the burdens assumed by public authorities in the prevention of disease and the care of the sick transcend even the visions of progress yet within the memory of the living.

Commensurate with other large-scale developments is the extension of preventive medicine by Americans to lands beyond the sea. Life is now as safe in the Canal Zone as anywhere. The Army Medical Department and the Public Health Service taught sanitation to the people of the Philippines. The Rockefeller Foundation has attacked hookworm infection and yellow fever all over the world, has given material aid to the organization of public health in Czecho-Slovakia, and is cooperating with the Italian government in the experimental prevention of malaria. These are first steps in the global or internationalized medicine of the future. William Charles Wells, Mme. Dejerine, William F. Dunbar had the honor of being born in the United States, but their accomplishments are part and parcel of English, French, and German medicine. Jacobi, Senn, Fenner, Knapp, Jacques Loeb, Osler, Novy, Noguchi, and other talented physicians of foreign birth have played a significant rôle in the development of American medicine. We may take pride in the fact that the most characteristic achievements of American medicine have been the work of physicians born within our shores, and that its progress has been in keeping with the poet's device—

Bidding the eagles of the West fly on.

CHAPTER THIRTEEN

EDUCATION

"When I first went to school, Webster's Spelling-Book was just supplanting Dilworth's . . . and the only grammar in use was 'The Ladies' Accidence,' by Caleb Bingham—as poor an affair as its name would indicate. Geography was scarcely studied at all; while chemistry, geology, and other departments of natural science had never been heard of in rural school houses. 'Morse's Geography,' which soon came into vogue, was a valuable compend of political and statistical information; but, having barely one map, would scarcely pass for a school geography now. The first book I ever owned was *The Columbian Orator* . . . a medley of dialogues, extracts from orations, from sermons, from speeches in Parliament, in Congress, and at the bar, with two or three versified themes for declamation. . . . The first large work that I ever read consecutively was the Bible, under the guidance of my mother, when I was about five years old."—HORACE GREELEY, *Recollections of a Busy Life,* pp. 45 f.

Education

By Charles H. Judd

THE typical school of the United States in 1830, and for many years afterwards, was a meagerly equipped frontier school. A few of the larger centers of population were able to maintain fairly satisfactory schools, but the small towns and villages settled by the pioneers as they migrated westward across the continent made the best shift they could to provide some education for the children. Schools were held in rough buildings of the simplest kind or even in private houses. The teachers were often men and women ill qualified for the duty of instructing children. The school year was short, seldom more than ten or twelve weeks in length. The average schooling of the ordinary citizen in 1840 is reported as 208 days—somewhat less than the number of days now required to complete the first two grades.

The primitive frontier school had one great virtue, which it passed on to later generations of American schools. It was democratic in spirit and in organization. It did not provide much education, but what it provided was open to the children of all the families in the community—both boys and girls.

In order to appreciate the heritage which the frontier school bequeathed to later times, one must understand how different the early American school was from European schools. In Europe, where society was divided into aristocracy and common people, there had long existed a dual school system. One branch of this dual system admitted only the children of the upper classes. The other branch furnished a very limited education to the children of the common people.

In the oldest American settlements along the Atlantic seaboard there was some tendency to imitate the European dual school system. Boston had very early its Latin school, which was quite distinct from the common school. The Virginians who could afford to do so sent their young people to England to be educated as members of the upper class should be. Some of the early colonies established what were known as "pauper schools."

Social distinctions of the type exhibited in the older parts of the country were, however, not typical. On the frontier, in the remote inland settlements, there was no possibility of a dual school system. There was no disposition to imitate Europe. Communities were small and were battling with the hard conditions of life on the borders of civilization. The kind of school which they developed reflected all the characteristics of their strenuous existence. It recognized no aristocracy and no pauper class.

When frontier towns developed into cities, and cities became wealthy and more populous, their schools increased the number of days of schooling for the average young citizen; they were housed in more commodious structures supplied with more complete equipment; they were conducted by better-trained teachers; they added the opportunities of a free high-school education to those provided by the elementary grades. With all these improvements, they continued to receive on an equal footing boys and girls of all social classes.

The frontier school of a century ago had a very narrow curriculum. It taught the three R's and very little else. It was what is called today an "elementary school," though at that time there were so few high schools that the contrast between elementary schools and high schools was not very much in people's minds. The school of that day was called a "common school" because it was open alike to all the children, or it was called a "district school" because the territory from which it drew its pupils was a political unit known as a "school district."

Such high-school education as was to be had a hundred years ago was provided almost entirely outside the public schools.

In 1840 there were less than twenty-five public high schools in the United States. Even a decade later most pupils who secured an education beyond the elementary level were registered in private and denominational academies. It is estimated that in 1850 there were 6,085 such institutions.

THE RISE OF THE HIGH SCHOOL AND THE COLLEGE

Slowly after the middle of the nineteenth century and with increasing rapidity after the Civil War, the public high school became an important part of the educational system. A crucial legal decision rendered in the city of Kalamazoo, Michigan, in 1873, established the right of the high school to a share in the school funds of the community. In this decision Judge Thomas M. Cooley gave clear expression to the fundamental attitude of the American people favorable to free higher education for all the people. After reviewing the history of Michigan's educational policy, he said:

> If these facts do not demonstrate clearly and conclusively a general state policy, beginning in 1817 and continuing until after the adoption of the present constitution, in the direction of free schools in which education, and at their option the elements of classical education, might be brought within the reach of all the children of the state, then, as it seems to us, nothing can demonstrate it. We might follow the subject further and show that the subsequent legislation has all concurred with this policy, but it would be a waste of time and labor. We content ourselves with the statement that neither in our state policy, in our constitution, nor in our laws, do we find the primary school districts restricted in the branches of knowledge which their officers may cause to be taught, or the grade of instruction that may be given, if their voters consent in regular form to bear the expense and raise the taxes for the purpose.

One cannot read this excerpt from Judge Cooley's decision without realizing that it brings to its natural consummation the movement which began when the American people departed from the traditions of Europe and opened a single school offering to all children whatever education the community could afford to provide.

It was many years after high schools became numerous before people took advantage in any large measure of the opportunities which these schools offered. By 1890 the total number of pupils in high schools had not reached 300,000. It was not until the early years of the twentieth century that the registration in high schools began to increase rapidly. At the present time 53 per cent of the boys and girls of high-school age in the United States, or a total of more than 4,250,000, are attending high school. European countries do not register 10 per cent of their young people in the schools corresponding to American high schools. This statement emphasizes once more the truth of the assertion that American education has been since early frontier days unique in the fact that it is open to all classes in the community. Furthermore, in the United States the high schools admit the sexes on equal terms.

It is not alone in the public high schools that this country offers equal opportunities of higher education to young people of all social groups. The colleges and universities of the United States are more generally attended by all classes of people than are the corresponding institutions of any other country in the world. Here again evolution has been slow. In 1830 colleges were few, and their students were very limited in number. Only young men attended, and most of them were preparing for the three learned professions—the ministry, law, and medicine. The percentages of graduates of the leading colleges of that day entering these three professions were 31, 27, and 11, respectively, only 31 per cent being distributed over all other callings.

The colleges have become so popular that today they register one out of every six persons of college age. Less than 25 per cent of the graduates enter the three learned professions; a large percentage go into engineering, business, and the other vocations which have in late years taken on the character of professions or semi-professions.

It is a long step in the history of education from the primitive school of 1830 to the American system of educational institutions of 1930, including more than 20,000 high schools and more than 900 colleges and junior colleges, enrolling in the aggregate more than 29,000,000 pupils and students, and employing approximately 1,000,000 teachers. Approximately a quarter of the people of the United States are attending schools and colleges, and this country spends annually on its educational system somewhat more than $2,250,000,000.

GENERAL EDUCATIONAL ADMINISTRATION AND LEGISLATION

One important reason for the development of free universal educational opportunities in the United States is to be found in the fact that from the earliest times control of schools has been largely in the hands of representatives of the local community. In 1830 state departments of education did not exist in even the most rudimentary form in most of the states, and there was no federal bureau of education. The first state board of education in continuous existence since its establishment was organized in Massachusetts in 1837. The local school trustee, appointed by the town meeting or chosen by the selectmen, was competent, so far as the law was concerned, to employ as a teacher virtually whomsoever he saw fit. The curriculum of the school was determined by law only so far as the most fundamental subjects were concerned. The curriculum as actually administered depended on the equipment of the teacher and on the approval of the local community. Very often it was limited in scope. School revenues were small. They were derived in some cases from so-called "rate bills"—that is, from payments made by parents having children in school. Otherwise, they were derived from public sources. Whatever the source, they were administered by local authorities.

Major changes have taken place in school administration during the century that has passed, but the principle that schools are under the immediate control of the community continues to be accepted throughout the American educational system.

Here again we come upon a sharp contrast with the educational systems of Europe. Most of the countries of Europe have Ministers of Education, who are part of the central government and are endowed with large powers. European schools

EDUCATION

are under centralized control to a degree never thought of in the United States. A century ago there was no agency whatsoever in the federal government which concerned itself with education. In 1867 Congress created a Department of Education, which was changed in 1868 to a subordinate office of the Department of the Interior. This office has never had the powers possessed by European Ministers of Education. It has been primarily a statistical bureau, reporting on the activities of the school systems of the states.

Congress has from time to time, however, brought the federal government into relations with the public schools by making grants of land or money for the promotion of education in the states. In some cases conditions have been attached to these grants which virtually control their use, but even when direct supervision over expenditures was assumed, as in the laws of 1914 and 1917, the federal government did not attempt to take charge of the general educational system of the country.

The states have been increasingly active since 1837 in supervising and directing the schools in their territories. The example of the state of Massachusetts in creating a state board of education in 1837 was followed in succeeding years by the other states. Today every state in the Union has an officer in charge of education. Many of the states also have boards of education. These state authorities exercise certain powers over the local school districts. Thus, in all the states except Massachusetts power to license persons to teach is vested in the state department of education. In thirty-six states no agency other than the state department can issue teachers' licenses. The reason for control of licensing by state departments is that local districts are too small to undertake the training of teachers. Normal schools have accordingly been organized by the states to supply teachers for local districts. It is natural and proper that the states should exercise the licensing power as a correlate of their training activity.

In addition to creating state departments of education, legislatures have taken steps to enlarge the scope of education in the areas under their jurisdiction. They have increased the requirements of the laws compelling communities to conduct schools. In many states the school year is defined by statute as two hundred days in length. Laws have been enacted in all the states compelling parents to send their children to school. The first compulsory school-attendance law was passed in Massachusetts in 1852. It prescribed that children must attend school between the ages of eight and fourteen for twelve weeks in the year, six of which must be consecutive. Since 1852 all the other states have followed the example of Massachusetts in passing attendance laws. The majority of the states now require children to attend school up to the age of sixteen. Four states require attendance up to eighteen years of age.

Compulsory school-attendance laws were not vigorously enforced until the decade 1880-90. Indeed, there are some parts of the United States where enforcement is lax even today, but, in general, it may be said that the last half of the past century witnessed the establishment of the principle that the state has sufficient interest in its children to compel parents to take advantage of the opportunities offered by the public educational system.

A further respect in which state legislatures have been active is in the prescription of subjects to be taught or not taught in public schools. In general, it may be said that legislation with regard to the curriculum has not been especially intelligent. Small groups of citizens interested in promoting the use of schools for some particular purpose, such as teaching children so-called "patriotism," or in excluding certain doctrines, such as biological evolution, can secure the passage of a law coinciding with their views. Not uncommonly the schools fail to act in accordance with the law, but nothing is done about their failure because people do not take seriously the prescriptions of the legislatures.

The fact that the states have increased their supervision over schools has not changed the fundamental principle referred to in an earlier paragraph. The duty of conducting schools under the law has remained in the hands of local trustees. State

EDUCATION

governments and the federal government contribute support, and the states direct in certain general ways the operations of their educational systems, but local authorities continue to exercise a high degree of autonomy.

LOCAL ADMINISTRATION

The result of local control is that great variations are to be found in the schools of this country. There are obvious advantages in local control for a community which is progressive and willing and able to provide ample funds for the equipment of its schools. But districts lacking financial resources and districts deficient in enterprise sometimes suffer from the large degree of local control which they possess. The United States has all kinds of schools, ranging from those that are excellent to those that are very inferior in quality.

Schools have changed greatly in their internal organization since 1830. At that date the vast majority of the schools of the United States were ungraded, one-room schools, presided over by a single teacher. The single teacher grouped the pupils for purposes of instruction as he or she saw fit. Sometimes, usually not more than once a year, a committee, which included the preacher and the trustees, came to the school to inspect the work. This visit provided the only supervision that there was of the work of the pupils and the teacher.

As communities expanded in size, the number of pupils increased, and it became necessary to organize them into classes. Each class was taught by a teacher who specialized in a single level or grade of instruction. Not only so, but many towns found it necessary to open more than one school. A new need made itself manifest—the need of a school officer whose duty it was to keep all the teachers in the system working in harmony. In the year 1837 the two cities of Buffalo, New York, and Louisville, Kentucky, appointed such a coördinating officer with the title "superintendent of schools." Other cities followed the example of Buffalo and Louisville, and today every city school system and many systems in smaller towns have superintendents. In the larger cities the superintendents have staffs the members of which assist in the supervision of the schools.

With the appearance of school superintendents, a number of problems of school administration have arisen that were unknown in the days when the school had only one teacher. Perhaps the most conspicuous of these problems is that of determining who shall select teachers. The superintendent, as an educational expert and as the officer responsible for the direction of teachers after their appointment, naturally regards it as desirable that his judgment should be of prime importance in the selection of teachers. On the other hand, local trustees and boards of education often hold that, because all contracts for the payment of teachers must be approved by representatives of the public, these representatives should make the selections. Other problems in the solution of which there is often a clash of authority are the choice of textbooks and the determination of plans for school buildings. Wherever financial questions arise, the traditional authority of trustees is likely to assert itself and to claim precedence over the authority of experts who came into the school system long after trustees.

The fact that the relation of the superintendent to the school trustees is uncertain and often difficult of adjustment has led to the adoption in many states of laws defining more or less fully the powers of the expert school officer. The trend is steadily in the direction of conferring on the superintendent control of all administrative details and leaving to lay trustees only the determination of general policies.

THE CURRICULUM

Schools have expanded during the past century not only in the number of pupils registered and in the organization of the instructional staff but also in the range of subjects taught. Nothing determines the character of a school so completely as do the subjects which it teaches. The subjects taught in the public schools in 1830 were of the most rudimentary grade. Reading, which included spelling, was limited in any single year to one reader. Arithmetic was very largely made up of drill on the tables and practice in solving problems by the rule of three. Penmanship was taught in very formal drills, which cultivated little fluency.

The simple curriculum described in the foregoing paragraph has given place to one which includes a wide range of subjects. One by one new subjects of instruction have been added until today the elementary-school curriculum includes numerous informational subjects, such as geography, history, and science. Courses in the manual arts have also been added, with a view to cultivating skill of hand. It was hardly thought necessary a century ago to develop manual skills in school because every child was occupied at home with work that supplied ample opportunity for the training of the hand. The inclusion of manual arts in the school curriculum was a direct response to the new demands arising from city life, where children have little or no opportunity for work at home. Courses in play and in hygiene are among the latest additions to the curriculum. One can well imagine the critical attitude which would be taken toward these new subjects by an observer transported from a school of 1830 to a school of 1930.

A comparison of the curriculums of present-day higher schools with the curriculums of similar schools of earlier times reveals differences even more impressive than those which one finds when one examines the lists of subjects taught in elementary schools. The college and the preparatory school of a century ago administered strictly classical curriculums. The ancient languages, including Hebrew, and the mathematical sciences were the staple contents of the curriculums of higher institutions during the first three-quarters of the nineteenth century. To be sure, some of the academies offered courses outside the classical-mathematical curriculum, but, in general, the statement holds that the staple subjects were the ancient languages and mathematics. Obviously, then, the narrow curriculums of the early colleges and academies had to be greatly extended to satisfy the students who entered institutions of higher education in large numbers during the years following 1880.

The list of subjects reported by the United States Office of Education as given in the high schools of 1890 is limited to nine. The list of subjects offered in high schools today includes vocational subjects, many branches of natural science and social science, modern languages as well as the classics, and numerous other subjects, such as art, physical education, stenography, typewriting, and commercial arithmetic. Altogether, one can find as many as 250 courses in the high schools of the present as contrasted with the nine subjects reported as given in 1890. Evidently, education is adapting itself to popular needs.

What is true of the high schools is equally true of the colleges. They have expanded in many directions. Indeed, there are special colleges for many of the interests which were not represented at all in the educational programs of 1830. For example, there are colleges of commerce, agriculture, engineering, and dentistry. Present-day colleges offer courses that provide for the broadest culture and for the narrowest specialization.

INTRODUCTION OF NEW EDUCATIONAL UNITS

With the enlargement of the curriculum there has come in recent years a breaking down of the sharp distinction between the different divisions of the school system. The elementary school is no longer limited in its curriculum to rudimentary subjects. It includes today some of the subjects which a generation ago were thought of as belonging to the high school. In like manner, subjects formerly taught in the college are now included in the high-school curriculum.

These shifts in the curriculum have been facilitated by the improvements which have been made in methods of teaching and by the lengthening of the school year. It is possible under present conditions to give a more complete elementary education in six years than was formerly given in eight years. The result is that the seventh and eighth grades are detaching

EDUCATION

themselves from the elementary school and are becoming part of the new unit of the school system known as the "junior high school." In like fashion, the upper years of the high school and the lower years of the college are uniting to make the so-called "junior college."

The new units thus appearing in the school system are indications of a new and broader conception of popular education. Opportunities for acquaintance with the higher branches of learning are now being offered to pupils at an earlier age than ever before. Such opportunities would not be offered if people continued to think of a common education as satisfactorily provided in the elementary grades. The older pattern of an eight-year elementary school followed by a high school attended by a few highly selected pupils served well enough in a day when the school furnished only a small part of a child's training and when only a limited education was deemed necessary. Modern civilization demands a new pattern of school organization.

The exact form which will ultimately be taken by the new pattern now in process of development is difficult to foresee. In some school systems the elementary school is made up of six grades; the junior high school, of three grades; the senior high school, of three; and the junior college, of two. This type of organization is described as the 6-3-3-2 plan. In other school systems there are other plans, such as the 6-4-4 and the 6-2-4-2 plans of organization. One fact is clear—whatever the details of the final pattern of organization, it will include some higher education as a part of the schooling of all pupils.

THE TRAINING OF TEACHERS

The teachers of a century ago were entirely innocent of professional training. They were, for the most part, indigent persons who thought of themselves as caretakers rather than as educators. An old soldier who could not perform the labors of an ordinary farm hand, a widow who was obliged to do something to maintain herself, or a young man who had not yet entered upon his profession was employed to drill the children for a few weeks and to keep them in order as best he or she could. The school which was conducted in the summer was attended only by little children because the older boys and girls were engaged in work on the farm. A woman teacher was thought of as equal to the task of conducting this school. In the winter, on the other hand, it usually required a man to keep order in the school. Older boys and girls, released from the duties of the summer, came to school quite as much for the enjoyment of social life as for the purpose of intellectual improvement. Many are the stories of rough encounters in the winter schools when teachers and pupils engaged in physical competition to discover who was to receive chastisement.

There were no special public institutions devoted to the preparation of teachers in 1830. It was popularly assumed that anyone who had reached mature years and had any training whatsoever could give instruction to children. The day was to come when this slight opinion of the art of teaching was to be corrected. As early as 1823 it occurred to one man, Samuel Read Hall, that a special seminary for the training of teachers would improve schools. Accordingly, he opened a private seminary of this type in Concord, Vermont. Some years later Horace Mann and some of his progressive contemporaries, especially Charles Brooks and James G. Carter, persuaded the legislature of Massachusetts to establish a public normal school, the first institution of its kind in America.

This first public normal school, opened in 1839, was located at Lexington and was afterwards moved to Framingham. It was patterned on the model of the teacher-training institutions of Prussia. It had no connection whatsoever with the colleges; it was an institution of secondary grade and followed the methods of teaching common in the elementary schools. Its curriculum consisted of recitations and drill exercises in the three R's, in geography, and in some of the subjects taught in the academies of that day. This curriculum was supplemented by some instruction in the art of teaching.

Normal schools did not develop at all rapidly. It was not until after the Civil War that they began to flourish. The ag-

EDUCATION

gregate amount spent for such schools in the United States in 1870 was $202,000, less than many states now spend on a single institution. The rapid increase in interest in the professional training of teachers in recent years is very impressively attested by the fact that in 1920 the incomes of all teacher-training institutions in the United States amounted to $31,000,000 and in 1928 to $70,000,000. Many of the institutions which were formerly classified as normal schools rank today as teachers' colleges. They have four-year curriculums and grant degrees.

Not only is professional training provided for teachers by greatly enlarged normal schools and teachers' colleges, but liberal-arts colleges and universities very commonly offer, especially since the beginning of the present century, courses for the training of teachers, supervisors, and administrators of all ranks.

SCHOOL FINANCE AND EQUIPMENT

The salary of a teacher in 1830 was not attractive. Usually a part of the teacher's compensation was secured by "boarding around." The salary, such as it was, was frequently collected in part from the families whose children went to school. The time had not yet arrived when schools were to be supported by taxation. Indeed, we read of stubborn resistance in many localities to the imposition of taxes for the maintenance of schools. It was argued by people who had no children in school that it was not just to extract money for the support of schools from property-owners who gained no direct benefit in their families from the services of the school.

The present-day attitude, which is that the education of citizens is essential to the preservation of the state, is the product of a social evolution that has gone on rapidly during the past half-century. It would be quite impossible to maintain the well-equipped schools of the present time and to provide the salaries paid to teachers today if it were not for the participation of the whole community in raising the school revenues. Yet it required the most strenuous endeavor on the part of the friends of school taxes to persuade legislatures to pass laws providing such revenues. During the years from 1840 to 1860 many legislatures imposed school taxes and then repealed the laws which they had passed. One writer, commenting on the struggle for tax-supported schools, compares the political excitement attending the debates over school taxes to the partisan fervor aroused by the slavery issue.

The history of school support is illuminating in view of the current discussions with regard to school costs. These costs have mounted very rapidly in recent years. One of the reasons for the increase is that since the World War the salaries of teachers have been appreciably advanced. A second reason is that high-school education is more expensive than elementary education and the growth in public-school attendance in recent years has been in the high schools. From 30 to 40 per cent of all the expenditures of American cities is devoted to the support of public schools. The burden on taxpayers is today very heavy, and complaints are heard in many quarters against the high cost of schools. As a result, there is a disposition to curtail expenditures.

It does not seem probable that the remedy for the present situation will ultimately be found either in retrenchment or in a return to private payments for school privileges. The battle for well-supported, tax-supported schools has been so completely won that there is little probability of a renewal of that particular contest. However, there is a new contest on at the present time—the contest for reform of the antiquated taxing systems of the states. It seems quite certain that property taxes will be increasingly replaced by income taxes. Education affects property comparatively little, while it affects incomes very materially.

It is not only on account of increased salaries for teachers and the development of high schools that the cost of education has mounted. The physical equipment of schools today is wholly different from that of a century ago. The school buildings of that date were of the simplest and most inexpensive type. The pupils' seats were often nothing more than benches

EDUCATION

hewn out of rough timber. In many of the early schools the only desk supplied for the pupils was a shelf projecting from the wall on three sides of the room. The windows were often closed with paper made semi-transparent by being greased with lard. There were no blackboards or equipment such as globes and maps. There were no pencils or steel pens. One of the most common duties of the teacher was to sharpen the quill pens with which the pupils practiced their penmanship. Paper was expensive and was carefully conserved. Much of the ciphering was done on birch bark or on sand boards. Books were few in number and were handed down from generation to generation. Often the teacher found that the members of his class in a given subject were supplied with books by different authors, inherited from their older brothers and sisters or even from their parents. Such practices as requiring conformity to text-book adoptions and supplying text-books at public expense never occurred to anyone in those early days.

Sanitary equipment was of the most primitive type. Drinking-water was brought into the school-room in pails, and a common dipper was used by all the thirsty children and their teacher. Ventilation was hardly thought of, and hygienic lighting was impossible because of the construction of the building.

Playgrounds were not regarded as essential parts of the educational equipment. Nature was kind to the children of those early times and provided them with much free space. The existence of play space was, however, a fortunate accident rather than a result of educational planning. The grounds around school buildings were often quite unkempt. In a biographical note Colonel Francis W. Parker tells of a school in Illinois of which he took charge in the late 'fifties. He says that the yard was rank with weeds and quite unusable as a playground.

Such conditions are to be explained in some measure by the lack of financial resources in early communities. The schools, like the roads, were developed only so far as was absolutely necessary to serve primitive community needs. The lack of money is not, however, the full explanation. The fact is that people did not value education very highly. The life of the farm and of the home required very little "book learning." Most of the experiences which prepared for successful adult life were supplied outside the school.

It was not until the transformation of American life through the development of machines and the migration of people from rural areas to cities that the need of an extended education began to be recognized by the popular mind. Education is prized today because the range of a person's contacts is so broad that he cannot be successful unless he can see far beyond his immediate environment. When the coal and iron industries of Pennsylvania began to supply on a large scale the materials for a new and rapidly expanding machine civilization, as they did in the decade 1880-90, the individual citizen found himself dependent on remote parts of the world for his food and clothing and for the raw materials used in industry. He then sought for himself and for his children the type of training that would prepare for success in the new world opened by transportation and the modern devices of communication. Social necessity led to the erection of the well-equipped schoolhouses of modern times. The desire for broad knowledge helped to bring expensive maps and globes into classrooms and led to the equipment of science laboratories.

METHODS OF TEACHING

The better training of teachers, the enrichment of the curriculum, and the other changes recounted up to this point have been accompanied by the adoption of methods of teaching and discipline altogether different from those which prevailed a century ago. In earlier days instruction was aimed chiefly at the cultivation of the memory. Discipline was harsh, consisting of frequent resorts to corporal punishment and petty forms of torture, such as requiring an offending pupil to stand for a period of five minutes or more with his arm stretched at full length, holding a book or some other weight in his hand. There was much of the Puritan attitude in the schools of the first half of the past century. Pupils were thought

EDUCATION

to be naturally perverse and in need of the salvation born only of adversity.

A general charge which can be made against teaching in the early schools is that it was formal in the extreme and therefore barren of desirable results. One example will make clear what is meant by this statement. When Horace Mann became the secretary of the Massachusetts State Board of Education in 1837, he found the schools teaching reading by the A-B-C method, which had come down from the days of Greece and Rome. This method was thoroughly intrenched in the leading text-book in use in American schools of that day—Noah Webster's speller. This book begins with the alphabet and then proceeds through meaningless combinations of letters, such as "ab," "ac," and "ad," to simple but wholly detached phrases, such as "he is" and "to go." Mann reported that eleven out of twelve of the pupils in the Massachusetts schools did not understand what they were reading. He pointed out that the A-B-C method is wrong in principle in that it does not begin with the element of the printed page which has meaning, namely, the word. No reading can be really educative, he held, unless it appeals to the interest of pupils.

Reforms in the methods of teaching reading came slowly even after Mann had called attention to the inadequacy of traditional procedures. A strong influence in the right direction was exerted by William H. McGuffey, who prepared a series of readers filled with interesting stories. It is said that McGuffey read the selections collected for his readers to the children in his neighborhood and retained or rejected stories according as he found children interested or not interested. Furthermore, the grading of the materials in his readers was made to depend on their acceptability to children of different ages.

When one compares the interesting, well-illustrated, and well-printed books for children now available in abundance with Noah Webster's speller, one realizes that there has been a complete change in the attitude of schools toward child nature. The new books are prepared with a view to attracting and holding attention. They draw on the experiences of pupils and on the best folk lore of the race. The Puritan austerity and the meager formalism of early education are entirely gone.

Another way of illustrating the change which has taken place in methods of teaching is to comment on the use of the word "recitation" in describing school exercises. If one could go back and observe a school in session in, say, 1865, one would find a teacher with book in hand reading part of a sentence and then pausing and requiring some member of the class to reproduce verbatim the last part of the sentence. "Recitation" as used in American educational history means literally an exercise in reproducing from memory what has been learned from a text-book.

This statement is revealing because it calls attention to two facts. The American school has always based instruction on text-books, and the American school of earlier days emphasized memorizing. The trend in modern schools is toward less formal methods of teaching. In sharp contrast with the exclusive use of text-books is the adoption of methods that bring pupils into direct contact with the objects about which they study. Where they use books, they are trained in library methods of study. They use collections of books rather than single text-books in preparing their lessons. They are encouraged to extract from many sources different views and different accounts bearing on the topics assigned. Furthermore, they are required to find ways of expressing in their own words the information they have secured. In short, the day of the formal recitation has passed, and instruction is aimed at the cultivation of initiative and intellectual independence.

In the fields of high-school and college education the abandonment of the recitation method is even more marked than at the lower levels of schooling. One influence that has contributed perhaps more than any other to produce a change in the spirit and methods of instruction is the introduction of the sciences. The sciences make large use of the laboratory method, which is altogether different from the text-book method. The student who sees the movements of a mechanical appliance in the physics laboratory or the student who sees chemical processes take place in a test tube is not able to depend on formal memory when he is asked to describe what he witnessed.

It would, of course, be contrary to the facts to assert that there is no formal teaching today or to say that the memoriter recitation is unknown. As long as human nature is at all disposed to find an easy way to accomplish intellectual tasks, there will be relapses into the method of memorizing by rote. The point to be emphasized is that the modern school recognizes memorizing as a form of mental achievement less worthy of cultivation than independent and comprehending formulation of personal experience.

Much of the change in the organization of schools and methods of teaching is due to the fact that present-day education is guided by scientific studies far more than was the education of former times. The science of psychology has revealed the ways of the mind in learning. The scientific measurement of school results has made it possible to compare methods of teaching and to select those productive of the best results. The last few decades have seen a vigorous and successful effort to establish a science of education and to guide all school practices by the results of this science.

One of the pioneers in the scientific study of educational problems was G. Stanley Hall. In 1884 he organized in Johns Hopkins University a department of psychology. He emphasized in all his studies the mental traits and abilities of children. He founded what he regarded as a new science. He called it "child study." With the aid of parents and teachers he collected facts about the stages of children's development as exhibited in such traits as their preferences for toys and in their tendencies to tell lies before they are fully aware of what is demanded by the canons of veracity. He studied physical growth and the periods in the lives of children when interest first appears in books and reading. He used the information thus secured to overthrow traditional practices and eliminate formal routine from school work.

Another vigorous contributor to educational reform based on scientific studies was John Dewey. He organized an experimental school that differed from ordinary schools in that it emphasized concrete experience and action. Dewey called attention to the fact that society had changed from a predominantly rural mode of life to one predominantly urban. Therefore, he argued, education must provide a substitute for farm life. The school must have laboratories and shops and kitchens. Class exercises must supply opportunities for activity. Dewey pointed out that the type of education which he advocated is in complete harmony with the psychological nature of children as well as with the history of society at large.

SCIENTIFIC STUDIES OF EDUCATION

The new movements in education resulting from the writings of Hall and Dewey took on a more strictly scientific character, with the beginning of the present century, through the invention of methods of measuring the results of school work. The first notable contribution to educational measurement was made in 1897. In that year J. M. Rice, who had been publishing for some years in the *Forum*, of which he was editor, articles describing the leading school systems of the country, conceived the idea of making an exact comparison of the success of the various school systems through tests of children's achievements. He secured the consent of a number of superintendents to the administration of a test requiring children to spell a list of words. He published the results secured in an article entitled "The Futility of the Spelling Grind." He found that school systems which taught spelling by widely different methods and for very different numbers of minutes a day stood exactly the same on his test. A few years later he published the results of tests in arithmetic. These results were, in general, like the results in spelling—very discrediting to the school practices of the times.

After Rice published the results of the first tests of pupil achievement, the measurement movement in education spread until it now covers all phases of school work. There are available today great collections of scores on nearly every subject taught in the schools. These "standard scores," as they are called, make it possible for the individual teacher to compare the achievements of his or her pupils with those of thousands of other pupils.

The possibility of making exact scientific comparisons of school results led in 1910 and the years following to general surveys of school systems. Tests and surveys have been supplemented by laboratory studies of the learning processes which take place under varying conditions and in different fields of knowledge. Methods have been devised for determining the general intelligence of individuals. The results of different administrative procedures, such as the organization of pupils into large and small classes, have been accurately compared. The effects of nutrition on school work have been investigated. The laws of physical and mental growth have been ascertained.

The science of education is one of the important contributions of America to modern civilization. The freedom to experiment, always characteristic of American schools, and the absence of domination by any central governmental authority, supplied the conditions favorable to the development of this science. The effort to rescue American schools from formalism and inefficiency stimulated its cultivation. The results of scientific studies are so significant and have done so much to improve education even in the short time during which they have been available, that there can be no doubt that in the future schools will continue to modify traditional methods of teaching and traditional systems of organization under the guidance of carefully planned and executed investigations.

CHAPTER FOURTEEN

THE ARTS

"Our houses and towns are like mosses and lichens, so slight and new; but youth is a fault of which we shall daily mend. This land, too, is as old as the Flood, and wants no ornament or privilege which nature could bestow. Here stars, here woods, here hills, here animals, here men abound, and the vast tendencies concur of a new order. If only the men are employed in conspiring with the designs of the Spirit who led us hither, and is leading us still, we shall quickly enough advance out of all hearing of others' censures, out of all regrets of our own, into a new and more excellent social state than history has recorded."—RALPH WALDO EMERSON, address on *The Young American* before the Mercantile Library Association, Boston, 1844.

The Arts

By Fiske Kimball

ART is not a field in which "progress"—as in material civilization—can ordinarily be said to exist, movements in this or that direction being rather like swings of a pendulum. In a new country, to be sure, there may be a progress in the first introduction of the arts and the effort to bring them abreast of contemporary technical standards in the Old World, but after that has been achieved the merits of art, in a new country as in the old ones, rest on the quality and frequency of individual creative acts.

The history of American art thus falls into two periods of different character—one of colonial implantation and conquest of artistic means; the other of independent mastery in the free seeking of artistic ends. The first by no means necessarily terminated with the Colonial period in politics, nor did it end simultaneously in the different arts. The attainment of creative independence in them has indeed been a long and unequal process, not yet complete in all.

THE COLONIAL HERITAGE

In the art of the first phase of the Colonial period the effort was one of conservation—to preserve, in the struggle for bare existence in a primitive environment, as much as possible of the civilization the colonists had known at the time they left it. They looked backward to what they remembered, to the frame cottages of England in the time of Elizabeth and James, which were there already being replaced by stone and brick, to the medieval survivals which were there giving way to the academic forms of the Renaissance, to such portraits as those of Shakespeare rather than to the incoming style of Rubens and Van Dyck.

We do not find in the arts and crafts of seventeenth-century America much that was native to the new continent or evoked by it—the old frontier did not create new types or borrow much from the Indian. The first shelters were like the "frail houses" of poor English folk, of poles, reeds, and earth, of wattle and palisades. The log house, which showed itself best adapted to the forest wilderness, was an importation by the Swedes of their ordinary type of dwelling at home. These themselves were abandoned as soon as possible for the steeply gabled houses of frame, in which the Middle Ages had a last afterglow. In some ways time brought even a recession, as the first generation of limners and craftsmen of foreign birth and training died out, and were replaced by apprentices who had learned their trade here without a glimpse of London or Amsterdam.

From such artistic stagnation or retrogression the opening of the eighteenth century saw the beginnings of an awakening. New fashions from Europe began to filter in—in the newly founded colony of Pennsylvania, in the colonial capitals with their succession of governors from England, in the maritime towns of the North, on the tidewater plantations of the South with their direct marketing in London.

It was in architecture, an art useful and necessary as well as "fine," that the approximation to English contemporary style was best achieved. The finer Colonial mansions were not more backward, in relation to the metropolis of empire, than the houses of small gentry in the English shires. Their columned doorways and carved chimneypieces were copied from the plates of those English manuals by which each new fashion of building was popularized. Their simplicity of form, their symmetry and proportion, their fine craftsmanship, make them objects of beauty, however closely imitative their style. More characteristically American were the ordinary farmhouses of different sections—long-roofed and clapboarded in New England, gambreled and wide-eaved about New York, whitewashed and of stone in Pennsylvania, narrow and high-chimneyed in Virginia—fresh in artistic character just in the degree to which they ignored or lagged behind the current fashion and embodied adaptations to local climate, customs, and materials.

Painting, too, began to approach more respectably the contemporary foreign standards. As in England itself, the portrait was almost the sole product in demand. The supply was provided by such newcomers from abroad as Hesselius, Smibert, Bridges, Woolaston, Feke, and Blackburn, with varying degrees of capacity or incapacity, in inherited formulæ admitting little individuality. From amid this school of Colonial portraiture emerged two men of native birth and ability, Benjamin West in Pennsylvania and John Singleton Copley in Boston.

West, of Quaker upbringing almost in the wilderness, excited the marvel of his contemporaries by precocious gifts which soon became legendary. Sent to Rome by generous patrons, he absorbed the tradition of historical painting from Mengs and Battoni and took it to London, where it made a fortune by contrast with the native school of portraiture. A classicist before David in France, a precursor of romanticism, both in his rich color and in his choice of literary and biblical subjects of dramatic character, West won the favor of the King and succeeded Reynolds as president of the Royal Academy. Henceforth his importance for American art lay in his example, in the influence of his style, and in his unfailing help to the American students who flocked to him in London.

Copley, with only the teaching of the local painters of Boston in his youth, and an unconquerable determination to profit by every resource available in the way of prints and paintings there, raised himself by literal faithfulness of observation to a very notable competence in his art, evidenced by his "Boy with a Squirrel," sent to the exhibition of the Society of Artists in London in 1762. A whole gallery of portraits of Boston worthies and their ladies, as sincere in their delineation of character as laborious in their elaboration of costume, followed down to the eve of the Revolution, when Copley, a Loyalist, sailed for England. The greater facility and breadth of subject-matter in his later work abroad were poor compensation for the loss of his admirable primitive quality.

INDEPENDENCE AND THE ARTS—CLASSICAL REVIVAL

In contrast with the Colonies, which, as we have seen, sought approximation to conditions in England, the American Republic from its foundation strove to break with England and establish its independence in artistic matters also. Architecture and the representative arts alike derived a new stimulus from the Revolution and the establishment of a national government. The alliance with France took Americans to the Continent; the opportunities afforded by the young nation attracted artists from abroad, among whom the refugees from the French Revolution were conspicuous. French relationships were thus influential, but they were not decisive. The fundamental influence was the sophomoric analogy to the republics of Greece and Rome, giving an unexampled force to the classical revival already beginning abroad and driving it to extreme consequences unacceptable to the conservatism of older countries.

In architecture, buildings of novel type were to be created to meet the needs of popular government with its democratic and humanitarian program. Not only the Capitols of state and nation but the prisons and asylums called for new types of building. The heroes and events of the struggle for independence were to be commemorated in monumental form.

The decisive work, determining the evolution of style, was the Virginia Capitol (1785), which preceded that of the nation as well as those of the other states. Jefferson, then Minister to France, had already shown his interest in public no less than private architecture, and had evidenced his academic and classical allegiance. Approached to secure a design for the Capitol of his native state, he proposed as a model the Maison Carrée at Nîmes, finest of surviving Roman temples. Although he consulted the French architect Clérisseau as to the

THE ARTS

details, the initiative was his own. Abroad the form of a temple had already been imitated in garden structures, but no example on a large scale, or divided for practical uses, had yet been attempted. The Madeleine in Paris was not begun until 1807. Jefferson thus anticipated Napoleon by a score of years in his appeal to the sanction of the classic. The great portico of the Capitol at Richmond became a frontispiece to the architecture of the young republic, which was turned in a monumental and classical direction.

For the new Federal City of Washington (1791), the French engineer L'Enfant made a plan based on that of Versailles, with avenues radiating from the principal sites across a gridiron of minor streets. The design for the Presidents' house, the White House, was made by an Irish builder, James Hoban, who closely followed English academic precedents. For the Capitol, a French architect, Stephen Hallet, first proposed the scheme with a dome and wings, finally embodied in the prize-winning design of the English amateur William Thornton, which adopted the low dome of the Roman Pantheon.

In domestic architecture the classical influence took two forms. In the North the delicate proportions and detail developed abroad by the brothers Adam were adopted by such designers as Charles Bulfinch of Boston, Samuel McIntire of Salem, John McComb of New York, and Philip Hooker of Albany. In the South, under Jefferson's influence, the massive portico, Palladian in proportion, was the most conspicuous element. The final step of making the dwelling also in the form of a temple was taken by Jefferson at the University of Virginia (1819-25). Here the rectangular pavilions housing the professors and their schools contrasted with the circular rotunda, as models, Jefferson said, of "cubic" and of "spherical" architecture. The whole group, monumentally and symmetrically disposed, was unified by the long colonnades of the dormitories, the recurrence of red wall and white portico.

Long before this a Greek revival had begun in the train of the Roman. Its pioneer in America was an English architect, Benjamin Henry Latrobe, who came in 1796. In his marble

Bank of Pennsylvania (1799), he adopted the Ionic order of Erechtheum; in his design for the Bank of the United States (1818), he went to the extremity of imitating the Parthenon, a decade or more before the European derivatives such as the Valhalla at Regensburg. Jefferson created for him the post of Surveyor of Public Buildings, and he was for many years in charge of the work at the Capitol. The old halls of the Senate and of the Representatives (now the Supreme Court and Statuary Hall) are the chief remains of his activity there. The cathedral in Baltimore (1805 ff.) severely classical, opened a new era in religious architecture. In his office was formed a whole generation of pupils, of whom the chief were Robert Mills and William Strickland. Among Mills' works were two notable monuments to Washington, that in Baltimore, an immense Greek Doric column, preceding the similar monuments to Wellington in Great Britain; that in Washington a vast obelisk, long the highest of human structures, which in its grandeur and simplicity of form well matches the character of the subject. These men dominated the period to the eve of the Civil War.

In painting, the Revolution widened the sphere of interest beyond the family portrait, which had hitherto been almost the sole employment of the artist in America. The personages and episodes of the war were alike to be represented and perpetuated on canvas. The task fell to pupils of West, attracted by his success and encouraged by his kindness—Charles Willson Peale, John Trumbull, and Gilbert Stuart.

It was Trumbull, who had himself been a Revolutionary officer of rank, who conceived the undertaking most broadly. With him the admirable individual portraits of the actors in the Revolutionary drama, executed from life both in America and abroad, on a miniature scale, were but means to his ultimate end—a series of great historical compositions. Early studies on a small scale for some of these, "The Battle of Bunker Hill" and the "Death of Montgomery before Quebec," had considerable merit. The commission from the government in 1817 to fill four panels in the national Capitol

THE ARTS

came too late. Hand and eye were alike failing, and the final results little answered the great expectations.

Gilbert Stuart, who in 1793 returned to live in America after considerable success abroad, had no such ambitions but far more genius. He painted freshly with rich color in the best traditions of British portraiture, at once with a fine sense of characterization and with an air of distinction and refinement. His portraits are the patents of the aristocracy of the Republican Court as those by Copley are of Colonial gentility. The first five Presidents were all among his sitters; his Washingtons, together with the sculptured likeness by Houdon, created the popular image of the father of his country.

After his death in 1828, Thomas Sully, Henry Inman, and John Neagle worthily continued the line of early portraiture to the middle of the nineteenth century.

The rise of romanticism about 1830, in America as abroad, loosed the bonds of tradition and of convention, for both good and ill. The supremacy of the classic in architecture weakened to admit the Gothic and other styles. The exclusive devotion to figure-painting relaxed in favor of an appreciation of landscape. The characteristic American scenery of broad valleys gave the subject-matter of what came to be known as the Hudson River school, in the canvases of Thomas Cole and Asher B. Durand and their followers, panoramic in composition, minute in execution.

NEW CONTACT WITH EUROPEAN CURRENTS

The Civil War, with the industrial transformation which accompanied it, brought to a close the cultural period of the early republic, and inaugurated a new era, with new artistic forces and personalities. Once more a stream of young men went to study in Europe. The brilliance of France under the Second Empire drew the majority to Paris, where the École des Beaux-Arts, with its unrivalled discipline and emulation, was at the height of its influence, and where, outside the schools, the new movements of realism and impressionism in painting were achieving their first triumphs.

In architecture, where the new historical knowledge brought on a "battle of the styles" here as well as abroad, the first result was a fresh series of revivals, led by the men returned from Paris.

Henry Hobson Richardson inaugurated a bold but short-lived attempt to find in the rugged masonry of the Romanesque an expression of the youth and energy of the new continent. His Trinity Church in Boston, his town halls and libraries, his Marshall Field warehouse in Chicago, were widely imitated in works which reflected less the spirit than the superficial features—the broad low arches, the towers picturesquely disposed. Richard Morris Hunt sought in the rich forms of the Valois châteaux an investiture for the town and country houses of the new "millionaires" of the Gilded Age. John Carrère and Thomas Hastings, in their great Florida hotels, were stimulated by colonial history to follow the buildings of the Spanish Renaissance.

More far-reaching in its scope was a neo-classical movement led by Charles Follen McKim and Stanford White. The beginnings were modest enough, but they had a vitality which came from contact with native tradition. The example of the "Queen Anne" movement in England led these men first to study and imitate the early American work, inaugurating a Colonial revival which, as time went on, acquired strength from various local traditions. The example of one of their associates, Joseph Morrill Wells, led them to Italian models, as in the Boston Library, and thus toward Rome and Greece, which had inspired the government buildings of the republic a century before.

At the Chicago World's Fair of 1893 the adherents of classicism, under the leadership of Daniel Hudson Burnham, gave in the Court of Honor a striking object lesson of the value of monumental dignity and harmony of style. Columned façades masked the sheds of steel and "staff," evoking a phantasm of tranquil magnificence. The buildings of McKim and of Charles Attwood most admirably realized the classical idea of form, which was deeply stamped on the consciousness of the

THE ARTS

whole nation. The great academic groups created in the 'nineties by McKim and White for Columbia University and New York University recalled alike the lessons of the Court of Honor and of Jefferson's plan for the University of Virginia.

In 1901 the makers of the Chicago Fair were called to advise on the development of the city of Washington. In spite of the grave and seemingly irreparable errors made there in the nineteenth century, they boldly undertook the restoration and extension of L'Enfant's plan, and returned to the classic traditions of the early Republic. A monumental site on the main axis was assigned to the Lincoln Memorial, in the execution of which McKim's idea of a great peristyle of the Greek Doric order was worthily carried out by his pupil, Henry Bacon.

The Chicago Fair also furnished the first example of a rich use of monumental sculpture, and gave play to the group of Paris-trained sculptors, of whom the leader was Augustus Saint-Gaudens. They adopted both the realistic handling and the allegorical allusiveness of their French masters of the 'seventies. Saint-Gaudens' monuments with their broad calm settings by Stanford White—many commemorating heroes of the Civil War: Farragut, Lincoln, Sherman, Robert Gould Shaw—were the characteristic works of this era. His inscrutable brooding figure for the Adams memorial in Washington is perhaps the finest of his works.

American painting was likewise vivified by new contact with French example and teaching.

William Morris Hunt, who had had a solid academic training under Couture, was the first to feel the influence of the romanticists and realists outside the official school. He became a friend and supporter of Millet, and worked with the men of Barbizon whose subjects and treatment influenced his. Their subjective vision of landscape, suffused with the feeling of the artist, he first brought to America on his return in 1855. There was little appreciation either for his landscapes or for his excellent figure paintings, and a large mural commission, for the Capitol at Albany, came only on the eve of his death. He was thrown back on portraiture, and on teaching, where his influence was salutary. It was he who first turned the stream of later students toward Paris.

George Inness, who had begun with panoramic scenes of the Hudson River type, learned from the work of the Barbizon masters the charms of intimate landscape, of mood and atmosphere. He studied the wood-interiors of autumn, the masses of spring bloom, the glow of evening and calm of twilight. His later canvases from the 'nineties are of saturated misty color, dissolving any firm structure. Homer Martin, who had begun under the influence of Corot, derived new inspiration from the French impressionists whose work he came to know in the 'eighties. Without abdicating his individuality, which was of a melancholy, almost tragic turn, he adapted his technique to the new method of color division, rendered in a delicate minor key.

John LaFarge, himself born of French parents in New York, studied with Hunt and abroad, less in the schools than independently, copying the old masters and making his own experiments. His early observations of effects of light in the landscape led him to results anticipating, in part, those of the French impressionists. His color was rich, and relatively high in key. His decorative sense and feeling for materials drew him into decorative work, his first large commission being the interior of Richardson's Trinity Church in Boston. For windows he was called on to design, he created a new type of glass, opalescent and shaded, which he used, with little opaque painting, in windows which represent one of the few independent developments of this art since the Middle Ages.

The relation to European developments is closest in three American painters who spent their lives mainly abroad: James McNeill Whistler, John Sargent, and Mary Cassatt. Whistler's fruitful early relationships were with Courbet and Manet; he made one of the *Salon des Refusés* of 1863. In England, where he then established himself, he was the protagonist of the new cause of *art pour l'art* in his "symphonies" and "nocturnes." Deprived of the stimulating environment and emulation of Paris, he was drawn by the influence of Japanese art into what proved to be a bypath in the evolution of Western painting, removed from the voluminous solidity of his great French contemporaries. Thus it was also with Sargent, who, too, left the Paris of Manet for London, and whose brilliant facility as the fashionable portrait-painter of his time was purchased at the expense of solid construction and serious attainment. Rebelling against the routine of portraits, he undertook mural commissions in the Boston Library and Museum, and painted for his own pleasure the multitude of landscapes of his later years. Mary Cassatt, friend and disciple of Degas, exhibited with the Impressionists in their early showings of the 'seventies, and shared their sure rich color and light tonality. Her favorite studies of children with their mothers and nurses were devoid of the sentimentality of minor artists of the period, and she remained steadfastly devoted to purely artistic ideals.

THE STRENGTH OF THE AMERICAN NOTE

Stronger and more native was the art of certain men who worked in America, the most important of our painters—Winslow Homer, Thomas Eakins, and Albert Pinkham Ryder.

Homer, robust and masculine, inaugurated American realism. Beginning as an artist at the front during the Civil War, he painted the soldier, the Negro, the hunter and guide of the north woods, the toilers of the sea, the sea itself. In his matter as in his manner there is something stern and elemental, something of fundamental power. The bare direct treatment, whether in oil or in water color, is of great technical competence, but this competence is in the service of a searching and orderly vision, a living art.

Thomas Eakins had the severe academic schooling of Paris in the studios of Gérôme and Bonnat, but his own strong personality determined the character of his work. He brought to the problems of form and movement the deepest and most intense study. It led him to the dissecting-room, where the surgeons he was afterwards to paint so remarkably became his friends, to the boxing-ring, to the swimming-hole, to the river with its oarsmen. He participated in the scientific studies of

Muybridge to determine the action of animals in motion. The horse, their chief subject, he dissected and modeled as well as painted. Nowhere else, as in his canvases, do we see so fully the life of America in the later nineteenth century, a life in which he himself participated quietly but richly. We see the hunters in the marshes, the sailboats on the river, the fishermen drawing and mending their nets, the cowboy with his guitar, the girl at the piano in the stuffy interior of the time. These scenes and types were not sought for their picturesqueness, nor with an intent to illustration, but were represented as the episodes and acquaintances of his various ordinary experience. In his passion for truth Eakins did not neglect form, which he viewed, indeed, as a higher truth of nature—the reduction of the complex to the simple. The realist was also truly an artist.

More purely individual was Ryder, a poet and mystic whose fantasies may be regarded less as late offshoots of romantism than as expressions of deeply personal feeling, reacting on appearance. His elements are few and simple—a sail by moonlight with wind-swept cloud; two figures, as the Christ and the Magdalen, in elementary juxtaposition—but with these, by just relation of shapes and tones, prevailingly somber, he evokes a deep and powerful emotional response. His world is less real than supernatural; his poetry, of another sphere.

Although the men just discussed lived and worked on, before the turn of the century new problems and new personalities came into play.

THE ARTS OF THE INDUSTRIAL AGE

Beneath the surface, currents of architecture in the later nineteenth century, in America as in Europe, swelled a mighty tide of new forces flowing from the industrial revolution. Materials essentially new were developed—iron and steel, plate glass, ferro-concrete. New types of building were called into being—for manufacture, business, and transportation. The effort to find a new and appropriate form for these, a "modern style," was spurred by the patriotic cry for an "American style," product of the new continent.

THE ARTS

European thinkers by 1870 had formulated the demand that the form of every technical product be a function of use and material, that architecture be bounded by imitation of no style, antique or medieval. This demand, still little satisfied abroad, fell on willing ears in the great American West, above all in Chicago, where all was yet to be created, and where the forces of materialism at that moment were perhaps more unrestrained than anywhere on the globe. Fortunately, there were in Chicago also men of artistic vision: Burnham with his dream of vast order and organization; Louis Sullivan with his passionate worship of nature, his freshness and consistency of thought, his deep poetic impulse. It was Burnham who clarified the requirements of the new type of business building; it was Sullivan who first gave it a crystalline artistic form.

The tall office-building, the "skyscraper," evoked on preferred sites by economic advantage, unhampered in individualistic America by legal restriction in the public interest, was made practicable by two inventions—the elevator, and the metal frame, supporting first the floors only, then the exterior wall as well. The development, begun in New York with self-supporting walls of masonry, was taken up in Chicago, which developed the steel frame and hung the walls upon it. The conception of a frame so used was perhaps first reached by Leroy Buffington. Its structural evolution was accomplished in the Home Insurance Building (1884-86) by William L. B. Jenney, and the Tacoma Building (1887-88) by Holabird and Roche. An artistic crystallization was first attained in the Wainright Building at St. Louis (1890), the decisive work of Sullivan.

The artistic essence of the steel office building lay for Sullivan in the thought, "It is lofty," and he wrote: "It must be tall, every inch of it tall. The forces and power of altitude must be in it, the glory and pride of exaltation must be in it. It must be every inch a proud and soaring thing. . . ." His Wainright Building with its multiplied and accented vertical lines was the embodiment of this conception. In its treatment Sullivan sought to exemplify his dictum, "Form follows function"; he rejected historical style and detail, and allowed the novel problem to suggest its own solution. Wall surface was abandoned for a system of tall piers and recessed spandrels of terra cotta casing the steel to protect it against fire. A delicate surface ornament removed any suggestion of self-supporting masonry. The design, however, was more than a mere mechanical expression of structure. In simplicity and unity of form, deeply felt by its creator, it rose into the realm of authentic creative art.

Sullivan's genial artistic conception became a formula to many architects who were far from having Sullivan's convictions and who compromised with historical style. Thus Cass Gilbert in the Woolworth Building, for many years the highest, took his suggestion from the Gothic, with its multiplied and extended verticals. Even within the classical camp, where colonnades of conventional proportions were applied to the upper and lower stories of the façade, the shaft of the building between was often marked by shallow vertical strips. At the height of the classical reaction, about 1910, this element of expression also was abandoned and the curtain of masonry was again treated merely as if it were a self-supporting wall, crowned by a great overhanging cornice.

In sculpture, by the turn of the century, the influence of Rodin was making itself felt in a freer, less academic treatment, seen, for instance, in the work of George Grey Barnard. His large groups at the Pennsylvania Capitol in Harrisburg, his "Two Natures," his "Hewer," have a power both of conception and of form, rare in American plastic art.

In painting, in the years about 1900, the influence of French impressionism was predominant. A multitude of painters, particularly painters of landscape, concerned themselves with the problems of light and atmosphere, with greater or less dependence, adopting the color division and the high key of Monet and his followers. Although the level of competence was higher than ever before in America, one cannot say that from it emerged any individual figures of the importance of those of the previous generation.

THE ARTS

On the eve of the Great War certain new tendencies and new figures began to be apparent.

Sullivan's chief disciple, an architect of powerful creative impulse, Frank Lloyd Wright, developed a characteristic style in his early houses about Chicago—houses low and widely ramified, each room separately projected in a multitude of parallel planes, with long banks of windows beneath a broadly overhanging cornice. His church at Oak Park was a significant study of monolithic cast concrete, and of unity of cubical forms. His building for the Larkin factory inaugurated a reform in industrial architecture, its broad windows extending from pier to pier, its exterior form enriched by the projecting masses of stairs and ducts. Wright welcomed the machine and welcomed the advent of new materials, inventing systems of construction with the greatest fertility and using each as the basis for a new artistic creation. His influence was deeply felt both in Japan, where he built the Imperial Hotel at Tokio, and in central Europe, where his name became one to conjure with. At home his general recognition was more gradual, and was finally achieved only when the ideas of the modern school in Europe began to be familiar.

In the design of the skyscraper, the New York zoning law of 1915, with its provision for setting back the upper stories, opened the eyes of architects to the possibilities of composition of mass. The slender towers of the Singer and Woolworth buildings had already pointed the way skyward. Now the towers, ever higher, were bastioned about, making transition from the cube below. As this lost its independence the heavy cornice was abandoned; the mass became organic. Partly because of the survival of historical eclecticism, partly because of conservative provisions in the building law, the organic quality did not equally extend to the surface, which retained many survivals of masonry construction. Wright's genial designs, exploiting glass and copper, have not come to execution. It may scarcely be said that the problem of the skyscraper has again, as in its infancy, evoked a masterpiece.

The revolution in European painting effected by Cézanne

and the post-impressionists was long in being felt in America. Maurice Prendergast, who returned full of enthusiasm for the work of its leader, and whose own canvases, of independent personality, show similar tendencies in form, was slow in achieving even a small recognition. The photographer Alfred Stieglitz furnished a center where European works of the school were shown and gave encouragement and support to a group of American pioneers who did notable work, especially to the water colorists John Marin and Charles Demuth. The Cubists and *Fauves* were first brought strikingly to public attention by the Armory exhibition of 1913, and an increasing number of American artists became interested in their experiments in form, which have given the basis for the chief body of creative work in America since the Great War.

In architecture the deeper problems of the moment go far beneath superficial considerations of "style" and are rooted rather in town planning and housing, imbedded as they are in the foundations of economic and social organization.

The dazzling achievement of the skyscraper, the amenity of suburban bourgeois life distract attention from the ruthless exploitation of urban land and from the drabness of speculative jerry-building for all below the executive and professional level. The vicious spiral, of land-values based on unrestricted heights, and of heights necessary to pay return on inflated land-values, has brought a congestion of central areas almost beyond the capacity of sidewalks, if not of streets. As the spiral still rises with increased elevator speed and further technical developments, the multiple levels of traffic envisaged by architects are little more than what already exist, under and above ground, at the busiest points, such as the Grand Central in New York.

The measures so far taken in these matters, by architects and engineers as well as by public bodies, can be regarded as little more than palliatives of conditions which have resulted and are still resulting from rank, chaotic growth. Certain attempts have been made, to be sure, to direct this growth in more orderly fashion—as in certain provisions of the Plan of

Chicago and of the Regional Plan of New York—but it is clear that architects and speculators alike are mastered by their own creation, and public authorities seem powerless to make really effective intervention.

The superficial problem currently posed in architecture is the acceptance of the "international style" based on the initiative of Le Corbusier, Gropius, Oud, and other continental designers, who have lately achieved a position of general leadership in the modern movement at large. Whether American architects can make significant contributions within this international style, or whether by adopting it they would sacrifice an opportunity of reëstablishing their own leadership on other lines, is for the future to decide.

Sculpture languishes at the moment between the empty formality of neoclassicism in the official art of the Roman prizemen and the experiments in abstraction of the disciples of the modern French school, still unacceptable to the public taste which governs monumental commissions. Freedom for creative effort can be scarcely found except in small decorative works—the statuettes of metal or fired clay, multiplied in "editions" sufficient to compensate the sculptor, and adapted to the domestic interior where one may be artistically adventurous without public responsibility. There can be no question, however, that the swing away from realism and toward abstract form is one fundamentally favorable to plastic art, for which nineteenth-century romanticism with its cult of mysterious vagueness, its corresponding preference for literature, painting, and music as modes of expression, had such injurious consequences. We may look, as the twentieth century advances, for a vital revival of sculpture.

The triumphant victory of anti-realistic tendencies in painting all over the world would seem to make certain that, in American painting also, the next generation will give prominence to the problems of form. No doubt the subject-matter of the American scene, as with John Sloan, Reginald Marsh, Thomas Benton, and James Chapin today, will furnish material for a body of work racily characteristic of nationality in that regard—and the overwhelming flow of population to the cities will doubtless determine the predominant character of types. As in most periods of supremacy of form, figure-painting may be expected to predominate, although the landscape and its buildings remain to be seen with other eyes.

There seems no prospect that the primacy of France in determining the direction of painting, a primacy which has now continued unchallenged for more than a hundred years, will be soon replaced—indeed, never was the *school of Paris* so all-absorbing in its attraction of the young artistic forces of every country. In coming movements abroad, as in those of the past, we may expect Americans to have a place. It is scarcely to be supposed, however—any more than in the past—that those who abdicate their national heritage and individuality there will be the strongest of our painters. These will rather be found, as hitherto, in the lonesome geniuses working here, inevitably of their artistic time, but fulfilling in relative isolation their individual destiny.

CHAPTER FIFTEEN

LITERATURE

"The loftier masters, though their technical power and originality, their beauty of form, strength of flight, music and variousness of rhythm, are full of interest and instruction, yet, besides these precious gifts, come to us with the size and quality of great historic forces, for they represent the hope and energies, the dreams and consummation, of the human intelligence in its most enormous movements. . . . For these we need synthetic criticism, which, after analysis has done its work, and disclosed to us the peculiar qualities of form, conception, and treatment, shall collect the products of this first process, construct for us the poet's mental figure in its integrity and just coherence, and then finally, as the sum of its work, shall trace the relations of the poet's ideas, either direct or indirect, through the central currents of thought, to the visible tendencies of an existing age."—JOHN MORLEY, in his essay on Byron.

Literature

By John Erskine

THE history of a literature can be written either in terms of individual masterpieces or in terms of the social, philosophic, and economic conditions in which they were produced. American literature in the last hundred years is rich in masterpieces, but neither Americans nor Europeans, when they review the contributions of our authors in this period, stress primarily the quality of the books; they are rather disposed to debate the conditions which influence writing in the United States, to discuss the background of the work rather than the work itself.

American literature, therefore, has perhaps not yet received the recognition that it deserves as pure accomplishment, but the instinct of reading folk the world over to use it as a starting-point from which to examine American life, is probably a sound instinct, which far-sighted writers would not protest against, even though it postpones for a while the kind of attention a writer primarily desires. The forces at work in American society are probably more significant for the moment than any book; they are setting the conditions in which the books of the future, the unquestionably American books, will be written; perhaps they are bringing about a strange new set of conditions in which no books will be written at all, but the literary man will address his audience through the film, or through some yet undiscovered medium. In any case, it is the distinction of American literature, a distinction of which the American writer is proud, that our novels, our poetry, our criticism, have been pretty intimately tied up with our life, and if readers in general are more concerned with

LITERATURE

the life to which our books are so intricately attached, it is not because the books are mean, but because the life is of dramatic and startling importance.

Thirty years ago in our schools and colleges the course in American literature was usually presented somewhat apologetically, as a dangling appendage to English literature. The implication was that our writers were provincial and second-rate. In the startling upheavals of our modern world, the stature of many of the older American writers has grown fast, and our literary values have been redefined, so that unless we are desperately untuned to the taste of our day, we are now finding it easy to rate our best writers well up among those who thirty years ago gave distinction to other lands. Our respect for American authors of the elder day has undoubtedly been increased by the reception our contemporary writers are receiving throughout the whole world. We have our voices today in international literature, expressing us as we are. A philosopher or a moralist might wish we were different, but for our literature, for the way in which we are getting ourselves portrayed, there is praise abroad, and no need at home to apologize.

A COMPARISON OF THE OPENING AND CLOSING DECADES

The achievement of American literature during these hundred years becomes clearer if we glance at some of the books of the opening decade, and of the closing. In 1833 Longfellow published *Outre-mer*; in 1835 the poet Drake published the *Culprit Fay*; in 1836 the modern and important phase of American literature opened with Emerson's little book *Nature*, containing the germ of his own philosophy, and perhaps also the philosophy of Whitman and of most American writers to this day. In the same year Whittier published *Mogg Megone*. In 1837 Hawthorne published the first series of *Twice Told Tales*, the volume with which his noble reputation as an artist began. The following year Fenimore Cooper published some of his irritated and irritating writings about Europe, not yet having found the subject-matter that made him world-famous. At the same time Poe published *The Narrative of Arthur Gordon Pym*. In 1839 Longfellow published *Voices of the Night*; in 1840 Cooper came to his own in *The Pathfinder*, and Poe published the *Tales of the Grotesque and the Arabesque*. In 1841, a marvelous year, appeared the *Deerslayer* by Fenimore Cooper, Emerson's *Essays*, first series, and Longfellow's *Ballads and Other Poems*. In 1842 Cooper published *The Two Admirals* and *Wing and Wing*, and Hawthorne brought out his first and second series of the *Twice Told Tales*. In 1843 appeared Cooper's *Wyandotte*, in 1844 the second series of Emerson's *Essays*, in 1845 Poe's *Raven*.

One is tempted to record other years in which the product for the point of view of permanent world literature was remarkable. The next three years, for example—1846, giving us Hawthorne's *Mosses from an Old Manse*, and Melville's *Typee*; 1847, giving us Emerson's *Poems*, Longfellow's *Evangeline*, and Melville's *Omoo*; 1848, giving us Lowell's *Biglow Papers*, first series, his *Fable for Critics*, his *Vision of Sir Launfal*, and Poe's *Eureka*. One is tempted to mention also 1850, in which appeared Emerson's *Representative Men*, Hawthorne's *Scarlet Letter*, Melville's *White Jacket*, Whittier's *Songs of Labor*; 1851, which saw Hawthorne's *House of the Seven Gables*, Longfellow's *Golden Legend*, Melville's *Moby Dick*; and 1852 in which appeared Melville's *Pierre*, Hawthorne's *The Blithedale Romance*, and Mrs. Stowe's *Uncle Tom's Cabin*.

A country that produces year by year books of this quality is doing its share for literature in general. If only a few titles are here named, many of them by the same authors, it is because time has sifted out these few; critical opinion in other lands as well as in ours would agree as to their importance. But only a rash critic would attempt to make the same selection for the writers of the present hour. The number of books produced in America today is vast, the army of writers is prodigious, and though out of the mass one can safely name a very respectable list who have an audience everywhere, who are seriously championed by thoughtful critics, and who are translated into many languages, we cannot guess who will still be in vogue fifty years from now.

Omitting dates and titles for lack of space, and apologizing in advance to those who are not mentioned here, we must in any case name, among the novelists, Sinclair Lewis, Willa Cather, Edna Ferber, Ellen Glasgow, Fannie Hurst, Edith Wharton, Zona Gale, Margaret Barnes, Gertrude Atherton, Dorothy Canfield Fisher, Booth Tarkington, Sherwood Anderson, Upton Sinclair, James Branch Cabell, Joseph Hergesheimer, Thornton Wilder, Ernest Hemingway, Floyd Dell, Louis Bromfield, and among the writers of short stories, Wilbur Steele and William Faulkner.

Among the poets we have Edward Arlington Robinson, Edgar Lee Masters, Robert Frost, Carl Sandburg, Edwin Markham; Vachel Lindsay, lost to us since these lines were first written, and Joyce Kilmer, a costly sacrifice in the World War; an extraordinary group of women—Edna St. Vincent Millay, Sara Teasdale, Leonora Speyer, and two, who though recently dead, belong vitally to our time, Amy Lowell and Elinor Wylie. Louis Untermeyer and Jessie Rittenhouse, accomplished verse-writers, furnish us with notable anthologies of contemporary and older poems.

In criticism we have an increasing group of highly trained essayists and philosophers who have made scholarship intelligible even to the general public, who write not only with competence, but with brilliance and beauty, and who from various angles have gone to the heart of our problems in art and in life. It is difficult to know where the list should begin—certainly we should not forget to mention William James, who, though already some time in his grave, still sets his mark upon us; and as certainly we should name John Dewey, who, modestly disclaiming literary pretense, has made his courageous experiments and explorations count heavily in our literature as well as in our education. George Santayana, for decades only a visitor among us and now residing abroad, taught us, especially in his early books, how closely the philosopher and the artist can be allied.

Paul Elmer More maintains in the midst of our machine age a faith in ancient classics and in older modes of thought matched only by his astounding erudition. Irving Babbitt, fine scholar, great teacher, an ardent lover of life, has rallied a hopeful band of cultured young men in the cause of what is called the New Humanism, the outline of which is not clear, but whose impulse is to assert once more the enduring values of life, with special protests against the kaleidoscopic shifts of our time.

Henry L. Mencken, the brilliant editor of *The Mercury*, has brought every weapon of wit, ridicule, and controversy to bear upon the smug complacences incidental to democratic prosperity. Himself the most kind and human of persons, he has struck ruthlessly at stupid narrowness, at lazy intellectual habit, at the moral timidity which is first cousin to hypocrisy.

The late Stuart Sherman, somewhat of the school of Paul Elmer More and Irving Babbitt, but with a marked personality of his own, discussed literature and its tendencies with special reference to our younger generation. His interest in youth and his concern for the directions it takes was perhaps sharpened by his own contacts with young folk in the classroom. But he was one of the first of our critics to feel the change coming over our social life, the innovations eagerly practiced by the young, if not first invented by them, the unconventional deviations for which Judge Ben Lindsey in his addresses and in his books has given intelligent and sympathetic explanation, but which Stuart Sherman contemplated with a grave and reasoned doubt.

The late W. C. Brownell occupied a unique place in our criticism. He followed no school and remained aloof from the literary moments of his time, but he was richly endowed with taste, the rarest of critical equipments, and he was profoundly read in the literature of the world, especially of modern France. His comments upon American literature, therefore, were unusually challenging in matters of taste, of style, of decorum, and his books seem likely to stand as the most serious and most subtle examination of contemporary writing.

Joseph Wood Krutch has brought to literary criticism a subtle and penetrating use of the modern psychology, and Van Wyck Brooks has compelled us, by the charm of his quiet reasoning, to see the esthetic lack, even at times the esthetic poverty, in American life—not in order to remove our pride in the past, but to awaken our conscience for the future. Agnes Repplier has maintained for decades a position of the highest distinction and honor as clear thinker, brilliant and kindly wit, subtle understander of human nature.

Irwin Edman, poet by temperament and philosopher by trade, speaks for the American generation which now, as the dust of the World War settles, looks about to get its bearings—much as Randolph Bourne, lonely genius, spoke for those American youths who were critically minded just before the war and during its early years. Carl Van Doren, scholar, novelist, and master of criticism, represents the enterprise and the sanity of the Middle West, and stresses for us the point of view of that section which in all probability will exert increasing control upon American opinion.

J. E. Spingarn, the most scholarly of our poet-critics, advocates an esthetic based on Benedetto Croce, but strongly personal and American. Kenneth Burke and Max Eastman, among the most subtle and brilliant of our prose writers, have redefined the nature of poetry and of the poetic experience.

In the early years of the century under consideration, American drama was not important. With all due respect to the theatrical glooms, gaieties, and romances which from the point of view of the practical stage made Harrigan and Hart outstanding, and in another vein Weber and Fields, or in a different kind Bronson Howard or Clyde Fitch or Augustus Thomas—it is not unfair to say that the American playwrights have contributed little to literature. They left us, that is, few plays that anyone now cares to read, and almost as few that any producer cares to revive. The stage today, by contrast, can name a large number of playwrights whose literary ability is extraordinarily high—such a dramatist as Sidney Howard or George Kelly or Philip Barry; and one, at least, Eugene O'Neill, who seems already a classic for the printed page as well as for the stage.

This brief and superficial glance at the activities of the decade beginning, and at the decade ending, the span between 1833 and 1933, leaps too lightly over important names in between, some of which will be mentioned in later paragraphs to illustrate other points, and some of which should at least have the tribute of a backward glance. Our poetry is richer for the work of George Edward Woodberry, accomplished writer, inspiring teacher, and for the work of William Vaughn Moody, fine and devoted torch-bearer on the threshold of our contemporary renaissance in verse-writing. Miss Harriet Monroe, as poet and as founder and editor of the Chicago magazine, *Poetry*, has provided for young talent a platform and a forum which, thanks to her devoted zeal, continues in unimpaired vigor after twenty years.

There will be occasion to speak more fully of Mark Twain and of Bret Harte, but here we ought to remember William Dean Howells, our most delightful prophet of realism; Hamlin Garland, who combined the realistic method with a romantic temperament and a love for our early national history; Stephen Crane, dramatic reporter of the colorful things in our life; Henry James, beautiful mind, willfully literary; George W. Cable, dreamy romancer of New Orleans; Joel Chandler Harris, who told us of Uncle Remus; Thomas Nelson Page, painter of the old South; Mary E. Wilkins, painter of New England; and O. Henry, epigrammatic romancer of the great city.

Eugene Field and James Whitcomb Riley have their permanent title to remembrance not only in their understanding of children, of whom and for whom they wrote, but in their shrewd and quaint perceptions of life which, as time goes on, seem obviously in the American vein.

If such names as Oliver Wendell Holmes, David Thoreau, Joaquin Miller, John Burroughs, Josh Billings, William Gilmore Sims, and dozens of others, are here passed over, along with such unique apparitions as the posthumous reputation of Henry Adams—the reason is that this article is brief, and its purpose is not to provide a bibliography of American writers during this period, but merely to suggest some broad points of view from which the achievement of that period in literature is significant.

THE INFLUENCE OF EUROPEAN CRITICISM

The product of our literature, then, if we consider the writers separately and their best books one by one, is impressive, yet few of us, even in America, are content to accept that product merely as art. We ask rather whether it is just the kind of product which the world should expect from America; either we think our books are too much like those written in Europe, or we wish they were more European; either we yearn for subject-matter peculiar to our world, or we regret that the author has specialized in it.

In fact, no great nation has ever been so uneasy about its literature as the United States. We are obviously not Europe, but Europe has just as obviously got on our nerves. No season goes by without a protest in some journal against the unfair attitude of foreign critics toward American writers, nor without sharp rejoinders from equally patriotic Americans who feel that the foreign critics are not severe enough. The irritation is maintained by visiting authors from across the Atlantic who, while they are with us, point out the defects of our books, or indict us more elaborately in whole volumes after they have returned home. A large part of the American reading public were fascinated by such criticisms, even when their feelings were hurt, and the visiting lecturer from abroad, who did not like us, has a way of returning, to dislike us some more.

Of course the visitor does not intend to be patronizing, and the American whom he irritates is neither a provincial imitator nor a frontier wild man. To the European America, in its literature as in other manifestations, means change, a threat against the habits he loves; he is free to like it only when it deals with scenes and manners which never occurred in Europe, and which therefore provoke no comparisons. To the American the European comment even when it seems unjust is a reminder that our country is still proving itself and has not yet arrived.

Between life in Europe and life here there is a real difference, not in kind but in degree; the European therefore begins by expecting from our books a picture of something he knows, but he ends by coming on something unfamiliar. Perhaps there is nothing in American social or economic life, the germs of which did not exist in the old country, but America has pushed to a conclusion what Europe was content to leave as a premise. Europe honors an engineer like Eiffel for his achievement in putting up one tall structure, but America builds many skyscrapers, of which the European is slow to approve. Europe sees the greatness of a genius like Pasteur, but in America it is easy to secure Pasteurized milk. It is fair to say, as Europe does say, that we have organized on a vast scale the life we inherited from the old country, and by so doing we have changed certain values to which the European reader was and is accustomed.

With us, to take at once the central example, business has a dignity which it does not yet enjoy abroad. We don't mind being shopkeepers if the shop is large enough and if we have plenty of customers. We'd rather see the eldest son run an automobile factory than hold a commission in the army. The learned professions carry with us no prestige in themselves over any other honorable employment, and practically all employment is honorable. And instead of paying respect to the aristocracy of birth and place, we like the man who begins at the bottom and works his way up. These views come from our history, and they affect the plot of the American novel and the emphasis which it gives to character.

On the other hand, however, art is in its nature the most traditional of pursuits, tending to avail itself of the medium and of the subject-matter already familiar and understood. The rapid changes of American life during these hundred years has embarrassed the American writer by depriving him of an adequate language. He rarely can speak through those images or facts around which in his youth he built up an emotional context; most of his readers have had other experiences, and have attached their emotions to other symbols.

Some of our writers, like Edith Wharton, meet the difficulty by declining to write at all about their own day, preferring to report as quaint news the days one can be sure of, since they are dead. This is not the same thing as using old characters to express universal ideas; Edith Wharton writes American, not universal or cosmopolitan, novels; she is quite as American as Booth Tarkington or Sinclair Lewis or Edna Ferber. She merely specializes in what has become, or can be made to seem, antiquarian. Superb artist though she is, she has chosen not even to try to interpret the America of the hour in which she writes. What she really interprets is not America at all, but the confusion of the European mind, trying to find in the United States a saving resemblance to the old country.

THE DECLARATION OF INDEPENDENCE

This retrospective attitude, flourishing even today among our writers, was well developed a hundred years ago, and Ralph Waldo Emerson unfolded his philosophy and began his literary career fighting it. His first little book, *Nature*, 1836, bade us to stand on our own feet, to cease building the sepulchers of the fathers, to take ourselves for better or worse, to build every man his own house. "Our Fathers," he said, "enjoyed an original relation with the universe—why should not we?" The address on the American Scholar, the Divinity School Address, the Essays, merely elaborated this doctrine of integrity, of self-reliance, this faith in the sources of life which frees us from imitative or servile relations to the universe.

Emerson seems to have spoken for those American writers who proved themselves our greatest—or at least who have commended themselves to the world as our legitimate—product. Thoreau was his colleague, and in some sort his disciple;

Walt Whitman was his supporter, the most powerful transmitter of his ideals, all the more convincing because more deeply immersed in that American world which Emerson counseled Americans to accept. Whitman had the prophetic gift; by a singular instinct he emphasized those aspects of the American scene that were destined to take on importance after he was gone; and he refused to attach too great importance to aspects that have since proved but temporary. Men in the street around him were saying that America was to be a worker's world, a machine world, a popular and therefore a vulgar world. Instead of trembling at the prospect, he sang of work and machines, of mass production and vulgarity, of arts reformed in the popular rather than to the aristocratic taste, of manners turned informal and spontaneous. Some readers paused to question his metric system, as though one should polish a river or put style on a volcano. Today he is an adequate spokesman for the toiler in Russia, busy with his astounding experiment, as for us at home, needing to be told what to do with the civilization we have made.

The sense of the gulf between America and Europe was present in writers so differently gifted as James Russell Lowell and Mark Twain. Lowell's brilliant paper on "A Certain Condescension in Foreigners" is still the best statement of the American attitude toward the patronizer from abroad—and it still makes the foreigner wonder why Lowell, and we, are sensitive. Mark Twain's *Innocents Abroad* repeated the Emersonian doctrine in the key of ridicule and irreverence. Few books have done so much to widen the Atlantic Ocean. Our tourists took courage from its pages, and learned to laugh at the absurdity of outgrown things.

But in *Tom Sawyer*, and especially in *Huckleberry Finn*, Mark Twain once more illustrated the Emersonian doctrine, this time on the great scale of creative literature, no less serious because it evoked deep humor. We had previously had no comparable pictures of common American life—if you choose, of vulgar life; and in these books the author taught the world

LITERATURE

to see in this phase of humanity elements of tragedy and of beauty.

Before Fenimore Cooper began his Leatherstocking series, he was defending the States from European criticism, and carrying the war into the enemy's country with more vigor than tact—pausing in his labors to say, from time to time, a few hard things of America on his own account. His stories of the woods and the prairie were his answer to the challenge to be oneself; this poetical aspect of America was entirely different from anything in the old country. Here, embroidering the contacts with nature and with frontier society which he had enjoyed in his youth, he discovered his unique genius as poet and landscape-painter. His fame is world wide. Of recent years he has been underrated, but unjustly. At his best he is one of the most truly creative writers the Western World has produced. The literatures of Europe would have difficulty in offering a match for such masterpieces as the *Deerslayer, The Last of the Mohicans,* and *The Prairie.*

Kindred to Cooper in spirit, less creative in his subject-matter but like him a poet, Francis Parkman wrote the early history of Canada in a series of volumes following the scholar's method, and yet making the appeal of an inspired epic. *The Jesuits in North America, The Conspiracy of Pontiac,* and *Montcalm and Wolfe* sound in their prose the football of destiny, the march of empire; they preserve great characters in vivid portraits; they set the scene, the continent of forests and mountains and lakes, on a majestic scale. They are among the noblest of our books.

Today the American novelist or poet is still conscious of the cultural rift with Europe, still wishing to bridge it over, or to accept it, in Emerson's fashion, and be frankly American. Sinclair Lewis is a satirist, but his books are a faithful report of at least that phase of American life which he has chosen for the moment to discuss. His portrait of us on the surface of our speech, our manners, our mass conduct, is disagreeably true. Edna Ferber draws a similar portrait, with sentiment and good nature, but still realistically. Theodore Dreiser works on native material with a devoted seriousness, squeezing out the last drop of grim truth, and ignoring somewhat those apparitions of truth which are radiant. Whether or not these three writers consider themselves Emersonian, they pursue their art with that integrity which he preached; they are of their place and time, and having written its story with honest power, they deserve the attention now paid to them everywhere.

American literature would be far poorer if we had to give up the occasional utterance of men in public life, who in moments of stress have struck into the national tradition words, phrases, even brief speeches—which remain the possession of us all. Theodore Roosevelt had a genius for inventing American phrases, or for turning known phrases to a new American content. "The Strenuous Life," "The Big Stick," "The Muckraker" have currency among us because of the way he used them. To a certain extent Woodrow Wilson had the same talent, though his most unforgettable contributions, like "Too Proud to Fight," were often unlucky. The contributions in this kind, of Abraham Lincoln, are altogether the most important, having a beauty and charm so inspired, in his letters as in his speeches, that many critics are persuaded he was a poet misplaced, a genius essentially literary. The art of letters has little to boast of more admirable than the Gettysburg Address and the tremendous letter to Horace Greeley.

THE QUEST FOR MATERIALS IN AMERICAN LIFE

The continued attempt of American literature to find its material in American life has raised among the literary the pertinent question whether there are not several kinds of American life, whether the states of the Union, the sections of the country, do not present variations in the type as wide as can be found from country to country in Europe. The Great American Novel, such questioners admit, might theoretically stand for all of us, but meanwhile it would be prudent for each novelist to confine himself to the limited scene in which he is thoroughly at home.

Consciously following this idea, many of our novelists and poets have specialized in a given locality. Robert Frost is the poet of New England, as George W. Cable was the novelist of New Orleans. But the sectional idea in novel-writing expands naturally into the reporting of all social pictures which to the general reader will have the appeal of news. Bret Harte's stories gave to the east and to Europe a report of the mining society in the West—a report which, whether accurate or not, had immense vitality and would now be difficult to revise. Writers like Margaret Deland, in *Old Chester Tales*, have chosen quaint areas, social back-waters, in which to study the quiet depths of human nature. Others, like O. Henry, have specialized in the great cities. Whether the emphasis is on the unusual scene or on the human nature it contains depends on the temperament of the writer, but for the general public it is the news value of such books that attracts.

This has always been true of novel-writing; the novel by definition is a record of society, a bulletin, as it were, of manners, morals, and etiquette. A hundred years ago, when the novel was not yet firmly established in Western literature, shrewd men and women doubted its permanent place as a literary form, on the ground that, since it dealt with what was, after all, nothing but news, it would come to an end as soon as the news had all been reported. The world has now been pretty well supplied with novels which report the doings of the rich and the sufferings, the toils, and the consolations of the poor. If the novelist in Minnesota now gives us a faithful bulletin of the farmer's life in that region, the result makes us think of the many similar bulletins already received from France, Germany, Russia, Norway, or Sweden. The American novelist who reports on the farm is not imitating the European novelists; farm life everywhere is in essence the same. When Sinclair Lewis wrote *Main Street*, the resemblance between his powerful book and Flaubert's *Madame Bovary* was at once noticed. The life in small towns is no longer new, except to those readers who have missed the earlier bulletins. And the life in various sections of the country is not unfamiliar, thanks to G. W. Cable, Owen Wister, James Lane Allen, Thomas Nelson Page, Irving Bacheller, Irvin Cobb.

The novelist, therefore, who remains loyal to the reporting ideal must, like any other journalist, go out and get a story. In recent years, since the United States is already well covered, there has been a tendency to report on Central America or South America, as in Thornton Wilder's *Bridge of San Luis Rey,* or in Mrs. Stone's *The Heaven and Earth of Doña Elena.* There has also been a tendency toward the autobiographical novel, since the personal career of each one of us has a fair chance of containing peculiar and original elements—original, that is, in the outward setting, if not in the human principle.

Theodore Dreiser is perhaps the chief of those who write biographically and autobiographically. His stories are well documented; in detail they report accurately things which have happened in the United States. His early novel, *Sister Carrie,* in the opinion of many readers his masterpiece, encountered bitter prejudice hard to understand today. It was thought to deal with subject-matter unsuitable for reporting in polite society. A novelist of Dreiser's character is too passionate for truth to be stopped by such opposition, but perhaps the opposition engenders an over-zeal for the kinds of truth objectionable to the tender-minded. In any case, the school of biographical novelists who have flourished among us in recent decades is supplying a remarkable record of our times, a record, on the whole, trustworthy, so far as the outward picture is concerned. Where the writer has attempted explanations of motive in the contemporary psychological style, one suspects that posterity will do some blue-penciling.

The growing search for new material to report on has led the American novelist into a rich field, the life of the immigrant peoples among us. In the United States the immigrants have added themselves, for the most part, to the classes economically poor. Our new arrivals are usually laborers and workers who may bring with them a cultural inheritance and a hunger for the arts, but who rarely are in position at first to contribute to American arts, and who almost never make a recognizable contribution to the arts of the country from which they came. The lives of these national or racial units during the period of their amalgamation with earlier arrivals in America, are picturesque always, tragic sometimes, sometimes humorous. If a novelist arises from among themselves, he has his material ready to hand, in the fortunes of his people, but more often it is a sophisticated writer from a racial element already established in America who seizes on this good material—an artist, for example, like Fannie Hurst, who in her numerous short stories and novels reports, with the skill of a trained and sympathetic observer, the lives of the new elements in process of becoming like the rest of us.

The colored race in the United States has made remarkable contributions to our literature, remarkable if judged even by absolute standards, without allowance for the handicaps of the Negro. Booker T. Washington, in his work for the industrial and social advancement of his people, was not more successful than William Edward DuBois or Paul Lawrence Dunbar or James Weldon Johnson in their labors for the intellectual and artistic progress of their race. DuBois was a sensitive thinker, Dunbar a true poet; Weldon Johnson is a poet, a scholar, and a statesman. What the Negro will eventually do in our literature will surely be important, if young poets like Langston Hughes are an indication of what is to come.

But meanwhile the books about the colored people which have made the widest impression have been written by the whites. *Uncle Tom's Cabin* seems permanently imbedded in the racial question; Joel Chandler Harris gave us our classic insight into the Negro folk-lore; Roark Bradford carries us farther in this field, and Marc Connelly, in his unique "The Green Pastures," building on the work of Joel Chandler Harris and Roark Bradford, adds his own dramatic sense to create what the American public has recognized at once as a spiritual monument to the colored race in America. Du Bose Heyward in *Porgy*, both as novel and as play, and Julia Peterkin in *Scarlet Sister Mary* and other novels, have reported on

Negro manners, modes of thought and feeling, with a poignant beauty arising partly from the material and partly from the talent of the writers. The Negro is still news among us.

The Irishman once was, and the German, but they have now ceased to be. In California the Chinese, and later the Japanese, were news. Oliver La Farge in *Laughing Boy* recently demonstrated that the Indian and his problem still appeal to us as in the days when Helen Hunt Jackson wrote *Ramona*. A novelist in search of fresh material to report on will discover new racial elements when we have grown saturated with these, but no doubt the possibilities of this kind of novel will eventually prove limited.

Not all writers of fiction, however, care to accept the formula of the novel. A story need not deal primarily with the behavior of society in a given time or place—it may rest rather upon a universal plot through which march typical characters, appealing to general emotion and illustrating broad principles of conduct.

The writers of stories before Richardson and Fielding were novelists, if at all, only by accident. Many of them told their stories in verse. The critic who wishes to be subtle can distinguish between those parts of Chaucer which are poetic and those portions which are novelistic; he can point out that in the Iliad we have a romantic story with the emphasis placed upon an ethical problem, and only incidentally upon the picture of manners; but in the Odyssey, in spite of the adventurous plot, the attention to manners is considerable. Probably one can write no good story without evoking by the way the kind of picture in which the novel, by definition, specializes.

A number of American writers, avoiding the pure novel for temperamental or other reasons, have given us the sort of romance that in other ages would have been told in verse. In the late years of the nineteenth century there was a vogue for historical novels, perhaps energized by the popularity of Lew Wallace's *Ben-Hur,* but developed out of various aspects of American history by Mary Johnston, Meredith Nicholson, and others. The paradox of the historical novel is that it avails itself of the poetry which can be extracted from the past, but gives only a weak simulacrum of ancient manners. After all, one can't report very fully on what one has not lived or seen for himself. In recent years the fiction-writers who give us poetry rather than a report of manners increase in number. A unique figure among them is James Branch Cabell, subtle thinker and sensitive stylist, who in a series of astonishing books, the most famous of them *Jurgen*, has adapted the manner of Rabelais to a fanciful discussion of human fortunes and foibles. He is in some respects as purely the artist as any of our writers, and if he seems detached from our daily life, he compensates for this aloofness by studying with relentless penetration certain weaknesses in our character. He is at heart, it would seem, an austere moralist, but he is also a poet enamoured of all beauty, and making a richer sensuous response to life than has been the tradition of American letters. If his audience is at all limited, it is perhaps because he is in manner neither very cheerful nor very sad. The American reader likes one extreme or the other—the gloom of Dreiser or the optimism of Harold Bell Wright.

Somewhat in the tendency illustrated by Cabell, the late Frances Newman wrote two remarkable books impossible to omit from even a brief survey of our literature. Her first novel *The Hard Boiled Virgin*, most unluckily titled, led ignoble readers to expect salacious entertainment. They found instead a difficult style, carefully wrought but slightly monotonous in structure and cadence. What such readers probably did not find was a sensitive record of a fine nature reaching out for wider knowledge—a record containing perhaps as high a percentage of truth to life as any book written in our time. The second novel, *Dead Lovers Are Faithful Lovers*, reinforced the impression made by the first among discriminating readers.

This kind of book is as characteristic of American life as any realistic novel could be. We have written our best poetry, perhaps, in prose-narrative of this kind, which moves, as much as books can, in the realm of pure art, and which indicates that

Americans, like other people, have a large preoccupation with dreams and with general ideas. It may be that Hawthorne's *House of Seven Gables* gives a picture of New England, that his *Blithedale Romance* represents the Brook Farm experiment, that his *Marble Faun* describes Rome, but what Americans have loved in these volumes is what makes the *Scarlet Letter* a masterpiece—the portrayal of human nature on a large scale, with the suggestion of the mystery, the shadows, and the allure of life. Edgar Allan Poe, in his prose as well as his verse, seems to be moving through a region of higher mathematics rather than inhabiting any earthly land, yet his romantic abstractions express one side of the American character. Nowhere else, perhaps, would a romancer have been quite so abstract as this American in the administration of President Polk, out of sympathy with the drift of national policies and the interests of society around him, but not of a temper to strike back as Lowell struck in the *Biglow Papers*. Poe's instinct was to take refuge in the mind. His comment upon contemporary life lies partly in the completeness with which he turned his back on it.

Herman Melville, in *Omoo* and *Typee*, reported on faraway societies, and to that extent was a novelist, but poetry and philosophy were his true fields, and in *Moby Dick* he gave us what many think the most colossal achievement of American literature in pure art. One who knows us well, however, would surely recognize the American characteristics even in *Moby Dick*. Melville, the friend and neighbor of Hawthorne, could not escape our moralizing tendencies, not even when he fixed his imagination on remote and vast stretches of water and told the feud between the mad sailor and the fabulous whale. Like Poe, he comments on American life by the omissions he makes from his picture. His resolute silence about contemporary events and customs is itself a kind of record.

In recent years the reputation of Emily Dickinson has risen enormously. The complete publication of her letters and the reissue of poems by her niece, Martha Gilbert Dickinson

Bianchi, who brilliantly carries on the family talent, have furnished occasion for numerous studies of this recluse poet, and biographies more or less speculative have tried to reconstruct her private affairs. Her electric kind of wit, her often poignant insights into experience, her strangely casual and sometimes irreverent dealing with what her countrymen consider sacred things, her tone of comradeship with the solar system, have startled modern readers into believing that she is perhaps our greatest poet. It is the fashion to say also that she is one of our pure artists, moving in the upper ether, uncontaminated by temporary incidents. Whether or not her detachment from life is a merit, those may debate who care to. It might be simpler to point out that she was not nearly so detached as she seemed, that her detached manner was a protection not infrequently set up in New England and in America generally, a blind behind which the hypersensitive would shield themselves. The audacities of her verse also are perhaps the nervous indication of what an exaggerated seclusion had done to her. They are not unfamiliar in American manners everywhere, especially in quiet villages.

Perhaps the poetry revival which began in the second decade of the twentieth century was only an illustration of the quest for new material. Edgar Lee Masters gave us *The Spoon River Anthology*, a striking picture of what to many readers was an unknown section, a picture which might well have been a novel. Amy Lowell excelled in exotic themes, few of her poems being on American subjects. Robert Frost limited himself to his special account of New England, and Sandburg turned to poetic account aspects of the machine age not yet fully developed in our literature. As the movement spent its force and died, the figure that chiefly emerged, Edward Arlington Robinson, was occupied with Old World themes quite as much as with American subjects. He was writing, in other words, about whatever he chose, letting his Americanism take care of itself.

The record of our literature, then, is rich in individual authors and books. It seems to show also that from the beginning until now our writers have spent much thought upon their proper mission in the world—too much thought, perhaps. They have tried to decide whether to maintain European traditions in poetry and fiction, or whether to invent new forms suitable to a new country, or whether to let themselves go, secure in their sincerity and trusting to the new country to force upon them the suitable new forms. The majority of writers today are in this last class. Our literature is definitely committed to be itself, and the excitement of social and economic upheavals at home and through the world at large seems calculated to supply authors here as well as elsewhere with new and important inspirations.

THE BUSINESS OF PUBLISHING

Something should be said, however, about the publishing conditions in the United States, in their origin geographical and economic. The size of the country has made the distribution of books difficult. Postal and other facilities have made the circulation of magazines comparatively easy. American literature has become, therefore, definitely journalistic, with certain consequences to the quality of our writing.

The *Atlantic Monthly* was established in 1857, *Harper's Magazine* a few years earlier, *The North American Review* still earlier. Among recent magazines we have nothing comparable to these dignified publications except *The American Mercury,* and Henry Mencken's sprightliness would for some Americans render impossible any mention of his magazine in the same breath with the *Atlantic.* The modern magazine is a quite different kind of vehicle, a purveyor of high-class news, a medium not for literature but for advertising. The advertising depends upon the circulation, the circulation is maintained by blurbs and bally-hoo, each number being quite different from the last, and every front page announcing the acquisition of the writer most in the public eye, or an article from some person temporarily talked about.

The tendency, therefore, among these magazines is to buy reputations rather than writings, to recruit the popular author rather than to hold him up to his best standard. The novelist who has written a best seller is not infrequently invited to contribute to the magazines his next story, sight unseen. The magazine may not be able to use the story when it comes in, but in most cases it will print the thing, no matter how disappointing.

Because the modern magazine is a vast business, and the vital spot is the advertising, maintained by the circulation, the modern editor naturally tries to play safe. He not only wants big names, but he wants his contributors to furnish him with the kind of writing for which they are popular.

He assumes, in other words, that literature is not an art, expressing the author, but a business, supplying a demand; he assumes that a public responsive to one "Hamlet" would be glad to come back for six more.

His assumption is of course quite wrong, and the magazines of large circulation which follow this policy are in constant danger of losing their readers. The moving pictures make the same assumption on an even larger scale, with the same result. In time the editors and the picture magnates will learn that the theme well expressed is a theme to avoid, and that there is no such thing in art as demand and supply. There was no demand, in advance, for Shakespeare's work or Homer's, and there is no demand for imitations of them. In other words, the editors and the picture directors will some day recall that though they are in business, they are dealing with an art.

RESTRAINTS ON THE ART OF WRITING

Meanwhile the effect upon our writers is dangerous. A novel which sells one hundred thousand copies is a best seller, even in a population of a hundred million and more; for the distribution of books is with us shamefully inadequate. A hundred thousand copies will bring the author no more than forty thousand dollars, but the serial publication of his book, if he is well known, will yield him at least fifty thousand dollars, often a much larger sum, and he will have the book profits besides. The magazine can promise him several million readers.

But the magazine editor probably will not publish the book quite as the author wrote it; for reasons of editorial policy, this will be omitted, and that will be, if not added, at least suggested to the author in strong terms. Magazine publication, therefore, means for the author a loss of freedom, and if he insists on the integrity of his art, it will cost him between fifty and seventy-five thousand dollars. Most novelists don't insist.

The novelist can sell his story to the movies and reach a still larger audience—and what writer would not address a large audience if he could? But in the pictures his freedom will be still further curtailed, and he is fortunate if he can recognize anything in the product eventually reaching the screen except his name. The plot of his story will be changed, the characters reorientated, and another title will be substituted. The temptation is great to resign oneself to these conditions, especially since the films are often excellent, even if they have nothing to do with the books on which they are supposed to be founded. But an author should be able to speak his message in his own way to his readers. In America today, the wider his audience the more his message will be altered, against his will.

If it were not for this condition, books might go out of fashion, and we might enjoy all our literature in newspapers, magazines, and pictures. But so long as the writer has more freedom of speech in books than in the other mediums, books will remain, even if authors have to pay for publication.

Some writers would say that no account of our literature during the past hundred years would be complete without mention of censorship, direct or indirect—of the attempts of prejudiced people to suppress thoughts they disapprove of. Others would add that the account should deal with the decline of literary style in America, the spread of colloquialism and vulgarity.

But as to the first point, it is fair to remember that all nations have a bad record in censorship, all peoples have among them some stupid and narrow-minded, and the problem is not peculiar to us. In fact, our record, bad as it is, is better than the average. We often try to suppress books, but we rarely succeed. We were unappreciative of Poe, but we didn't try to put him in jail, as the French tried to put Baudelaire. Some of us were annoyed at Sinclair Lewis's *Main Street*, but we did not arrest him for indecency, as the French arrested Flaubert, on the appearance of *Madame Bovary*.

And as to the vulgarity—there is something to be said for it. Our art represents us. If we are vulgar, then our art should be vulgar, too. It is useless to expect literature to serve as a screen between the reader and what we are. Too often art has become refined by moving beyond contact with the common man, as though the common man were hopelessly contaminating. In the end it is better for literature to belong to us all, an honest index of our condition. This American ideal of art was well stated by Lincoln Steffens in his *Autobiography*, in his comment on the moving pictures: "This new machine mass art cannot develop for its own sake; it is so tied to democracy that it cannot rise to its obviously potential heights without lifting and being lifted by the human race."

American literature, like the pictures, can wait patiently for its potential heights, provided it remains closely tied to the daily life of the human race.

Index

Abbot, Charles Greely, 300
Abbott, Edith, quoted, 276-277
Abel, John J., 308, 353
Abt, Dr., 342
Accuracy, economic effects of, 79
   the craftsman and, 78-80
Acheson, Edward Goodrich, 307
Act of 1921, 192
Adams, Abigail, 208
Adams, Henry, 407
   quoted, 86
Adams, John Quincy, 86
Adams, Roger, 306
Adams, Walter S., 298
Addams, Jane, 274
Adler, Alfred, quoted, 289
Administrative efficiency, the quest for, 213-216
Adrenal glands, isolation of the secretion of, 308
Aedes mosquito, 347
Aërial exploration, 319
Aërodynamics, principles of, 52
Africa, American scientists in, 319
Agassiz, Alexander, 314, 350
Agassiz, Louis, 313, 314, 318
Agramonte, Aristides, 312, 345
Agricultural colleges and stations, 133-136
Agricultural coöperatives, 139-140
Agricultural Market Act, 219
Agricultural organizations, 136-138
Agricultural research, 130-132
Agriculture, before the rise of, 123
   butter-fat test invented, 134
   colleges and stations of, 133-136
   commercialized, 129
   dark side of, 142-145
   decline of workers in in the United States, 36
   effect of specialization on, 128-130
   government protection of, 219
   Hatch Act, 133

Agriculture— (*continued*)
   historic system of in the United States, 123-126
   improvement in crops, 135
   in Egypt, 123
   invention and science as applied to, 131
   livestock improvement, 135
   loss through erosion, 144
   markets and prices, 138-139
   mechanical invasion of, 126
   Morrill Act, 133
   need for foresight and planning, 145-147
   plant pests and disease, 132-133
   progress of, 123
   research in, 130-132
   revolution in, 37
   the depression and, 130
   Virgil and, 130
   waste of fertile soils, 142-145
Aircraft, transportation by, 51
Airplane, the, 24
Air travel, the future of, 51
Airways, opening of the, 109-114
Aitken, R. G., 298
Albany Post Road, 92
Albec, Dr., 343
Allegheny Observatory, 296
Allen, E., 308
Allen, James Lane, 414
Allison, Fred, 306
Alter, Dr. David, 301
Alternating current, 29
Altgeld, Governor, 343
America, adoption of universal education in, 12
   an agricultural nation, 124
   archeological researches in, 320-321
   chemical laboratories in, 63
   development of the steam-engine in, 28
   discovery of, 9
   early surgery in, 329
   extension of the suffrage in, 12

426   INDEX

America—(continued)
    fewer locomotives today than in 1911 in, 48
    growth of a leveling freedom in the Agrarian West in, 12
    low-cost handling of raw material in, 45
    machines that make machines imported, 73
    pioneering spirit in, 71
    rediscovery of, 320-321
    the most highly mechanized social organism in the world, 26
    war on suffering, disease, ignorance, and misfortune, 224
American Academy of Science (Boston), 294
American achievement transcends best of Greeks, 16
American Association for Labor Legislation, 251
American Association for Old Age Security, 225
American Association for the Advancement of Science, 324
American banking, present position of, 195
American bill of fare has changed, 48
American bourgeois, the, 13
American Chemical Society, 305, 324
American civilization, 13
American colonies, and the concept of progress, 11
    a united and independent power, 11
    Great Britain and the importation of machinery into, 71
    idea of progress in, 9
American culture, 20
American Farm Bureau Federation, 137
American Federation of Labor, 154, 167, 169, 239
American finance and Eastern financiers, 173
American industrial centers huddle near coal-mines, 28
American industrial pioneering, 70-72
American industry, changes in, 80
American Journal of Science, 294
American life, actualities of, 12
American literature, 420
American locomotive, development of the, 46
American long-haul transportation, 47
American Medical Association, 324, 339, 350
American medicine, beginnings of, 326-330
American Mercury, The, 420
American Museum, 324
    expeditions, 320
American note in the arts, 391-392
American Notes (Dickens), 100
American Philosophical Society of Philadelphia, 294
American Pleistocene, 321
American political economy, transformation of, 216-223
American surgeons, 328
"American System of Manufacture," 34
American Telephone and Telegraph Company, 119, 120
American wage-earner, the, 29
Americans, meaning of life for, 14-15
    native, attitude toward immigrants, 237-238
    pay large honoraria to Europeans, 15
America's first inhabitants, 321
Amiel, 13-14, 15
Amos, Dr., 349
Ancient books, slavish adherence to, 9
Anderson, Dr. Sherwood, 403, 547
Andrews, Edmund, 337
Anesthesia, 310-311, 334-338
Ångström, 301
Antarctic continent, discovery of, 319
Anthony, Susan B., 208, 256, 283
Anthropology, 322
Antisepsis, 328
Appendicitis, 342
Apperson, Elmer, 107
Appleby knotter, 43
Appointment of women to political positions, 286-287
Archeology, American exploits in, 319
Archimedes, 22
Architecture, American, 382, 384-386, 388, 392-395, 396
Archives of Ophthalmology, 344
Arc light, 26
Argive Heræum, excavation by Americans in, 320
Aristotle, 22
Arizona, the suffrage in, 208
Arkwright, Richard, 41
    spinning frame of, 38

INDEX 427

Armsby, Henry P., 309
Army Nursing Service, women in the, 287
Art, supply and demand in, 421
Art of writing, restraints on the, 421-423
Arts, the, in America, 380-399
Ashburn, Dr., 347
Ashford, Col. Bailey K., 346
Asia, American scientists in, 319
Association of Highway Engineers, 213
Association of University Women, 273
Astræa, 14
Astronomy, 295-300
Athens, the slave-owners of ancient, 9
Atherton, Gertrude, 403
Atlantic, the (balloon), 110
Atlantic cable laid, 117
Atlantic Monthly, The, 420
Attwood, Charles, 388
Atwater, W. O., 309
Audobon, 292, 293
Austen, Jane, 275
Austen-Leigh, James Edward, 275
Austin, 106
Australia, output of work per person in, 35
Austria, women's rights in, 283
Autobiography, Steffens', 423
Automatic screw-machine, 34
Automobile, the, 23-24, 37, 48, 49-50, 69, 74, 105-109
Ayres, Daniel, 336

Babbitt, Irving, 404
Babcock, Dr. Stephen M., 134, 399
Bach, 21
Bache, Alexander Dallas, 318
Bacheller, Irving, 414
Bacillus Welchii, 342
Bacon, 10
Bacon, Henry, 389
Bacteriology, 340-342
Baekeland, Leo H., 307
Bailey, Liberty Hyde, 316
Baird, John L., 59
Baird, Spencer Fullerton, 313
Baldwin, Matthias, 96
Ballads and Other Poems, Longfellow's, 402
Baltimore & Ohio Canal, 94, 95, 100, 104
Baltimore & Ohio Railway, 93, 94-95
Baltimore & Susquehanna Railroad, 96
Bandelier, A. F., 320

Banking and finance, 170-200
Banting, F. G., 308
Bard, John, 328
Bard, Samuel, 327, 329
Barker, Dr., 341
Barnard, George Grey, 394
Barnard College, 266
Barnes, Margaret, 403
Barnum and Bailey's "horseless carriage," 50
Barry, Philip, 406
Barton, Clara, 284
Basket-makers, 321
Battey, Robert, 337
Baudelaire, 423
Baudot, 56
Beacon Lights of Science, 111
Beard, George M., 343
Beaumont, William, 309, 326, 330-331, 333, 352
Beck, Emil, 352
Beck, T. R., 329
Bedford Cut Stone Company *vs.* Stone Cutters Association, 165
Beers, Clifford W., 322
Beginning of the Mechanical Transport in America, 107-108
Belgium, output of work per person in, 35
Bell, Alexander Graham, 21, 24, 25, 57, 71, 118, 303
Bell scissors reaper, 26
Bell Telephone Company, 119
Bell Telephone laboratories, 63, 64
Bellevue Hospital (N. Y.), 338
Benedict, Francis G., 309
Ben-Hur, 416
Benjamin, Judah P., 101
Bennett, Dr. H. H., 144
Bennett, Jesse, 328
Benton, Thomas, 397
Benz, 24, 49, 106
Bernard, Claude, 344
Besant, Walter, 239
Bessemer, Sir Henry, 25, 41
Bessemer steel, 25
Best, C. H., 308
Best Friend of Charleston (locomotive), 96
Bethlehem Steel Works, 36
Beth-Shan, excavation by Americans at, 320
Bianchi, Martha Gilbert Dickinson, 418

Big industry, 246
Bigelow, Henry J., 335, 336
Biglow Papers, Lowell's, 402, 418
Billings, John S., 332, 338-339
Billings, Josh, 406
Biological chemistry, 308-313
Biology and entomology, 313-317
Birkhoff, George David, 304
Bismarck, 282
Bitter Cry of Outcast London, 240
Bituminous coal fields, 89
Blackburn, 383
Blackstone Canal, 89
Blackstone quoted, 227, 257, 261
Blackwell, Elizabeth, 271
Bland-Allison Act, 183
Blast-furnace, the, 25
Blithedale Romance, The, Hawthorne's, 402, 418
Block printing, invention of, 54
Board of Tax Review, women in the, 287
Boat propulsion, adapting the steam-engine to, 44
Boaz, Franz, 322
Bobbs, John S., 336
"Bonanza" farms of the Far West, 44
Bond, W. C., 297
Booth, Charles, 239
Boss, Lewis, 298
Boston, early education in, 359
  philosophers arrive at, 235
  religion in, 235
  sanitary conditions in, 223
Boston & Lowell Railroad, 96
Boston & Maine Railroad, 104
Boston & Worcester Railroad, 96
Boston Post Road, 92
Boulton, 32, 88
Bourne, Randolph, 405
Bowditch, Henry Ingersoll, 334, 338, 345
Bowen, Joseph T., 241
Boylston, Zabdiel, 327
Bozeman, Dr., 337
Bradford, Dr., 343
Bradford, Roark, 415
Brainerd, Daniel, 326
Bramah, 25, 33
Brandeis, Judge, 230
  quoted, 165
Brashear, Walter, 329
Brehmer, Dr., 341
Bretonneau and intestinal lesions, 334
Bridge of San Luis Rey, Wilder's, 414

Bridges, 383
Bridges, C. B., 315
Bright's disease, 343
Brill, Nathan E., 347
British India, output of work per person in, 35
Broadcasting. *See* Radio
Bromfield, Louis, 403
Brook Farm experiment, 418
Brooks, Charles, 370
Brooks, Van Wyck, 404
Brooks, William Keith, 96, 314, 350
Brown, Rev. Antoinette, 273
Brown, Charles Brockden, 208
Brown, E. W., 299
Brown, Joseph R., 34
Brownell, W. C., 404
Brush, C. F., 23, 303
Bryan, William Jennings, 210
Bryant, Gridley, 93
Bryce, James, 60
Bryn Mawr College, 266
Bubonic plague, 348
Buckingham, 56
Budget system, 192, 216
Buffalo (N. Y.) school system, 365
Buffington, Leroy, 393
Buick, Herman, 108
Bulfinch, Charles, 385
Bullock, William, 55
Burbank, Luther, 351
Bureau of Agriculture established, 131
Bureau of Foreign and Interstate Commerce, women in the, 287
Bureau of Public Personnel, 213
Bureau of the Budget, installation of, 192
Burke, Edmund, 11
Burke, Kenneth, 405
Burnham, Daniel Hudson, 388, 393
Burnham, Sherburne Wesley, 298
Burrill, Thomas J., 311
Burroughs, John, 406
Burrows, 351
Burton, William M., 307
Bury, J. B., 9, 10
Bushnell, George E., 347
Butter-fat test, invention of the, 134
Byrd, Admiral Richard E., 319

Cabell, James Branch, 403, 417
Cable, George W., 406, 413, 414
Cable railways, 103
Cady, H. P., 307

INDEX

Cæsarian sections, 328
Calhoun, quoted, 205
California, civil service in, 212
  discovery of gold in, 47
  suffrage in, 208, 284
California Fruit Growers Exchange, 139
California Institute of Technology, 302
Camden & Amboy Railroad, 96
Campbell, W. W., 298
Canada, output of work per person, 35
Canadian banks, 176
Canal Zone, safety of life in, 355
Canals, early, 89-91
Cannon, Walter B., 313, 344, 352
Cannon-boring, 32
Capital investment in the hauling of wood or coal, 44
Capital punishment, reduction in crimes entailing, 229
Capitalism, 17
  rich profits of, 12
Capitalist, enrichment of by machinery, 66
  became a dictator, 149
Carleton, Mark Alfred, 316
Carrel, Alexis, 312, 351
Carrel-Dakin solution, 312
Carrère, John, 378
Carroll, Dr. James, 312, 343, 345, 346
Carnegie, Andrew, 41
Carnegie Institution of Washington, 321, 326, 349
Carnegie, non-magnetic, 319
Carter, Henry R., 345
Carter, James C., 370
Carthage, excavation by Americans in, 320
Carty, 63
Cascade Tunnel, 104
Cassatt, Mary, 390, 391
Castberg law (Norway), 262
Caste, barriers of, cut, 4
Castle, W. E., 316
Cather, Willa, 403
Catholics denied a share in government, 206
Catlin, George, 344
Catt, Mrs. Harriet Chapman, 283
Cattell, J. McKeen, 322
Cayley, 24, 52
Central banking system called for, 197-198
Central Pacific Railroad, 101

Central Park, arrest of first man who drove an automobile through, 50
Central-station system, 26
  the modern, 29, 30
Centralization, 231-232
Chamberlin, Thomas C., 299, 318
Chapin, Dr. James, 347, 397
Charlotte Dundas, 88
Chase, Salmon P., 181
Chaucer, 416
Chemical industry, German, 63
  coal tar, 63
Chemistry, 305-313
Cherry Mine disaster, 242
Chesapeake and Ohio Canal, 89
Chicago, industries in, 235
  gang violence in, 249
  Juvenile Court in, 243
  Juvenile Protective Association 243
  Mothers' Pension Act, 243
  necessities of factories in, 31
  political corruption in, 239
  settlements in, 240-244
  Visiting Nurses Association, 243
  World's Fair of 1893, 238, 388, 389
Chicago & Northwestern Railroad, 100
Chicago, Milwaukee, St. Paul & Pacific Railroad, 104
Chicago packers, 50
Child labor, 168
Children's Bureau, 286
Chilled-steel plow, invention of the, 42
China, output of work per person in, 35
  population of, 70
Chinese Exclusion Act of 1882, 157
Chloroform, administration of, 329, 335, 337
Cholera epidemic of 1832, 333
Chomel, 334
Christian heritage near the close of the thirteenth century, 7
Chromium alloy steel, 25, 37
Cincinnati College, 333
Circumnavigation of the globe, 9
City managership, 215
City planning, 146
Civil Service Assembly, 213
Civil Service Commision, 285
Civil service reform, 210-213
Civil War, American neurology cradled in the, 342
  bonds of the, 184

Civil War—(*continued*)
  finance preceding the, 186
  internal revenue system of, 187
  resources of the United States at the opening of, 179
  taxes during the, 182
  the cotton gin and the, 39
  the reaper during, 41-42
  value of currency at the close of the, 180
Civilization, America a guiding principle for a, 12
  American, 13
  beginning of, 15
  fatalism and, 7
  in the United States, 16
  is characterized by the buoyancy of youth, 6
  is only in its infancy, 10
  is on the threshold of time, 6
  issue of, 15
  material benefits of, 6
  rise of, 3
  the machine and, 66
Clark, Alonzo, 334
Clark, Alvan, 298
Clarke, Frank W., 305, 306
Class, barriers of, cut, 4
Classic, despotism of the, 9
Classical economy, doctrines of, 217-218
Clayton Anti-trust Act, 165, 218
Clemen, Dr. Rudolf A., xxii
Clermont, the, 45, 88, 89
Clevenger, S. V., 343
Coal, a potential energy, 27
  capital investment in the hauling of, 44
  cost of in manufacturing, 28
  cost of transporting, 28
"Coal oil," 49. *See also* Gasoline; Kerosene
Coast and Geodetic Survey, 318
Cobb, Irvin, 414
Cocaine injections, 337
Cohn, E. J., 308
Cohn heart disease, 349
Cohnheim, Dr., 338
Cold Spring Harbor laboratory, 316, 351
Cole, Rufus, 349
Cole, Thomas, 387
Cole, W. H., 352
Collective agreement in labor, 158-160
College of Physicians and Surgeons, 332
Collins, Dr. Joseph, 331

Collip, J. B., 308
Colonies, physicians in, 328
Colonial portraiture, 383
Colorado civil service in, 212
  the suffrage in, 208, 283
Columbia, District of, malarial fever in, 348
Columbia University, 389
Commercial revolution, 9
Commission government in cities, 214-215
Committee on the Cost of Medical Care, 354
Common things, natural science and the observance of, 9
Communicable diseases, control of, 340-342
Communication, 24-25, 86, 115-121
Communism, 17
Compton, Arthur H., 301
  spinning-mule of, 38
Compulsory school attendance, 364
Comstock, Anna B., 316
Comstock, John Henry, 316
Concept of progress, affiliated with democracy, natural science, technology, social amelioration, 3
  a gigantic intellectual outcropping, 9
  a gospel of futurism, 6
  and Western civilization, 3
  a philosophy of history, 6
  arouses lethargic Russia, 4
  assumes an indefinite future, 6
  conditions necessary to the flowering of, 9
  confronts obstacles with assurance, 4
  contends that mankind is advancing, 6
  critics and scoffers of, 16
  cuts across the barriers of caste, class, race, and nationality, 4
  decentralization of, 129
  democracy acting under, 13
  deserves thorough exploration and illustration in particulars, 4
  discussion of, 6
  dissolves the feudal institutions of Europe, 4
  disturbs the slumbers of the Orient, 4
  escapes the illusion of finality, 4
  European criticism of idea of progress in American civilization, 13
  finds a naked avowal in the United States, 4
  gives a clue of meaning to the rise of civilization, 3

INDEX 431

Concept of progress—(*continued*)
has established comforts and conveniences, 3
has demonstrated that symmetry and efficiency can be carried into modern life, 4
has lengthened the span of life, 3
has made famine obsolete, 3
has made knowledge popular, 3
has silenced or assuaged pain, 3
has stamped out disease, 3
has supplied sanitation to multitudes, 3
hotly attacked, 6
in America, 12
in the American Colonies, 9
in the hands of the demagogue, 4
in the Middle Ages, 9
in the nineteenth century, 12
is opposed to certain views of life, 7
it suggests faith of power, 4
its controlling interest, 6
kernel germinating in the heart of the, 17
may be employed to cloud issues, evade responsibilities, and justify cruelties, 4
offers a philosophy of individual and collective action, 4
plans for coming years, 6
rationality and, 18-19
regards forces of nature as potential instruments of humane purposes, 4
runs counter to the doctrine of fatalism, 6
shares the strength of universality, 3
superficial champions of, 17
survives friends and enemies, 4
unknown to the ancients and Middle Ages, 3
warmly defended, 6
what it can do, 3
what it implies, 3
what it is a product of, 9
Concrete History of Medicine, A, 325
Conduct, our fundamental attitudes toward, 5
Conference of State Sanitary Engineers, 213
Congress, women in, 287
Congress and the public school, 363
Conklin, E. G., 315
Conspiracy of Pontiac, The, Parkman's, 411

Convict labor, exploitation of, 168
Cooke, 56
Cooley, Judge Thomas M., quoted, 360
Coolidge, W. D., 302
and the incandescent lamp, 63
Coolidge tube, 352
Cooper, James Fenimore, 401, 402, 411
Cooper, Peter, 95
Cope, E. D., 318, 319
Copley, John Singleton, 383-384
Corinth, excavation by Americans in, 320
Corliss, George H., 23, 27, 28
Cornell University, 266
Corner, 351
Corporate farms, 142
Corporations, as manufacturers, 81
large-scale, 218
Corruption in cities, 214
Cotton, 39-40, 124
Cotton gin, the, 26, 38-40, 124
Cottony scale, 132
Cottrell, F. G., 307
Court decisions in labor cases, 163-164, 165
Bucks Stove and Range Co. *vs.* Gompers, Mitchell and Morrison, 164
Commonwealth *vs.* Hunt, 164
Hitchman Coal and Coke Co. *vs.* Mitchell, 164
Loewe *vs.* Lawlor, 164
Craftsman, the, and accuracy, 78-80
Craig, Dr., 347
Crane, Stephen, 406
Creditors, satisfaction of claims of, 226
Crile, Dr. George W., 337, 338, 352
Critics of progress, 16-17
Crocker, Charles, 101
Croly, Herbert, 214
Crops, improvement in, 135
Cugnot, Captain, 48
Culprit Fay, Drake's, 401
Culture, what we conjure up as, 21
democracy incompatible with, 13
"Currency reform," the old question of, 185
Curtis, George William, 211
Curtiss, Glenn H., 113
Cushing, Dr. Harvey, 337, 343, 352
Customs Court, woman judge in the, 287
Czecho-Slovakia, output of work per person in, 35
public health in, 355

432 INDEX

Dacks, the, 351
Da Costa, Dr., 341
Daimler, 24, 49, 106
Dalton, John Call, 343, 344
Dampier-Whetham, 131
Dana, James Dwight, 317, 318
Dandy, Walter E., 352
Danielson, 328
Dante, 16, 21
Darby, 25
"Darius Green." *See* Langley, Samuel P.
Darwin and evolution, 313, 314, 318
Davenport, Dr., 351
Davenport, Charles B., 316
Davis, William Morris, 318
Davisson, C. J., 301
Davy, Sir Humphry, 26, 335
Dead Lovers Are Faithful Lovers, 417
Debt repudiation, 192
Declaration of Independence and the consent of the governed, 207
Decline of Aristocracy in the Politics of New York, 201
Deere, John, 42, 128
Deerslayer, Cooper's, 402
Defoe quoted, 264
De Forest, 25
Degradation of the Democratic Dogma, The, 86
Dejerine, Mme., 355
Delafield, Dr., 341
Deland, Margaret, 413
Delaware & Hudson Canal, 89, 95
Dell, Floyd, 403
Demagogue, concept of power in the hands of the, 4
Democracy, and progress, 3
  extension of, 206
  incompatible with "culture," 13
Demodulation, 57
Demuth, Charles, 396
Department of Education, first, 363
Depression, world-wide, 1929, 250-251
de Schweinitz, George E., 344
Detmold, William, 336
de Tocqueville quoted, 203
Detroit Observatory, 296
Dettweiler, Dr., 341
Devignes, 24
Dewey, John, 378, 403
De Witt Clinton (locomotive), 96
De Witt Clinton (steamboat), 45
Dick, George F., 312
Dick, Gladys H., 312

Dickens, Charles, 100
Dickinson, Emily, 418
Dickinson, John, xii
Dickinson, Martha Gilbert, 418
Diesel, Rudolf, 23
Diesel engine, 23
Differences between American and European cars, 50
Dingley Act, of 1894, 187
  of 1897, 187
Diphtheria in Roxbury, Mass., 327
Disease, stamping out of, 3
Diseases of the Interior Valley of North America, 333
Dix, Dorothea, 256, 343
Dixon, Ryan Fox, 201
"Documents Relative to Manufactures," 277
Dodge Commission, 345
Doisy, E. A., 308
Dom Pedro, Emperor, 118
Doubt, Nirvana of, 4
Drake, Daniel, 49, 332, 333
Drake, Joseph Rodman, 401
Drama, American, 405-406
Draper, John William, 297
Dreiser, Theodore, 411, 414, 417
Drew, Daniel, 99, 351
Drift of rural population to industrial centers, 43
Drosophila Melanogaster, experiments on, 315
DuBois, William Edward, 415
Dudley, 329
du Motay, 26
Dunbar, Paul Lawrence, 415
Dunbar, William F., 355
DuPont laboratories, 64
Durand, Asher B., 387
Duryea, Charles E., 105, 106, 107, 108
Duryea Motor Wagon Company, 107
Dutcher, R. A., 310
Dutton, Clarence E., 318
Dutton, H. D., 318
Dynamo, invention of first, 55

Eakins, Thomas, 391, 392
Early epidemics, 327
Early traveling, 47
Earth, subjugation of the materials and forces of the, 3
East, E. M., 316
Eastman, George, 307
Eastman, Max, 405

INDEX 433

Eastman Kodak laboratories, 64
Eastwick, 96
Economics, influence of electricity upon, 60
Eddy, Walter H., 310
Edebohls, George M., 343
Edison, Thomas A., 21, 22, 23, 26, 28, 29, 30, 56, 59, 64, 71, 72-73, 303
Edman, Irwin, 405
Education, in the United States, 356-379
  popular, 232
  women and, 264-269
Effect of specialization on agriculture, 128-130
Efficiency engineering, 37. *See also* Scientific management; Taylor's system
"Efficiency" systems, 161
Egoist, The, 254
Egypt, agriculture in, 123
  carved gates of ancient, 19
Eiffel, 53
Eighteenth Amendment, 247
Eighteenth-century medicine, 326
Einstein theory, 301
E. K. Lab., 307
Electric cell, primary, invention of, 55
Electric communication, 115-121
Electric generator, the, 23
Electric power, centralization of, 129
  supplanting steam power, 129
Electric transportation, 103-105
Electrical Age, the, 29, 72
Electricity and economics, 60
Elementary Principles in Statistical Mechanics, 304
Elevators, mine, 24
  Otis, 24
  passenger, 24
Eliot, Charles W., 332, 338
Eliot, George, 275
Elsberg, 353
Emancipation, 205
Emancipation Proclamation, 206
  of the mind from the despotism of theology and the classics, 9
Emerson, Ralph Waldo, 380, 401, 402, 409, 413
  quoted, 4
Emmet, Dr., 337
Employers' liability, 168, 228, 229
Encyclopædia of Universal Knowledge (French), 10
Energy, generation of, 23
Engelmann, George, 314

Engine, high-pressure or non-condensing, 23
England, country developed the railroads in, 46
  Factory Acts in, 236
  leader in metal-working tools, 73
  match girl's strike in, 239
  Reform Acts in, 236
  slide-rest invented in, 33
  social legislation in, 237
  theory of socialism, 240
English common law, and rights of women, 256-257
  and strikes, 163
English monarchy, overthrow of, 202
Entomology, biology and, 313-317
Epinephrine, isolation of, 308
Equitable Life Assurance Society investigation, 245
Eretria, excavation by Americans in, 320
Erie Canal, 90, 93, 94, 97, 98, 115. *See also* New York State Barge Canal
Eskimos, seeking the oldest, 320
Essay on Mediæval Economic Teaching, An, 7 n.
Essays, Emerson's, 402
Ether, introduction of, 335
  transmission of radiation by the, 56
Ethiopian kings, discovery of tombs of, 320
Euclid, the validity of, 6
Eureka, Poe's, 402
Europe, feudal institutions of, 4
  loans to in the World War, 190-191
  Taylorism in, 37
  theoretical science in, 28
  views of life in in the Middle Ages, 7
European countries and superior minorities, 15
European criticism of idea of progress in American civilization, 13
European traditions in poetry and fiction, 420
Evangeline, Longfellow's, 402
Evans, Alice, 347
Evans, Dr., 353
Evans, Herbert M., 310
Evans, John J., 308
Evans, Oliver, 23, 27, 45, 48
Evils that can be eliminated, 7
Evolution, 314, 316, 317, 318
Excavation and exploration, 319-322
Experiment stations, 133-134

INDEX

Experimentation, in finance, 1830-1860, 172-179
  natural science and, 9
Experiments in Aërodynamics, 110
Exploration and excavation, 319-322
Eye, diseases of the, 342-345

Fabian Society, 240
Fable for Critics, Lowell's, 402
Factories, capacity of American, 195
  establishment of smaller, 80
Factory Acts (England), 236
  Illinois, 242
Factory inspection provisions, 168
Factory system, beginning of, 22
Factory town, trend away from the great, 80
Famines, emancipation of mankind from, 3
Faraday, 22, 55, 335
Farm Board, 251
Farm press, 135
Farmers' Alliance, 137
Farmers' Union, 137
Fatalism, the doctrine of, 6-7, 8
Faulkner, William, 403
Federal Employees Compensation Commission, women in the, 287
Federal Reserve Act of 1913, 185, 189
Federal Reserve banks, 189, 190
Federal Reserve Board, 189, 190
Federal Reserve system, 179, 183, 189, 197, 198
Federal Trade Commission, 218-219
Federation of Trades and Labor Unions of the United States and Canada, 153
Federations of Labor, state, 243
Feke, 383
Feminist movement, 227. See also Married women rights of; Woman suffrage
Fenger, Christian, 343
Fenner, 355
Ferber, 403, 409, 411
Ferenbaugh, Dr., 347
Fessenden, Reginald, 25, 57
Fevers and infections, the war on, 345-348
Fewkes, J. Walter, 322
Fibers, mechanization of the spinning and weaving of, 38
Field, Cyrus, 56, 117
Field, Eugene, 406
Field Museum of Natural History, 324
Fifteenth Amendment, the, 206, 207, 283

Finality, the illusion of, 4
Finlay, Carlos, 312, 345
First agricultural college, 133
First American train, 45
First Bank of the United States, 173, 177
First locomotive in the United States, 46
First state school board organized in Massachusetts, 362
Fischer, 351
Fisher, Dorothy Canfield, 403
Fisk, James, 16, 99
Fitch, Asa, 316
Fitch, Clyde, 405
Fitch, John, 24, 25, 34, 45, 86-87, 88, 89
Fitz, Reginald H., 342
Flaubert, 423
Fleming valve, the, 25
Fletcher, Robert, 339
Flexner, Abraham, 350
Flexner, Simon, 312, 349
Flint, Dr. Austin, 341
Flint, Austin, Jr., 344
Flower, Mrs. Lucy M., 243
Flying shuttle, the, 38
Folin, 353
Foot-and-mouth disease, 133
Ford, Henry, 33, 41, 49, 50, 108
Ford cars, 35, 50
Ford industries, 83-84
Ford Motor Co., 75
Foreign countries, "rationalizing" and recapitalizing, 196
Forest, Lee de, 303
Forest reserves, 221
Foster, Dr., radio sermon of, 61
Foster, Sir Michael, 338
Fourteenth Amendment, the, 206, 207, 208, 231, 285
Fowler, George R., 343
France, output of work per person in, 35
  return of the Bourbons in, 202
Francis, Edward K., 312, 347
Francis of Assisi, 16
Frankfurter, Felix, xxii
Franklin, Benjamin, 11, 292, 294, 302, 327-328
Franklin, Edward C., 305
Franklinism, 328
Franz, S. I., 323
Frazier, 353
Free soil parties, 206
French Revolution, 10, 202
French thinkers and the new philosophy, 10

INDEX 435

Freud, 323
Frick, George, 329, 344
Frontier schools, 359
Frost, Robert, 403, 413, 419
Frost, Wade H., 348
Fulton, Robert, 24, 45, 71, 87, 88, 89
Fundamental attitudes toward life and conduct, 5
Funk, Casimir, 310
"Futility of the Spelling Grind, The," 378
Future of Farming in England, The, 141
Future of the United States, 194-200
Futurism, gospel of, 6

Gaillard, Thomas, 337
Gale, Zona, 403
Galena & Chicago Railroad, 100
Galileo, 22
Gallic header of A.D. 70, 40
Galvani, 22
Galveston (Texas) reconstruction of government, 214
Gang violence, 249
Garfield, James A., 211
Garland, Hamlin, 406
Garrison, William Lloyd, 205, 208
Gas-balloon, 109, 113
Gasoline, 49. See also "Coal oil"; Kerosene
Gasoline tractor "combine," 43
General Electric Company, 250
General Electric laboratories, 64, 307
Generation of energy, 23
Geneva Medical School (N. Y.), 271
Gentry, Dr., 347
Geological Survey, 318
Geology, 317-319
Geraghty, 353
Gerhard, William W., 333, 341
German chemical industry, 63
Germany, medical ascendancy in, 338
output of work per person in, 35
Wöhler's synthesis of urea, 305
women's rights in, 282, 283
Germer, L. H., 301
Gibbs, Josiah Willard, 304
Gibbs, Oliver Wolcott, 305
Gilbert, Cass, 394
Gilbert, G. K., 318
Gilman, Daniel C., 332
Glasgow, Ellen, 403
Glover, Townsend, 316
Godkin, E. L., 211
Goebel, 26

Goforth, William, 332
Goldberger, Dr. Joseph, 310, 347, 353
Golden Legend, Longfellow's, 402
Gold standard, adoption of the, 184
Gold Standard Act, 180
Goldthwait, Dr., 343
Gomberg, Moses, 306
Gompers, Samuel, 149
Goode, George Brown, 314
Goodnow, Frank J., 214
Goodwin, Rev. Hannibal, 59
Goodyear, Charles, 64, 65, 71, 307
Gorgas, William C., 346
Gould, George M., 344
Government and law, 201-252
Governors mere figureheads, 213
Graf Zeppelin (airship), 114
Graham, E. A., 352
Gramme, 23, 55
Grand Canyon of the Colorado, 318
Grand Central Station (N. Y.), 104
Grand Trunk Railroad, 104
Grange, the. See Patrons of Husbandry
Gray, Asa, 24, 313, 314
Great Britain, and the importation of machinery into the Colonies, 71
certificates loaned to, 184
output of work per person in, 35
requirements of manufacturers in, 217
women's rights in, 282. See also England
Great Eastern, the, 117
Great Lakes a highway for wheat, 45
Great Northern Railway, 104
Great War, financing the, 185
industrial conditions in and after, 72
Great Western Turnpike, 92
Great White plague, 341
Grecian Remains in Italy, 319
Greeley, Horace, 412
quoted, 356
Green, Horace, 344
"Greenbacks," 180
"Green Pastures, The," 415
Greenwich House (New York), 244
Grinding machine, introduction of, 34
Gross, Dr. Samuel D., 331, 341, 343, 345
Group aspirations, our, 21
Group inventions, 63, 64
Gunn, Moses, 343
Gurney, Sir Goldsworthy, 49
Gutenberg, 33, 54, 55
Guthrie, Samuel, 329

436 INDEX

Gwathmey, Dr., 337
Gynecology, 337
Gypsum Cave (Nevada), exploration of, 321

Hadley, Philip B., 313
Hale, George, 296, 300
Hall, Asoph, 299
Hall, Charles Martin, 307
Hall, G. Stanley, 322, 347, 377, 378
Hall, Samuel Read, 370
Hallet, Stephen, 385
Hallidie, Andrew, 103
Halsted, William H., 337, 338, 343, 350, 352
Hamilton, Alexander, 172, 173, 217, 277
Hamilton, Frank, 343
Hammond, William A., 343
Hard Boiled Virgin, The, 417
Harding, Warren, 192
Hargreves, spinning-jenny of, 38
Harlem Railroad, 98. See also New York Central & Hudson River Railroad
Harper's Magazine, 420
Harrigan and Hart, 405
Harriman, Edward H., 102
Harrington, M. R., 321
Harris, Joel Chandler, 406, 415
Harris, Thaddeus William, 316
Harrison, R. G., 96, 351
Harrison's equalizing lever, 46
Hart, Edwin B., 310
Harte Bret, 406, 413
Harvard College, 332
Harvard College Observatory, 296, 297
Harvard University, 271-272
Harvesting, mechanization of, 38
  problem of the Romans, 40
Hastings, Thomas, 378
Hatch Act, 133
Hatshepsut, Queen, discovery of sarcophagus of, 320
Hay, O. P., 319
Hayes, Ellwood, 105, 106, 107, 108
Hayne, Senator, 203
Haynes, Elwood, 50
Hawthorne, Nathaniel, 401, 402, 418
Heaven, this world as a vestibule to, 7
Heaven and Earth of Doña Elena, The, Mrs. Stone's, 414
Hektoen, Ludwig, 313, 349
Hemingway, Ernest, 403
Henking, 351
Henry, Joseph, 22, 56, 118, 302

Henry, O., 406, 413
Henry Draper Memorial, 297
Henry Street Settlement (New York), 244
Herald, the (locomotive), 96
Heredity of man, studies on the, 316
Hergesheimer, Joseph, 403
"Heroic" inventions, the end of, 62-65
Herrick, Dr., 341
Hertz, 25
Hertzian waves, 57
Hess, Alfred F., 310, 353
Hesselins, 383
Heyward, Du Bose, 415
High-pressure engine, 27
Highway Engineers' Association, 213
Highways, 91-93, 221
Hill, James J., 41, 102
Hillebrand, W. F., 305
Hinckley, 96
Historical novel, the, 417
History, concept of progress a philosophy of, 6
History of History, 61
Hitchens, Dr., 347
Hoban, James, 385
Hodge quoted, 62
Hodgen, John T., 343
Hoe, Richard, 55
Hoe, Robert, 55
Hoff, Dr., 347
Hog cholera, 133
Holabird and Roche, 393
Holland, output of work per person in, 35
Holmes, Justice Oliver Wendell, 230, 311, 334, 340, 406
Holmes, W. H., 322
Home Economics Bureau, women in the, 287
Homer, Winslow, 391
Homestead Act, 124
Hooker, Philip, 385
Hookworm disease, 312, 346, 347, 355
Hoosac Tunnel, 104
Hoover, Herbert, 30, 250
Honoraria paid by Americans to Europeans, 15
Hopkins, B. S., 306
Horner, Dr. William E., 331, 341
"Horseless carriage." See also Automobile
Horse-power and manpower, 29
House of the Seven Gables, Hawthorne's, 402, 418
Howard, Bronson, 405

INDEX 437

Howard, L. O., 316-317
Howard, Sidney, 406
Howard University, 271
Howe, 24, 26, 63, 71
Howell, William H., 313, 344
Howells, William Dean, 406
Hrdlicka, Ales, 322
Hubbard, Gardiner, 119
Hubble, Edwin P., 297, 298
Huckleberry Finn, Twain's, 410
Hudson, C. S., 306
Hudson River Railroad, 98. *See also* New York Central & Hudson River Railroad
Hughes, David, 56
Hughes, Langston, 415
Hull House, 240, 241, 242, 243
Human beings, experience of, 5
Human labor compared with machine labor in agriculture, 43
Human welfare, requirements of, 6
Humane spirit in government and law, 223-231
Humanson, M. L., 297
Hunt, Reid, 26, 353, 389, 390
Hunt, Richard Morris, 388
Huntington, Collis P., 101
Hurst, Fannie, 403, 415
Hussey, Obed, 40
Hussey, W. J., 298
Hussey reaper, 26
Hutton, 317
Hyatt, J. W., 307
Hyatt, L. S., 307
Hygiene, 347

Ice Age, 318
Idaho, the suffrage in, 208, 283
Idea of Progress, The, 9 n.
"Ignoble," sweat of the, 17
Iliad, the, 416
Illinois, Cherry Mine disaster, 242
  civil service in, 212
  Factory Act in, 242
  humane spirit in, 223-224
  woman suffrage in, 284
Illinois Central Railroad, 100
Illinois Mining Investigation Commission, 242
Illumination, 25-26
Immigrants in the United States, 156
  treatment of, 238-245, 247-248, 249
Immigration, 237-238
  and organized labor, 156-157

Imprisonment for debt, abolition of, 226-227
Incandescent lamp, 22, 26, 28, 62, 63
Income taxation, 187, 188, 193, 194
Independent Treasury system, 177, 178. *See also* Sub-Treasury system
Indeterminate sentence, the, 230
Index Medicus, 339
India, population of, 70
Indiana, State Bank of, 175
Induction, discovered by Faraday, 55
Industrial age, arts of the, 392-398
Industrial classes, 71
Industrial laboratories, American, 63
Industrial management, science of, 75-77
Industrial revolution, 22, 30
Industrialization of old crafts, 23
Industry, a large public service, 67
  and agriculture, same workers in, 81
  and wealth, 67
  cheap electricity an essential of, 72
  chemical application to, 306-308
  decentralization of, 31, 80-81
  depends upon manufacture of power, 72
  growth of, 67
  machine and the capitalists, 66
  not a private prequisite, 67
  social objective of, 67
  status of wage-earners in, 158
  two great divisions of, 72
  women in, 276-281
Infantile paralysis, 349
Inheritance taxes, 193, 194
Inman, Henry, 387
Inness, George, 390
Innocents Abroad, Twain's, 410
Inquisition, cruelties of the, 17
Institute of History of Medicine, 350
Insulin, isolation of, 308
Interchangeable parts, 26, 33, 34, 37
International Harvester Co., 250
International Health Board, 347
International Iron Molders, 159
International Typographical Union, 159
Internal-combustion engine, 23, 49
Invention, as a social manifestation, 20-66
  contrasting methods of, 62
  impetus of, 32
  implies research, 62
  natural science and, 9
Inventions are composites, 62
Investors, overseas, 72
Irrigation, 221

438 INDEX

Italian government, the Rockefeller Foundation and the, 355
Italy, output of work per person in, 35

Jackson, Andrew, 207, 209, 216, 217, 221
Jackson, Helen Hunt, 416
Jackson, James, 334
Jacksonian democracy, 208, 210, 217, 221, 222
Jacobi, Abraham, 334, 342, 355
James, Henry, 406
James, Horatio G., 329
James, William, 322, 403
James River Canal, 90
Janeway, Dr. (the elder), 341
Japan, output of work per person, 35
the Tokugawa era of, 14
Japanese exclusion, 157
Jefferson, Thomas, 33, 294, 318
and architecture, 384-385
Jeffersonian school, doubters of the, 205
Jenkins, C. Francis, 59
Jenney, William L. B., 393
Jennings, H. S., 315
Jerusalem, 21
Jervis, 46
Jesuits of North America, The, Parkman's, 411
Jesus, teachings of, 17
Jews denied a share in government, 206
John Bull (locomotive), 96
John McCormick Institute for Infectious Diseases, 349
Johns Hopkins Hospital, 332, 338
Johns Hopkins University, 266, 271, 332, 377
Johnson, Andrew, 210
Johnson, James Weldon, 415
Johnson, Samuel W., 309
Johnson, Treat B., 306
Johnston, Mary, 416
Jones, John, 329
Jones, L. R., 316
Jordan, David Starr, 314
Journal of Experimental Medicine, 350
Journal of the American Medical Association, 339
Jukes, the, 351
Jurgen, 417
Juvenile Court, establishment of in Chicago, 243
Juvenile Protective Association (Chicago), 243

Kallikaks, the, 351
Kamm, Oliver, 308
Kansas, the suffrage in, 208
Kansas Pacific Railroad, 101
Kaye, flying shuttle of, 38
Keen, Dr., W. W., 342, 343
Kellogg Peace Pact, 232, 251
Kelly, George, 406
Kelly, Howard, 25, 350
Kelvin, Lord, 56. *See also* Sir William Thomson
Kendall, Arthur I., 313
Kendall, Edward C., 308
Kent, James, 227
Kentucky, State Bank of, 175
Kerosene, 49. *See also* "Coal oil"; Gasoline
Kier, Dr., 49
Kilborne, F. L., 311, 345
Kilmer, Joyce, 403
Kimball, Dexter, 35
King, John, 328, 329
King's College, 332
Kirjath-Sepher, exploration of, 320
Kish, excavation by Americans at, 320
Knapp, Hermann, 344, 355
Knights of Labor, 153
Knonecker, Dr., 345
Knowledge, attainment of exact, 6
made popular, 3
Koch, Dr., 338, 340
Koplik, Dr., 342
Krutch, Joseph Wood, 405

Labor, one hundred years of, 149
Labor movement, the, 149, 150-152, 153, 155, 157, 158
Labor-saving machines and devices, 26, 162
Labor unions, benefits, 154, 161
functions of, 157-163
historical development of, 152-154
legal status of, 163-166
social service of, 166-171
Ladd, George Trumbull, 322
Laënnec, 334
LaFarge, John, 390
La Farge, Oliver, 416
Lake iron-ore carriers, 45
Lane, John, 42
Langley, Samuel Pierpont, 24, 52, 110-112, 300
Langley Memorial Aëronautical Laboratory, 303

INDEX 439

Langmuir, Irving, 63, 307
Lanston type-setting machine, 54
Laplace nebular hypothesis, 299, 304
Large corporation in agriculture; its future, 140-142
Laryngology, 344
Lashley, K. S., 323
Last of the Mohicans, The, Cooper's, 411
Lathrop, Julia, 243, 286
Latrobe, Benjamin Henry, 385-386
Laughing Boy, 416
Lavassor, 106
Lawes, Sir John B., 130
Lazear, Jesse, 312, 345, 346
Lea, Carey, 305
Leavitt, Miss Henrietta S., 297
Le Bon, 49
LeConte, Joseph, 318
Legal-tender treasury notes, issuance of, 179
Le grand monde, 14, 15
Leidy, Joseph, 311, 314, 318, 344
Leland Stanford University, 271
L'Enfant designs the City of Washington, 385
Lenhart, C. H., 353
Lénoir, 23, 49
Letters and arts, women in, 275-276
Levene, Dr. P. A., 306, 349
Lewis, G. N., 304
Lewis, Sinclair, 403, 409, 411, 412, 423
Lewis, Warren, 351
Lewis and Clarke, 294
Liberator, 205
Libman, Dr., 341
Lick, James, 296
Lick Observatory, 296
Liebig, Justus von, 130, 329
Life, after death, 9
  our fundamental attitudes toward, 5
  purposes of the good, 3
  resignation as a philosophy of, 4
Life and Labor of the People, 239
Lilienthal, 24, 52, 53
Lillie, F. R., 315
Limitation of distance power can be transmitted by a steam-engine, 28
Lincoln, Abraham, 101, 125, 131, 206, 211, 377, 412
Lind, Samuel C., 306
Lindbergh, Charles A., 321
Lindsay, Vachel, 403
Lindsey, Judge Ben, 404
Linotype type-setting machine, 54, 62

Lippmann, Walter, quoted, 248-249
Lister, Lord, 340
Liston, Dr., 335
Literature, American, 399-423
Littell, Squier, 344
Little, C. C., 316
Little systems, the doom of all, 4
Livestock improvement, 135
Living, standards of, 70, 71
Llewellyn, Karl N., 230
Loans to Europe in the World War, 190-191
Locomotive, the, 23
  electric, 103
Loeb, Jacques, 315, 349, 355
Loevenhart, 353
Long, Crawford W., 310, 335
Long Island Railroad, 104
Longfellow, Henry W., 401, 402
Loomis, Dr., 341
Louis, 334
Louisville (Ky.) school system, 365
Lovett, Dr., 343
Lowden, Governor (Illinois), 214
Lowe, 26
Lowell, Amy, 403, 419
Lowell, James R., 402, 410, 418
Lowell, Percival, 299
Lowell Observatory, 296, 299
Lucin cut-off, 102
Luckhardt, Arno B., 310
Lumières, 59
Lyell, 317
Lynching, 232
Lyon, Mary, 256

Machine, the workman and the, 81-82
  and the capitalist, 66
Machine Age, 34
  craftsmen of the, 12
  idea of progress slow in taking form in the, 9
  other-worldliness of the Christian, 8
  religion in the, 21
  religious philosophers of the, 8
  social tension in the, 21
  the landlords of the, 9
  theology of the, 9
  tools, 25
  writers of the early, 7
Machine industry enrichment of the capitalists, 66
Machine process, true nature of, 68

INDEX

Machinery, American method of co-ordinating, 74
  and mass production, 12
  Great Britain prohibits importation of into American Colonies, 71
  improvements in, 68
  introduction of, 22
  labor-saving, 68
  labor-*serving*, 68
  man-displacing, 68
  profitable use of, 68
Machines, motor-in-the-head, 73
  that make machines, 31-37, 73
Macht, 353
Macleod, J. J. R., 308
Maclure, William, 317
MacCurdy, George Grant, 322
Madame Bovary, Flaubert's, 413
"Made in U. S. A.," 295
Madison, James, 203
Magazine publishing, 420-422
Magendie, 331
Main Street, Lewis's, 413, 423
Major industries in small towns, 31
Mall, Dr., 350, 351
Malta fever, transmission of, 347
Man, heredity of, studies on the, 316
  is a poor and miserable sinner, 7
Mankind, improvement of the lot of, 6, 9
  influence of the French Revolution on the fortunes of, 10
  is advancing, 6
  the allegiance of, 6
  use of science and invention by, 3
Mann, Horace, 328, 353, 370, 375
Manufacturing, cost of coal in, 28
  decline of workers in the United States, 36
Marconi, 21, 25, 57
Marin, John, 396
Marine, David, 353
Marine Hospital Service, 355
Mark, E. L., 314
Mark Twain, 237, 406, 410
Markets and prices, agricultural, 138-139
Markham, Edwin, 403
Married women, rights of, 227. *See also* Changing position of women
Married Women's Property Bill, 259
Marriott, McKim, 353
Marsh, Othniel Charles, 42, 318, 319
Marsh, Reginald, 42, 397
Martin brothers, 25
Martin, Henry Newell, 338

Martin, Homer, 390
Martin, Newell, 332, 344
Maryland, civil service in, 212
Mason, William, 96-97
Mass consumption, 74
Mass production, 12, 13, 22, 33, 34, 50, 66, 74, 80, 161, 170
Massachusetts, first school board in, 362, 363
Massachusetts banks, 175
Massachusetts General Hospital, 329
Massachusetts Sanitary Commission, 339
Masses, upward thrust of the, 15
Masters, Edgar Lee, 403, 419
Matas, Dr., 337, 343
Materialism, American civilization as the apotheosis of, 232
Material world, subjugation of the, 6
Maudsley, 25, 33, 37
Maury, Lieut. Matthew Fontaine, 319
Maxim, 52
Maxwell, Clerk, 56
Mayas, the, 321
Maynard, J. Parkers, 307
Mayo, Charles, 353
Mayo, James, 353
Mayo Clinic, 353
McBurney, Charles, 342
McClung, C. E., 315, 351
McCollum, Elmer V., 310, 353
McComb, John, 385
McCormick, Cyrus H., 26, 38, 40, 41, 42, 43, 71, 126, 127
McCormick reaper, 26
McCoy, George M., 312, 347
McDougall, William, 323
McDowell, Ephraim, 326, 328
McGuffey, William H., 375
McGuire, Hunter, 336
McIntire, Samuel, 385
McKim, Charles Follen, 388, 389
Mechanical equipment, development of, 72-74
Mechanical invention, wave of, 22
  in western Europe, 22
Mechanical reapers, 40
Mechanization of agriculture different from that of industry, 44
Mechanized society foreshadowed, 23
Medical Center (N. Y.), 354
Medical College of Ohio, 333
Medical education, advances in, 338-340
  origins of, 331-333
Medicine, 325-356

INDEX 441

Mees, C. E. K., 307
Meltzer-Lynn method, 353
Melville, Herman, 402, 418
Memoir of Jane Austen, 275
Mencken, Henry L., 404
Mendel, Lafayette B., 309, 312, 315, 350
Mendeleev, "Eka-iodine" of, 306
Meningitis, cerebro-spinal, 349
Mergenthaler, 54
Merriam, John C., 319
Metallurgy, 25
Metal-working machines, 25
Metropolitan Life Insurance Company, 31
Miami Canal, 90
Mihály, Denis, 59
Michelson, Albert A., 298, 300
Michelson-Morley experiment, 301
Michigan, woman suffrage in, 284
Michigan Central Railroad, 100, 104
Michigan Southern Railroad, 100
Middle Ages, views of life in Europe in the, 7
Middle Western Utilities Company, 30, 31
Middleton, John I., 319
Midgley, Thomas, Jr., 307
Mill, Henry, 24
Mill, John Stuart, 257, 261, 283
Millay, Edna St. Vincent, 403
Miller, Joaquin, 406
Millikan, Robert Andrews, 298, 301, 302
Milling machine, first, 25
Mills, Robert, 386
Mining, decline of workers in in the United States, 36
Minot, George R., 308, 350, 354
Missals, reproduction of, 55
Mississippi Valley, fevers infesting the, 333
  invasion of by machinery, 235
  religion in, 235
Missouri, Bank of, 175
  juvenile court in, 225
Missouri Compromise, the, 40
Missouri Pacific Railroad, 101
Mitchell, John K., 328
Mitchell, Weir, 336, 342-343, 344
Mitman, Carl W., quoted, 107
Moby Dick, Melville's, 402, 418
Modern factory, the, 73
Modulation, 57
Mogg Megone, Whittier's, 401
Monks, printing by, 55

Monotype typesetting machine, 54
Monroe, Harriet, 406
Montana, the suffrage in, 208
Montcalm and Wolfe, Parkman's, 411
Montgomery, 351
Moody, William Vaughan, 406
Moore, E. H., 304
Moore, R. B., 307
Moorehouse, Dr., 342
More, Paul Elmer, 404
Morgan, John, 328
Morgan, J. P., 16
Morgan, Thomas Hunt, 315, 350
Morley, Edward W., 305
Morley, John, quoted, 399
Morrill Tariff Act, 133, 186
Morris, William, 240
Morris and Essex Canal, 90
Morse, Samuel F. B., 21, 22, 24, 56, 63, 71, 115, 303
Morton, Thomas G., 337, 343
Morton, William T. G., 310, 311, 335
Mosses from an Old Manse, Hawthorne's, 402
Motion pictures, 24, 58, 59, 60, 61
Motor-bus, the, 51, 103
Motor-car, 103
Motor-in-head machine, 73
Motor-trucks, 48, 103
Mott, Lucretia, 282, 283
Mott, Valentine, 329
Moulton, F. R., 299
Mount Holyoke College, 266
Movable type, 54
Muller, H. J., 315
Multitude, the good life for the, 17
Municipal government, elevation of, 215-216
Municipalities, responsibilities of, 222
Munition-makers and patriotism, 4
Murdock, 26
Murphy, John B., 343, 354
Murray, 56
Muscle Shoals, 30

Napier, 106
Napoleon, 19
Narrative of Arthur Gordon Pym, The, Poe's, 402
National Academy of Sciences, 324
National Advisory Committee for Aëronautics, 303
National Bank Acts of 1863, 1864, 1865, 181-183

National banking system, 184
National Budget Act, 216
National Bureau of Standards, 304
National Civil Service Reform League, 213
National Conference of Social Work, 274
National Education Association, 274
National Federation of Settlements, 250
National Health Department, 355
National Institute of Health. *See* United States Public Health Service Hygienic Laboratory
National Labor Union, 153
National Monetary Commission, 189
National Research Council, 323, 324
National Road, the, 91, 93
National system of banking, development of, 1860-1900, 179-188
Nationality, barriers of cut, 4
Nations, experience of, 5
the fatalist and, 6
Natural science and progress, 3
and the concept of progress, 4
and the emancipation of the mind, 9
the development of, 9, 291-324
Nature, Emerson's, 401, 409
Naval Observatory (Washington), 295, 299
Neagle, John, 387
Negro, in industry, the, 281
the franchise and the, 207, 208
Negro writers, 415-416
Nerve-blocking, 337, 345
Neurology, 342-343
Nevada, the suffrage in, 208
Newberry, J. S., 318
Newcomb, Simon, 299
Newcomen mine-pump, 27
New England banks, 174
New England defies the laws of the federal government, 203
New England educational units, 368-369
New England town meetings, 244
New Haven Railroad, 104
New Jersey, civil service in, 212
women's rights in, 282
Newlands Act, 221
Newman, Francis, 417
New philosophy, French thinkers and the, 10
Newton, 21
New World, the concept of progress and the thinkers of the, 11

New York Bureau of Municipal Research, 214, 216
New York Central Railroad, 97, 104. *See also* New York Central & Hudson River Railroad
New York Central & Hudson River Railroad, 99
New York City, challenge of Standard Oil Company methods in, 245
Greenwich House in, 244
Henry Street Settlement, 244
investigation of the Equitable Life Assurance Society, 245
necessity of factories in, 31
sanitary conditions in, 223
school system, 365
skyscrapers of, 19
water supply of, 348
New York & Erie Railroad, 99. *See also* Erie Railroad
New York State, civil service law in, 212
"safety-fund" system of, 176
the suffrage in, 208
woman suffrage in, 284
New York State Barge Canal. *See* Erie Canal
New York University, 389
Extension Lecture Course, 244
Nichols, E. F., 301
Nicholson, Meredith, 416
Nightingale, Florence, 274
Nineteenth Amendment, 284
Nirvana of doubt and oblivion, 4
Nobel-prize men, first American, 30
Noguchi, Hideyo, 312, 349, 355
Norfolk & Western Railroad, 105
Normal schools, 363, 370-371
Norris, 96
North Dakota, grain elevators in, 222
North Pole, discovery of the, 319
Northern Pacific Railroad, 102
Northwestern University, 271
Norway, Castberg law enacted in, 262
Nott, Eliphalet, 89
Nott, John C., 345
Novelty, the, 89
Novocaine, discovery of, 337
Novy, George, 313, 355
Noyes, A. A., 305
Noyes, W. A., 305
Nullification, the doctrine of, 40
Nuttall, George H. F., 312

INDEX 443

Oblivion, Nirvana of, 4
O'Brien, George, 7 n.
Obscurantism, 324
Observations on the Continuous Progress of Universal Reason, 9
Obstetrics, 336
Oder, Dr., 341
O'Dwyer, Joseph P., 344
Odyssey, the, 416
Office of Education, women in the, 287
Ogden, John B., 100
Ohio, civil service in, 212
  the Negro in, 207
Ohio Wesleyan University, 296
Oil-well, first, drilled in the United States, 49
Old-age pensions, 169, 225
Old Chester Tales, Mrs. Deland's, 413
Old crafts, industrialization of, 23
"Old King Caucus," 209
Old World, secular supremacy in the, 10
Oliver, James, 42
Olynthus, excavation by Americans in, 320
Omoo, Melville's, 402, 418
O'Neill, Eugene, 406
"On the Equilibrium of Heterogenous Substances," 304
Ophthalmology, 344
Ophthalmometer, invention of the, 344
Ophthalmoscope, invention of the, 344
Opium in peritonitis, 334
Oregon, the suffrage in, 208
Organized labor and immigration, 156-157
Orient, disturbing the slumbers of the, 4
  fatalism in the, 6, 8
Oroya fever, 349
Orwin, C. S., 141
Osborn, Henry Fairfield, 319
Osborne, Thomas B., 309
Osler, Sir William, 325, 332, 338, 350, 355
Otto, John C., 23, 49, 106, 328
Output of work by the machines, 35
Outre-mer, Tennyson's, 401
Ovariotomy, 328
Overproduction, 42

Pacific railroads, rapidity of building, 101
Pacinotti, 23
Page, Thomas Nelson, 406, 414
Pain, assuagement of, 3
Paine, Thomas, 11

Painting, American, 382, 386, 389-392, 394, 396, 397-398
Palace of Delight, 239
Paleontology, invertebrate, 319
  vertebrate, 318
Panama Canal, the, 47, 221
Pancoast, Dr., 336
Panhard, 106
Panic of 1873, 183
  of 1907-08, 185, 189
Papish, Jacob, 306
Parker, Col. Francis W., 373
Parker, G. H., 315
Parkman, Francis, 411
Parlin, William, 42
Parole boards, 230
Parsons, Sir Gilbert, 23
Pascal, 10
Passport Division, women in the, 287
Pasteur, Louis, 311, 340
Patent laws, enactment of, 22
Pathfinder, The, Cooper's, 402
Patriotism, 4
Patrons of Husbandry, 137
"Pauper schools," 359
Pavloff, 331
Payne-Aldrich Act of 1909, 187
Peace Pact. *See* Kellogg Peace Pact
Peale, Charles Willson, 386
Pearl, Raymond, 316, 351
Pearl Street (N. Y.) power house, 29
Peary, Admiral Robert E., 319
Pease, Francis G., 298
Pediatrics, 342
Peirce, Benjamin, 304
Pellagra, 347
Pennock, C. W., 334
Pennsylvania Hospital founded, 328
Pennsylvania Railroad, 105
Pennsylvania Station (N. Y.), 104
Pepper, William, 342
Personal feelings, divorcing our, 22
Personality, the worth of, 16
Peru, mummified prehistoric bodies in, 320
Peruvian cities mapped from the air, 321
Pests and diseases of agriculture, war on, 132-133
Peterkin, Julia, 415
Peugert, 106
Philadelphia, a center for locomotive construction, 96
  as a leader in science, 293
Philadelphia Centennial Exposition, 118

444 INDEX

Philippines, sanitation in the, 355
Philosophies, our, 5
Phipps Institute for Tuberculosis, 349
Phipps Psychiatric Clinic, 349
Phonograph, the, 22, 62
Physical Geography of the Sea, 319
Physicians of seventeenth and eighteenth centuries, 327
Physick, 329, 344
Physics, 300-304
Physiology, advances in, 329-331
Pickering, E. C., 297
Pierre, Melville's, 402
Pilcher, 24
Pioneering, American industrial, 70-72
Pious frauds and religion, 4
Pirogoff, Dr., 335
Pisidian Antioch, excavation by Americans in, 320
Pittman Act of 1918, 184
Pitts, Hiram, 42
Pitts thresher, 42
Plagues, emancipation of mankind from, 3
Planting and reaping machinery, 26
Plow, first steel, 127
Poe, Edgar Allan, 402, 418, 423
Poems, Emerson's, 402
Poetry revival, 419
Poincaré, Henri, 6
Poland, output of work per person in, 35
Politics, women in, 282-290
Pollution of streams, 348
Pony express, the, 116
Population, increase in the growth in cities, 129
Populism, 208
Porgy, 415
Portraiture, Colonial, 383
Post, Wright, 329
Post-Civil War prices, 183
Post Office Department, extended functions of, 220
Potter-Bucky diaphragm, 352
Poulsen, Valdemar, 25, 57
Powell, John Wesley, 318, 322
Power, 27-31
 dependence of industry upon the manufacture of, 72
 faith of, 4
 steam, 72
Power-dispatcher, dominance of the, 30

Power houses of today, 29
Prairie, The, Cooper's 411
Prandtl, 53
Pre-Civil War finance, 192
Prendergast, Maurice, 396
President, the election of a, 209-210
Press double-octuple, 55
 double-sextuple, 55
 octuple, 55
 quadruple, 55
 rotary, 55
 sextuple, 55
 type-revolving, 55
Preventive medicine, 354-356
Prevost, François, 328
Pride and Prejudice, 275
Priestley, Joseph, 293
Prince, Morton, 323
"Principle of sovereignty," 192
Printing-press invention, United States the leader in, 54
Private life, the issues of, 5
Privilege, dreams of, 15
Probation for criminals, 229
Problems yet unsolved, 232
Production, and demand, 75
 of energy, centralization of the, 31
Productive workers, decline in, 36
Professions, women in, 269-274
Progress, concept of, 3-19
 critics of, 16
 the idea of, 1-19
"Progressive assembling," 50
Progressive party, 245
Prohibition movement, 248
Property qualifications, removal of, 206, 207
Prout, 330
Prudden, Dr., 341
Psychiatry, 322
Psychology, 322-323
 in education, 377
Public affairs, the issues of, 5
Public-health movement, 313
Public ownership, 221
Public utilities, the federal government and, 219-220
Publishing, the business of, in the United States, 420-421
Pueblos, 321, 322
Puerperal fever, contagiousness of, 311
Pullman strike, 242
Pupin, Michael I., 303
Pure-food movement, 309

INDEX 445

Quantity production *vs.* mass production, 74
Quarterly Cumulative Index, 339

Rabelais, 417
Race, barriers of, cut, 4
Radcliffe College, 266
Radiation, transmission of, by the ether, 56
Radio, difference between wire telegraphy and, 56
  how it became possible, 56
  in the United States, 60
  signaling, 25
  telephone, 57
  telegraphy, 57
Railroad brotherhoods, 155
Railway, beginnings of, 93-97
  development of, 97-102
Ramona, 416
Raritan Canal, 90
Rationality, 18-19
Raven, Poe's, 402
Read, Thomas T., quoted, 35
Realism, the growth of, 230
Recollections of a Busy Life, 356
Rediscovery of America, 320-321
Reed, Walter, 312, 345, 346
Reflections on the French Revolution, 12
Reform Acts (England), 236
Regional power pools, 30
Reid, William W., 343
Reiss, 24
Religion, 4
Religious beliefs, our, 5
Religious disabilities, 232
Religious qualifications, removal of, 206, 207
Remsen, Ira, 305
Renaissance, recovery of ancient learning in the, 9
Renault, 106
Reorganization of state governments, 214
Representative Men, Emerson's, 402
Reproduction, instrumentalities of, 3
Republican party and slavery, 206
  and the Gold Standard Act, 180
  and women appointees, 286
  born of the cotton gin, 40
Research, Institutions of, 348-352
Resignation as a philosophy of life, 4
Resumption Act of 1875, 180
Rhythm of progress and reaction, 202
Rice, J. M., 378

Richards, Theodore W., 306
Richardson, Henry Hobson, 388, 390
Riches an object of suspicion, 7
Ricketts, Howard Taylor, 312, 347
Rights of Man, 11
Riley, Charles Valentine, 316
Riley, James Whitcomb, 406
Rittenhouse, Jessie, 403
Ritter, William E., 315
Road Locomotive Act (England), 49
"Road locomotives," 49
Robbins, Robscheit, 354
Robertson, T. Brailsford, 308
Robinson, Edward Arlington, 403, 419
Rockefeller, J. D., 41
Rockefeller Foundation, 355
Rockefeller Institute for Medical Research, 312, 324, 348, 349, 351
Rock Island Railroad, 100
Rocky Mountain spotted fever, 347, 348
Rodgers, David I., 329
Roebling, John, 92
Rogers, W. G., 318
Rolling mill, invention of the, 25
Roman roads, 91
Romans, harvesting problems of the, 40
Romanticism, rise of, 387
Röntgen's discovery, 352
Room of One's Own, 288
Roosevelt, Theodore, 412
  and civil service, 237
Rosa, E. B., 309
Roseneau, 346
Rous, Peyton, 349
Rowland, Henry Augustus, 301, 356
Rowntree, 353
Royce, 106
Rubber, vulcanization of, 64, 307
Rumford, Count, 292. *See also* Thompson, Sir Benjamin
Rumsey, 45, 87, 88, 89
Rush, Dr. Benjamin, 293, 322, 327, 329
Rush Medical College, 326
Russell, Col. F. F., 345
Russell, Henry Norris, 29
Russia, fear of, 247
  arousing lethargic, 4
  output of work per person in, 35
  stores of wealth in, 70
Rustless metals, 73
Rutherford, Lewis Morris, 301
Ryder, Albert Pinkham, 391, 392
Ryerson Physical Laboratory (University of Chicago), 301, 303

446 INDEX

"Safety-fund" system of New York State, 176
Saint-Gaudens, Augustus, 389
St. John, Charles E., 298
St. Martin, Alexis, 309, 330
Saint-Pierre, Abbey de, 10
Salmon reaper, the, 26
Salvation Army, 240
Samaria, American excavation in, 320
Sand, George, 275
Sandburg, Carl, 403, 419
San Francisco, humane laws in, 224
Sanitation, 3, 347
Santa Fé Railroad, 102
Santayana, George, 403
Saranac Lake Sanitarium (N. Y.), 341
Sargent, John, 390, 391
Savage, Dr. Thomas S., 314
Savannah, the, 45
Say, Thomas, 316
Sayer, Dr., 336
Sayre, Louis R., 343
Scandinavian countries, women's rights in, 283
Scarlet fever, 312
Scarlet Letter, Hawthorne's, 402, 418
Scarlet Sister Mary, 415
Schoepf, J. D., 329
School of Hygiene organized, 350
Schools, attendance compulsory, 168
   curriculum of the, 366-368
   development of school books, 375-376
   difference between European and American, 358, 362
   early, in the United States, 358, 373-374
   finance and equipment, 371-374
   high cost of, 372
   methods of teaching in, 374-378
   money spent on in the United States, 362, 371
   normal, 370-371
Schurz, Carl, 211
Schwann, 330
Science, organization of, 322-323
Scientific management, 37. See also Efficiency engineering; Taylor's system
Scientific studies of education, 378-379
Scotch reaper, the, 26
Scott, W. B., 319
Scribner, 63
Sculpture, 389, 394, 397

Second Bank of the United States, 173, 174, 175, 176, 177
Secular supremacy in the Old World, 10
Security of employment, 161, 162
Seidell, Atherton, 310
Selden, George B., 105, 106, 108
Sellers, Coleman, 24
Semmelweiss, 334
Senators, election of, 210
Seneca Falls Convention, 282
Senn, Nicholas, 343, 355
Settlements, establishment and success of, 240-245
Seventeenth Amendment, 210
Seventy Years of Life and Labor, 149
Sewage and typhoid, 348
Seward Observatory, 296
Sewing machine, the, 26
Seybert, Henry, 305
Shakespeare, Edward O., 345
Shakespeare, W., 21
Shapley, Harlow, 297
Shattuck, Lemuel, 339
Shaw, Rev. Anna Howard, 283
Shaw Botanical Garden, 314
Sherman, H. C., 310
Sherman, Stuart, 404
Sherman Anti-Trust Law, 164, 165, 218
Sherman Silver Purchase Act, 184
Shifting of transportation from railway track to highway, 51
Shippen, William, 328
Sholes, Christopher, 24
Shotwell, Professor, 61
Shull, George H., 316
Siemens, 23
Siemens-Martin process, 25
Siler, Dr., 347
Silliman, Benjamin, 293, 294
Silliman, Benjamin, Jr., 306
Silver, free coinage of, 183
Silver "certificates," 184
Simpson, Dr., 335
Sims, James Marion, 336-337
Sims, William Gilmore, 406
Sinclair, Upton, 403
Sinner, man a poor and miserable, 7
Skinner, Richard C., 329
"Skyscraper," the development of the, 393, 395
Slater, Samuel, 26
Slavery, abolition of, 205-206, 232, 236
   defense of, 205

INDEX

447

Slavery— (*continued*)
   Emancipation Proclamation, 206
   the Constitution and, 205, 206
Slide-rest, the, 25, 32, 33
Sloan, John, 397
Slosson, Edwin E., 308
Slye, Maude, 351
Smallpox, epidemics, 327
   human inoculation against, 327
Small towns, industries in, major, 31
Smibert, 383
Smiles, Samuel, 64
Smith, Adam, 10
Smith, J. Lawrence, 305
Smith, Job Lewis, 342
Smith, Nathan R., 332, 334, 343
Smith, Theobald, 311, 345
Smith College, 266, 271
Smithsonian Institution, 302, 324
Smyth, Andrew Woods, 337
Snowshoe principle, application of the, 46
Social amelioration and progress, 3
Social and economic effect of automobiles, 50
Social disasters, emancipation of mankind from, 3
Social medicine, 354
Social service, women in, 274-275
Social tension, 21
Social transformation, the problems of, 234-252
Society of American Foresters, 213
Soils, science of the, 130
"Some Realism about Realism," 230 n.
Songs of Labor, Whittier's, 402
Sons of Vulcan, 159
Soubeiran, 329
South America, American scientists in, 319
South Carolina, and the existence of hell, 206
   nullifies acts of the United States, 203
South Carolina Railroad and Canal Company, 96
Southern Confederacy, weaknesses of the, 204
Southern Pacific Railroad, 102
Southwest, discovery of prehistoric inhabitants of the, 321
Spanish-American War, 345
Span of life lengthened by the idea of progress, 3
Speculation, 197

Spencer, R. R., 348
Speyer, Leonora, 403
Spingarn, J. E., 405
Spinning-frame, 38
Spinning-jenny, the, 38
Spinning-mule, 38
Spoils system, 210
Spoon River Anthology, The, 419
Stampfer, 24
Standard of living, the, 13
Standardization, 33
   of machines and processes, 37
   principles of, 50
Standard Oil Company methods, 245
Stanford, Leland, 101
Stanley, William, 303
Stanton, Elizabeth Cady, 208, 253, 256, 282, 283
State administrations, defects in, 213
State banks, 174, 182
State governments, enterprises carried on by, 221-222
Stead, William, 240
Steamboat, development of in the United States, 24, 45
Steamboating in the United States, 87-89
Steam-coach, experiments with, 48-49
Steam-engine, before the coming of the, 26
   first appearance of, 22
   invention of in its modern form, 23
   limitation of power transmitted by a, 28
   number operated in the United States in 1930, 28
Steam-tractors, 44, 48
Steam-turbine, invention of the, 23, 28
Stearns, J. B., 56
Steel-making, modern, 25
Steel plow, the first, 42
Steele, Wilbur, 403
Steenbock, Harry, 310
Steffens, Lincoln, 423
Stegomyia fasciata, 345
Steinhal, 56
Stephens, J. L., 320
Stephenson, George, 23, 41
Stephenson, Robert, 23
Sternberg, Surgeon-General George M., 311, 340
Stevens, Colonel (of Hoboken), 45, 85, 95
Stevens, George F., 344

448 INDEX

Stieglitz, Julius, 306
Stiles, Charles W., 312, 347
Stimpson, Louis R., 343
Stockard, 351
Stock-market financing, 198
Stone, 34
Stone, Lucy, 256, 265
Stourbridge Lion (locomotive), 95
Stove Founders National Defense Association, 159
Stowe, Harriet Beecher, 256, 402
Streams, pollution of, 348
Streeter, 351
Strickland, William, 386
Stuart, Gilbert, 386-387
Sturtevant, A. H., 315
Subjection of Women, 283
Submarine cabling, application of telegraphy to transoceanic, 56
Sub-Treasury system, creation of the, 177
Suffolk Bank, 175
Suffrage, extension of the in America, 12, 206, 232
  woman, 283
Sullivan, Louis, 393, 394, 395
Sully, Thomas, 387
Sumner, Charles, 236
"Superintendent of schools," 366
Supreme Court, U. S., and labor legislation, 208
  and social improvement laws, 246-247. *See also* Court decisions; United States Supreme Court
Surgeon General's Library, 338
Surgery, in early America, 329
  first books on, 329
Suspension bridges, Roebling, 92
Swan, 26
Swift, Dr., 349
Swiss Lake Dwellers, 38
Swiveling truck, the, 46
"Sydenham of America." *See* Rush, Dr. Benjamin
Syme, Dr., 335
Symington, William, 88
Syphilis in Boston, 327

Taft, William H., 192
Takamine, Jokichi, 308
Tales of the Grotesque and the Antique, Poe's, 402
Tariff, the, 218

Tariff Act of 1913, 187
Tariff duties, 187
Tariff taxation, 187-188
Tarkington, Booth, 403, 411
Taxes and education, 372
Taylor, Frederick W., 25, 36, 37, 307
Taylor, Dr. Graham, 242
Taylor's system, 37. *See also* Efficiency engineering; Scientific management
Teacher, 351
Teachers, training of, 369-371
Teaching, methods of, 374-378
Tear, J. D., 301
Teasdale, Sara, 403
"Technological unemployment," 35, 36
Technology, and industrial economy, 222
  and progress, 3, 4
Telegraph, 22
  beginnings of, 115-118
  first signals through space, 57
  how it became a social force, 56
Telegraphy, electric, 24
  Baudot system, 56
  Buckingham system, 56
  difference between radio and wire, 56
  Murray system, 56
  picture, 57-58
  printing, 57
  radio, 25
  Rowland system, 56
  telephone, 57
  two and four messages over a single wire, 56
Telephone based on same principles as telegraph, 57
  beginnings of the, 118
  in the United States, 60
  radio, 57
  stations in the world, 60
Telephony, 24-25
  radio, 25, 57
Telephotography, 57-59, 60
Television, 58-59, 60, 61, 120
Terman, Lewis M., 323
Texas fever, 133, 311, 345
Textile machinery, 26
Thacher, Thomas, 326
Theology, despotism of, 9
Thermodynamics, science of, 28
Thimmonier, 26
Thirteenth Amendment, the, 206
Thomas, Augustus, 405

INDEX 449

Thompson, Sir Benjamin, 293. *See also* Rumford, Count
Thomson, Elihu, 303
Thomson, Sir William, 56, 118, 344. *See also* Kelvin, Lord
Thoreau, David, 406, 409
Thorndike, E. L., 323
Thornton, William, 385
Thyroid gland, isolation of the secretion of the, 308
Time and motion studies, 37
Times, London, 339
Times, New York and appointments of women, 286
Titchener, Edward Bradford, 323
Tobacco coöperatives, 140
Tokugawa era in Japan, 14
Tolman, Richard C., 298, 304
Tom Thumb (locomotive), 95
Tom Sawyer, Twain's, 410
Tool steel, high-speed 25, 37
Torrey, John, 314
Tosillotome, invention of the, 344
Trachina, discovery of, 311
Trachoma, 349
Traditions, divorcing our, 22
Transmission, and reception of ideas, 54
 instrumentalities of, 3
Transportation, 44-52
 decline of workers in in the United States, 36
 improvements in, 80
 in the United States, 23, 86-114
 See also Airplane; Automobile; Electric; Motor-bus; Motor-car; Motor-truck; Steamboating; Traveling; Trolley-car
Transporting iron ore from the Mesaba district, 45
Transylvania College, 332
Traveling prior to 1830, 93
Trevithick, Richard, 23, 27, 48
Trolley-car, electric, 103
Trudeau, Edward Livingston, 341
Trumbull, John, 386
Trust, the vertical, 81
Tularemia, 312, 347
Tungsten, alloy steel, 25, 37
Turbine, the, 23, 28
Turner's Lane Hospital, 342
Turret lathe (Fitch's), 25, 34
Twain, Mark. *See* Mark Twain
Twenty Years at Hull House, 234
Twice Told Tales, Hawthorne's, 401, 402

Twilight sleep, 337
Two Admirals, The, Cooper's, 402
Type, printing from, 24
Typee, Melville's, 402, 418
Typewriter, the, 24, 62
Typhoid fever, 333
Typhus fever, 324
 difference between and typhoid fevers, 333, 334
 Mexican, 347
Tyson, Dr., 341

Uncle Tom's Cabin, 256, 402, 415
Understanding Human Nature, 289
Unemployment insurance, 250
Union, the, danger of disruption, 204, future of the, 232
 preservation of the, 203-205
Union Pacific Railroad, 101
Unitarians denied a share in government, 206
United Mine Workers of America, 164
United States, after the Civil War, 204
 a generation behind other industrial nations, 239
 agrarian democracy of, 217
 a hundred years ago, 20
 and the protective tariff, 239
 banks in the, in 1846, 171
 business depression in, 238
 Carolina nullifies acts of, 203
 central stations in, 30
 Civil service in, 237
 colleges in the, 360, 361
 commercial airplanes in, 54
 communication in, 86, 115-121
 concept of progress in, 4
 corporation farms in, 141
 currency in the, in 1846, 171
 dearth of labor in, 32
 decade beginning 1830, 234
 decline of workers in agriculture, 36
 decline of workers in manufacturing, 36
 decline of workers in mining, 36
 decline of workers in transportation, 36
 development of railroad in, 47-48
 development of steamboat in, 45
 development of steam-engine in, 27
 divisible wealth per capita in, 35
 early canals in, 89-91
 early epidemics in, 327

INDEX

United States—(continued)
  early material and spiritual conditions, 234-237
  early pioneers and the railroads, 235-236
  enmeshed in copper, 30
  equity and law united in, 225
  financial condition during the World War, 190-191
  first laboratory of experimental physiology in, 338
  first locomotive in, 46
  first oil-well drilled in, 49
  first practical steamboat in. See *Clermont, the*
  first women's convention in, 282
  freight rates in, 44
  high schools in the, 360-361
  historic system of agriculture in, 123-126
  home market of, 70
  immigrants in the, 156
  increase of industrial employees in, 239
  influence of the cotton gin on the informative period of, 39
  in front rank of industrial nations, 27
  inland waterways, 86
  invention of the slide-rest in, 33
  lacked mechanics, 33
  largest coal-owning country, 27
  leader in printing-press invention, 54
  mechanical invention in, 22, 23
  money spent on education in the, 362
  natural resources of, 70
  need for steamship and railroad in, 44
  newspapers in, 55
  nineteenth century in, 205
  number of engines operated in, in 1930, 28
  number of locomotives used to convey coal in, 28
  output of work per person in, 35
  overseas investors in, 72
  political corruption in, 239
  primary power plant of agriculture in, 44
  production of steam-cars in, 49
  progress of, 20
  radio in, 60
  railroad indispensable in, 45
  railroads developed the country, 46
  railways in, 28, 93-114
  resources of at the beginning of the Civil War, 179
  "safety-fund" system of, 176
  social reconstruction in, 237
  steamboating in, 87-89
  theoretical science in, 28
  transportation in, 86-114
  25,000,000 automobiles registered in, 50
  typical school of in 1830, 358
  women's rights in, 282
  young men's movement in, 240
United States Dispensary, 327
United States Exploring Expedition, 319
United States Geological Survey, 318
United States National Museum, 314
United States Office of Education, 267
United States Public Health Service, 312, 346, 347, 348, 355
United States Railway Mail Service, 119
United States Steel Corporation, 60
Universal emancipation, the genius of, 4
Universal scheme, the, 5
Universality, strength of, 3
University Extension Lecture Course, 244
University medical centers, 354
University of Chicago, 243, 271
University of Pennsylvania, first medical school founded at, 328
University of Toronto, 308
University of Virginia, Jefferson and the, 385
University of Wisconsin, 309
University Women, Association of, 273
"Unskilled labor" suffers most, 35
Untermeyer, Louis, 403
Ur of the Chaldees, excavation by Americans at, 320
Uses of electricity, 29
Utah, the suffrage in, 208, 283, 284
Utopianism and the concept of progress, 8-9

Vaccination, 327
Vacuum-tube, the, 25
Vail, Alfred, 115
Vail, Theodore N., 56, 119
Valve-gear, invention of the, 23, 27
Vanadium alloy steel, 25, 37
Van Buren and the sub-Treasury system, 177
Vanderbilt, Cornelius, 98-99
Van Doren, Carl, 405
Van Slyke, Dr., 349, 353

INDEX

Van Wagenen quoted, 111
Vassar College, 266
Vaughan, Victor C., 313, 345
Veblen, Oswald, 304
Vedder, Dr., 347
Vertical transportation. *See* Elevators
Vertical trust, the, 81
Vindication of the rights of women, 264
Virginia, Capitol (1785), Thomas Jefferson and the, 384-385
  education in, 359
  statute of 1776, 227
Virginian Railroad, 105
Vision of Sir Launfal, Lowell's, 402
Visiting Nurses' Association (Chicago), 243
Vitamins, 310
Voices of the Night, Longfellow's, 402
Volta, 22, 55
Voltaire, 21

Wage-earners, consuming power of, 69
  status in industry, 158
Wages, distribution of wealth through, 83-84
  no conflict between low costs and high, 68
Walcott, Erastus B., 336
Wald, Lillian, 274
Wallace, Lew, 416
War contractors, 236
War on agricultural pests and disease, 132-133
Warburton Anatomy Act (England), 329
Ware, John, 328
Warren, John Collins, 329, 335
War taxes, 186-189, 193
Washburn, Hon. E., 260
Washington, Booker T., 415
Washington, D. C., planned, 385
Washington, George, 89, 131
Waste of our fertile soils, 142-145
Water gas, invention of, 26
Waterhouse, Benjamin, 327
Waterwheels, 30
Watson, John B., 323
Watt, James, 21, 23, 27, 30, 32, 34, 40, 62, 88, 127
Wealth, and industry, 66, 67
  distribution of through wages, 83-84
  satisfying distribution of, 70
Wealth of Nations, 10
Webb Act, 218
Weber and Fields, 405

Webster, Daniel, 99, 205, 223
  quoted, 203
Webster, Dr., 349
Weeks, John E., 344
Welch, William Henry, 312, 338, 340, 341, 350
Welding, developments through, 73
Wellesley College, 266
Wells, Horace, 311
Wells, Joseph Morrill, 388
Wells, William Charles, 337, 355
West, Benjamin, 383
Western civilization, 3
Western railway advance, 100-102
Western Union Telegraph Company, 29
Westinghouse, 29, 71
West Point (locomotive), 96
Wharton, Edith, 403, 409
Wheat, mechanization of the harvesting of, 38
  the Great Lakes a highway for, 45
Wheatstone, 56
Wheel of Equity, 137
Whipple, George H., 297, 308, 353
Whistler, James McNeill, 390-391
White manhood democracy accepted, 207
White, Maunsel, 25, 36, 307
White, Stanford, 388, 389
White, William A., 323
White House, design for the, 385
White Jacket, Melville's, 402
Whitman, Walt, 350, 410
Whitney, Eli, 25, 26, 33, 34, 36, 37, 38, 39, 40, 50, 71
Whitney, Willis R., 307
Whittier, John Greenleaf, 208, 401
Wilder, Thornton, 347, 414
Wiley, Harvey W., 309
Wilkes, Charles, U. S. N., 319
Wilkes expedition, 318
Wilkins, Mary E., 406
Wilkinson, David, 25, 33
Wilkinson, John, 32
William H. Wilmer Institute of Ophthalmology, 349
Williams, Robert R., 310
Williams, Whitridge, 350
Wilson, Alexander, 292, 293
Wilson, Edmund B., 314
Wilson, Woodrow, 187, 209, 221, 412
Winans, Ross, 96
Windsor, 26
Winthrop, John, 293, 295
Winton, Alexander, 108

INDEX

Wisconsin, civil service in, 212
Wistar, Caspar, 329
Wister, Owen, 414
Withington, Charles, 42
Wollstonecraft, Mary, 264
Woman suffrage, 208-209. *See also* United States and various states
Woman's Rights Convention in 1848, 208
Women, and education, 264-269
and English common law, 256-257
changing position of, 254-290
colleges for, 265-269
enfranchisement of, 206. *See also* various states
in Congress, 287
in industry, 276-281
in politics, 282-290
in the Board of Tax Review, 287
in the Federal Employees Compensation Commission, 287
in the Home Economics Bureau, 287
in the Office of Education, 287
in the Passport Division, 287
in the professions, 269-274
New York *Times* and appointments of, 286
protective legislation for, 168
rights of married, 227
Women in Industry, 276
Women's Bureau, 285
Wood, Henry Wise, 55
Woodberry, George Edward, 406
Woodhouse, Dr. James, 293
Woodruff, L. L., 315
Woods Hole, 314
Woolaston, 383
Woolf, Virginia, 275, 276, 288, 289
Workman, the, and the machine, 81-82
Workmen's compensation, 229
World, this, as a vestibule to heaven, 7

World Court, United States and the, 251
World War, aircraft in, 54, 114
James Bryce on the, 60
loans to Europe in the, 190-191
preventive medicine in, 354
revenue conditions which preceded, 185
taxes during, 193
transformation, 1900-30, 188-194
use of Carrel-Dakin solution in, 312
World's Fair, Chicago, 238 (1933), 252
Wright, Frank Lloyd, 395
Wright, Harold Bell, 417
Wright, Orville, 24, 52, 53, 112-113
Wright, Wilbur, 24, 52, 53, 112-113
Wyandotte, Cooper's, 402
Wylie, Elinor, 403
Wyman, Jeffries, 314
Wyman, Morrill, 334
Wyoming, Territory, woman suffrage in, 283, 284

X-rays, discovery of, 352

Yale College, 332
"Yankee experiment" in democracy, 203
"Yankee ingenuity," 72
"Yellow dog contracts," 165
Yellow fever, 311, 312, 327, 345-346, 349, 355
Yerkes, Robert M., 323
Yerkes Observatory, 296
York, the (locomotive), 95
Young, Charles A., 299
Young, Ella Flagg, 274
Young, John R., 330
Young men's movements, 240
Yucatan, expedition into, 320

Zeitgeist of the eighteenth century, 27
Zeppelin, Count Ferdinand von, 113